GUANYLATE CYCLASE

Guanylate Cyclase

Edited by

RAMESHWAR K. SHARMA

The Unit of Regulatory and Molecular Biology
Departments of Cell Biology and Ophthalmology
SOM and NJMS
University of Medicine and Dentistry of New Jersey
Stratford, NJ
USA

Reprinted from *Molecular and Cellular Biochemistry*, Volume 230 (2002)

Kluwer Academic Publishers
Dordrecht / Boston / London

Distributors for North, Central and South America:
Kluwer Academic Publishers
101 Philip Drive
Assinippi Park
Norwell, Massachusetts 02061 USA
Telephone (781) 871-6600
Fax (781) 681-9045
E-mail <kluwer@wkap.com>

Distributors for all other countries:
Kluwer Academic Publishers Group
Distribution Centre
Post Office Box 322
3300 AH Dordrecht, The Netherlands
Telephone 31 78 6392 392
Fax 31 78 6546 474
E-mail <services@wkap.nl>

Electronic Services <http://www.wkap.nl>

Library of Congress Cataloging-in-Publication Data

A C.I.P. Catalogue record for this book is available from the Library of Congress

ISBN 0-7923-7682-8

Printed on acid-free paper

Printed in the United States of America

Molecular and Cellular Biochemistry:

An International Journal for Chemical Biology in Health and Disease

CONTENTS VOLUME 230, Nos. 1 & 2, January (II) 2002

GUANYLATE CYCLASE
Dr. Rameshwar K. Sharma

Molecular and Cellular Biochemistry **230:** 1, 2002.

Preface

The guanylate cyclase signal transduction field started about 40 years ago with the discovery of cyclic GMP in rat urine. It, however, encountered strong turbulence. From the beginning, the identity of guanylate cyclase itself was wrapped up in mystery. It did not resemble its predecessor enzyme adenylate cyclase. The painstaking ensuing research by many laboratories has started to indicate many intriguing features of this signal transduction field, which are unique to itself. One of those features is that the guanylate cyclase enzyme exists in two major forms: membrane-bound and soluble. Both forms are independently regulated by distinct mechanisms. The soluble form is emerging to be an important transducer of the neurotransmitter signals. A fascinating signal transduction story is evolving with the membrane guanylate cyclase form. This form is a transducer of both the peptide hormone signals and the Ca^{2+} spikes arising within the sensory neurons. Thus growing from its initial skepticism, now the field occupies a position of major prominence in the signal transduction area. This volume focuses on the most current avenues of Guanylate Cyclase research. It contains sixteen chapters written by the leaders in the field. The first chapter is an historical account of the evolution of the membrane guanylate cyclase field. It is followed by five chapters, which cover the peptide hormone receptor research. Chapters seven through ten deal with Ca^{2+}-sensitive membrane guanylate cyclase. Chapters eleven and twelve identify the congenital retinal diseases that have been linked with a specific Ca^{2+}-sensitive photoreceptor guanylate cyclase. A membrane guanylate cyclase with unusual features has been discovered in unicellular organisms, this has been discussed in chapter 13. Chapters fourteen and fifteen discuss the biochemistry and physiology of the soluble form of guanylate cyclase. And chapter sixteen discusses the status of soluble guanylate cyclase in the retina.

Rameshwar K. Sharma
Guest Editor, Stratford, USA

Molecular and Cellular Biochemistry **230:** 3–30, 2002.

Evolution of the membrane guanylate cyclase transduction system

Rameshwar K. Sharma

The Unit of Regulatory and Molecular Biology, Departments of Cell Biology and Ophthalmology, SOM and NJMS, University of Medicine and Dentistry of New Jersey, Stratford, NJ, USA

Abstract

Almost four decades of research in the field of membrane guanylate cyclases is discussed in this review. Primarily, it focuses on the chronological development of the field, recognizes major contributions of the original investigators, corrects certain misplaced facts, and projects its future trend. (Mol Cell Biochem **230**: 3–30, 2002)

Key words: membrane guanylate cyclase, atrial natriuretic factor receptor guanylate cyclase, rod outer segment membrane guanylate cyclase, calcium-sensitive membrane guanylate cyclase

Introduction

The present day status of the mammalian membrane guanylate cyclase signal transduction field is based on four eureka findings: (1) The discovery of cyclic GMP and the existence of the membrane guanylate cyclase enzyme in the biological systems; (2) the demonstration that existence of this enzyme is independent of its soluble form; (3) the discovery of the first membrane guanylate cyclase, which showed that the membrane guanylate cyclase enzyme is a surface receptor; (4) the current realization that all membrane guanylate cyclases are not surface receptors, some of them are regulated by the intracellularly generated Ca^{2+} spikes in the sensory neurons.

These findings have established that the membrane guanylate cyclase exists in multiple forms, and that it is a critical transducer of the signals arising both outside and inside the cell. In this newly developing phenomenon, the Ca^{2+}-dependent membrane guanylate cyclase signaling field also internets with the G-protein coupled receptor field. The studies with the pinealocytes show that an $\alpha_{2D/A}$-adrenergic receptor generated signal activates a Ca^{2+}-modulated membrane guanylate cyclase. Thus, the initial concept where the membrane guanylate cyclase transduced only the extracellularly generated hormonal signal is under revision. In the revised concept, the transduction field is wider than believed earlier. It is directly involved in the transduction of both the extracellularly and intracellularly generated signals, and, via Ca^{2+}, it may also be indirectly linked with the signals originating from the G-protein coupled seven transmembrane-spanning receptors. One of the most exciting developments in the field has been that it is beginning to define several new paradigms of Ca^{2+} signaling in the sensory neurons. With the present review, it is the intent of the author to walk the reader through the past almost four decades of this exciting area of research, to recognize the contributions of the original laboratories, to correct certain misplaced concepts, and to project its future trend. Some sections of the review have been drawn from earlier reviews and publications where this investigator has been an author or a co-author; in those cases, appropriate citations have been made. This review is totally focused on the mammalian membrane guanylate cyclases, a principal field of research by the author for over three decades.

Background

The term 'second messenger'

The modern usage of the term 'second messenger' is rooted in the original discovery of cyclic AMP [1]. Based on studies with glycogen metabolism in skeletal muscle, Suther-

Address for offprints: R.K. Sharma, The Unit of Regulatory and Molecular Biology, Departments of Cell Biology and Ophthalmology, SOM and NJMS, University of Medicine and Dentistry of New Jersey, Stratford, NJ 08084, USA (E-mail: sharmark@umdnj.edu)

land's group coined the term 'second messenger' in proposing the role of cyclic AMP in hormone action [2]. In this concept, the hormone was the first messenger. Its interaction with its receptor resulted in the production of the second messenger, which was cyclic AMP. This then acted as the hormone messenger in eliciting the biological response inside the cell. The principal features of this concept were envisioned to be applicable to all hormonal systems [3].

The term 'signal transduction'

At the time when the term second messenger was coined, no plasma membrane receptor was purified. Therefore, the biochemical nature of the receptor was not known. It was not clear whether the receptor and the adenylate cyclase – the enzyme catalyzing the formation of cyclic AMP from ATP – were separate entities or a combined entity, the adenylate cyclase being a subunit of the receptor. The concept, however, favored the receptor to be an integral part of the adenylate cyclase enzyme. The receptor was conceptualized, as defined by studies done in the 1930s, as being the critical component of a cell where the hormone interacted to produce a stimulus (For early studies on the receptor concept see refs [4, 5]). The term 'stimulus' was equivalent to the present term 'signal', which, then, according to the 'second messenger' concept, meant that adenylate cyclase was the means to convert the extracellular ligand-binding signal into the production of the intracellular chemical, cyclic AMP. Implicit in this concept was the understanding that the events leading to the transformation of the binding signal to the production of the second messenger occurred in the cell's plasma membrane.

Studies done in the 1970's by Rodbell's group showed that the transformation of the hormonal signal into the production of the second messenger required GTP. The term 'transduction' was introduced to indicate the GTP-dependent step [6], which converted the hormonal binding signal into the formation of cyclic AMP. The transduction process envisioned two steps: (1) at the extracellular surface, the signal (hormone) bound to its receptor; (2) within the interior of the cell, GTP bound to G protein and produced the second messenger, cyclic AMP (These early studies are reviewed in refs [7–9]). The groups of Gilman and Birnbaumer then defined the detailed sequence of the transduction steps (reviewed in refs [10, 11]). Two types of G proteins, one stimulatory, termed G_s, and the other inhibitory, termed G_i, were purified and characterized [12, 13]. G_s mediated the stimulatory and G_i, the inhibitory transduction process. The nature of the receptor defined specificity of the hormone. The adenylate cyclase signaling system, therefore, represented a three-component system, consisting of three distinct proteins – the receptor, the G protein, and the adenylate cyclase. Subsequent studies showed a similar type of operation for the phosphatidylinositol-signaling pathway where cyclic AMP did not act as the second messenger [14, 15]. Here, the hormonal signal converted phosphatidylinositol-4,5-bisphosphate (PIP_2) into two separate second messengers, diacylglycerol and inositol triphosphate (IP_3). IP_3 signal transduction then split into two pathways: one mobilized the intracellular Ca^{2+} pool, and the other resulted in the activation of protein kinase C [reviewed in: 16, 17].

In this manner, these and other complementary studies from many other laboratories established the mechanism of two major G protein-linked signal transduction pathways – cyclic AMP and phosphatidyl inositol (reviewed in ref. [18]). Both of these pathways were three component signal transduction systems and in both cases the receptor component, where the signal originated, was a seven transmembrane-spanning protein. These signal transduction pathways provided the means of converting the extracellularly-generated hormonal signal into the production of the intracellular second messengers. In turn, these second messengers initiated a cascade of chemical reactions, which ultimately translated the signal into a cell-specific biological response.

Cyclic GMP and guanylate cyclase activity

Guanylate cyclase is an enzyme that catalyzes the transformation of GTP into cyclic GMP. Following the discovery of cyclic AMP, Ashman *et al.* identified cyclic GMP in rat urine [19]. Six years later cyclic GMP [20, 21] and guanylate cyclase were detected in all tested-tissues [21, 22]. Following the framework formulated for the second messenger role of cyclic AMP, various groups conceptualized cyclic GMP also as a second messenger of certain hormonal signals (reviewed in ref. [22]). The early studies observed some features, which were unique to the guanylate cyclase enzyme and were not shared with its counterpart enzyme, adenylate cyclase: (1) The adenylate cyclase existed exclusively in the membrane-bound form; the guanylate cyclase was mostly soluble [23–25]; (2) The adenylate cyclase activity was manifested freely in the particulate fractions; particulate guanylate cyclase activity was hidden and was manifested only when the particulate fractions were treated with detergents [21, 26, 27]; (3) The preferred cofactor of adenylate cyclase was Mg^{2+}; of guanylate cyclase, it was Mn^{2+} [21, 23–25]; (4) The adenylate cyclase responded to the hormonal signals in broken cell preparations [28], the guanylate cyclase did not [29]; 5) GTP stimulated adenylate cyclase; ATP, the counterpart of GTP, inhibited guanylate cyclase [21].

The primary conclusions of these studies were that: (1) in contrast to the adenylate cyclase system, the guanylate cyclase system is not linked to hormonal signals; (2) the predominant form of guanylate cyclase is soluble; and (3) the biochemical features of the guanylate cyclase and the adenylate cyclase enzymes are different.

Thereafter, the guanylate cyclase field witnessed three dramatic phases. First there was a logarithmic surge, which lasted for about 5 euphoric years and the proposal for an intriguing 'Yin Yang' hypothesis. Then, came about a decade of twists and turns, when major laboratories denied the independent existence of the hormone-dependent guanylate cyclase. And finally, there was a characterization of the first member of the hormone-dependent membrane guanylate cyclase, which had an unusual feature of also being a surface receptor for the peptide hormone, ANF. This discovery resulted in the rebirth of the guanylate signal transduction field and also in the origin of a novel concept, which proposed that the membrane guanylate cyclases belong to the earlier known family of surface receptor protein tyrosine kinases.

Euphoria and 'Yin Yang' hypothesis

In the early 1970's great excitement was generated in the cyclic GMP signal transduction field when the original studies of George *et al.* [30] implicated this nucleotide as a biological effector molecule, mediating the effect of acetylcholine through adrenergic receptors. This implied that acetylcholine via adrenergic receptors activated the membrane guanylate cyclase. An intriguing concept was proposed in which the biological regulation of a cell is governed by the opposing activities of cyclic AMP and cyclic GMP [31]. It was named the 'Yin Yang' hypothesis [31]. The coinage of this Chinese term coincided with president Richard Nixon's trip to China. Given the political twist and novelty of the concept, the 'Yin Yang' hypothesis attained instant notoriety, a focal point of intense interest, and a subject of many influential publications [22, 32–34].

The 'Yin Yang' hypothesis proposed that a cell stimulus (hormonal signal) caused a rise of cyclic GMP, which was accompanied by a reciprocal decline in the level of cyclic AMP, and vice versa. The investigators cited numerous cases in support of this concept (reviewed in ref. [22]). The concept rested on two central ideas. One, it envisioned that the sole form of the guanylate cyclase is membrane-bound. Two, that this cyclase is reciprocally interlocked with the adenylate cyclase. Thus, the interlocked guanylate cyclase-adenylate cyclase was a 'bi-directional' transduction system. Activation of guanylate cyclase caused the concomitant inhibition of the adenylate cyclase, and vice versa.

The 'Yin Yang' concept did not survive experimental scrutiny. The principal reason was that its 'bi-directional' transduction concept was not found to be universal. Studies with the isolated tissues of liver [35–38] and kidney showed that the exogenously supplied cyclic GMP and cyclic AMP were almost equally potent in advancing glucose output, glycogenolysis, phosphorylase activation and the conversion of glycogen synthetase from the active to the inactive form [39]. Similarly, cyclic GMP mimicked cyclic AMP in causing hyperpolarization of the liver cell membranes [40, 41]. In addition to the liver and kidney, studies with the adrenal slices and the isolated fasciculata cells of the adrenal cortex also showed that both exogenously applied cyclic nucleotides mimicked each others activity in stimulating steroidogenesis [42–44]. This indicated that these signal transduction systems were unidirectional.

Twists and turns

In the period spanning 1974 to 1984, the guanylate cyclase field folded in utter confusion and contradictions. The monoclonal antibody studies showed that the membrane-bound and the soluble forms of the guanylate cyclases were structurally distinct [45], the biochemical and functional studies indicated that the differences were minor. For instance, both forms were not the direct transducers of any hormonal signal and both responded non-specifically to the common signals of hydroxyl radical, hydrogen peroxide, lipids and unsaturated fatty acids, oxidants [46–49], nitric oxide and a variety of other nitric oxide generating compounds such as nitrosoamines [50–52], including cigarette smoke [53]. It was proposed that the underlying mechanism of action for the nitrite-generating compounds was the same, via nitric oxide (NO) gas [50, 54, 55]. Additional evidence was provided that indicated that NO activated the membrane form of the guanylate cyclase at its catalytic site [52], and the presence of NO-dependent membrane guanylate cyclase was reported in almost all tested mammalian tissues ranging from peripheral to the central nervous system [56]. The response of the cyclase to NO varied, ranging from 3-fold stimulation in the liver to 14.7-fold in the rat cerebellum [50].

The intense feeling about the physiological irrelevance of the hormone-dependent membrane guanylate cyclase was extended to the cyclic GMP-dependent protein kinase. At the time, besides guanylate cyclase, this was the only other known cyclic GMP signal transduction component. It was reported that cyclic GMP-dependent protein kinase did not phosphorylate any protein with higher specificity than cyclic AMP-dependent protein kinase did. The implication was that in the instances where a stimulus (signal) generated cyclic GMP in the intact cell, it acted through a cyclic AMP transduction component instead of the cyclic GMP transduction component. Thus, in these cases the cyclic GMP signaling pathway was a subservient to the cyclic AMP transduction system (reviewed in ref. [57]).

Propelled by this state of mind, the original key investigators then revised the interpretations of their earlier findings where they had concluded that the membrane-bound and the soluble forms of guanylate cyclases were separate structural and functional identities. In the reinterpretations, those results were artifactually attributable to the 'assay conditions, other constituents in crude preparations, or possibly other factors' [58]. Typical conclusions on the guanylate cyclase field as summed up by the two leading groups are cited below. One group states:

> While the metabolism of cyclic GMP has many parallelism to that of cyclic AMP, it is becoming quite clear that the two systems have distinct differences and that the functions of one system may not necessarily provide clues to the function of the other. In retrospect, the apparent parallelisms of the two systems hindered the originality of approaches to the cyclic GMP system and may, in fact, have impeded some of the developments in the area. Certainly, many early experiments with guanylate cyclase and cyclic GMP conducted in this laboratory and others were done based on previous experience and biases with the cyclic AMP system . . ., the properties of the soluble and particulate activities in tissues are quite different. . . . Minor differences have also been noted between particulate activities and in plasma and endoplasmic reticulum preparations. In spite of these significant physical and/or kinetic differences that suggest two or more forms of guanylate cyclase may exist in tissues, *it is not yet known whether or not the proteins or their subunits are different. The apparent differences may be attributable to assay conditions, other constituents in crude preparations, or possibly other factors* [58]. (Emphasis added).

The other group concludes:

> Considerable effort has been put forth to demonstrate an activation of the enzyme by hormonal or other physiological agents that promote the accumulation of cGMP in intact cells. To date, no specific direct effect of such an agent on guanylate cyclase activity *in vitro* has been convincingly demonstrated. This is a characteristic of the cGMP system that contrasts sharply with its cAMP counterpart; adenylate cyclase from disrupted cells as a general rule is activable by the agents that stimulate cellular cAMP accumulation. . . . *It is becoming evident that the greatest dissimilarity between guanylate cyclase and adenylate cyclase may be that the activity of the former is modulated indirectly with respect to the cellular action of hormones (and other cell stimuli) compared with the direct effect that hormones seem to have on the activity of adenylate cyclase* [29]. (Emphasis added).

Our present perception of the guanylate cyclase transduction field is that only the soluble guanylate cyclase responds to NO (reviewed by Russwurm and Koessling; Bellamy and Garthwaite, this volume). This response results from NO binding to the cyclase heme moiety. The NO is generated by NO synthetase from L-arginine. NO synthetase is regulated by Ca^{2+}/calmodulin. Thus, it is possible that this mechanism was observed earlier in the systems where intracellular rises in Ca^{2+} promoted cyclic GMP to increase (reviewed in ref. [59]). Earlier reports bearing upon the nitric oxide-dependent characteristics of the membrane guanylate cyclase/s in many mammalian tissues [50, 52, 60, 61], and in one instance for the completely purified form from the rod outer segments [62] are not compatible with the most recently established characteristics of the known members of the membrane guanylate cyclase family. None of these cyclases is NO-dependent (*vide infra*).

A few laboratories supported the second messenger role of cyclic GMP in hormone action

In direct contrast to the above dominant concept, which lasted almost to the mid 1980's and negated the role of cyclic GMP in hormonal signal transduction, a few laboratories, including this investigator's, supported the mediatory role of cyclic GMP in hormonal signal transduction (reviewed in refs [63–66]). Because the findings of these studies played an important role in refuting the 'Yin Yang' hypothesis and also in overcoming the prevailing bias against the cyclic GMP 'second messenger' concept, they are briefly discussed below.

In the formulation of the original 'second messenger' concept, Sutherland's group used the model system of liver for their investigations. The hormone under study was epinephrine and the process studied was glycogenolysis [2, 67]. In almost the same time frame, Haynes' group showed the involvement of cyclic AMP in the adrenocorticotropin (ACTH)-induced process of steroidogenesis [68, 69]. This group used the model system of adrenal cortex. The marker steroid was corticosterone in the rat or cortisol in the human, produced by the fasciculata cells of adrenal cortex (the reader is referred to several excellent early reviews on this topic [70–74]). Subsequently, studies with the two model systems of liver and adrenal cortex complemented each other in advancing the second messenger role of cyclic AMP in hormone action. To qualify as a second messenger, Sutherland's group laid down the following set of rigorous guidelines [75]. (1) The hormones should be capable of stimulating adenylyl cyclase in broken cell preparations from the appropriate cells, while hormones which do not produce the response should not stimulate adenylyl cyclase. (2) The hormone should be capable of increasing the intracellular level of cyclic AMP in intact cells, while inactive hormones should not increase cyclic AMP levels. It should be demonstrated that the effect on the level of cyclic AMP occurs at dose levels of the hormone, which are at least as small as the smallest levels, which are capable of producing a physiological response. The increase in the level of cyclic AMP should precede or at least not follow the physiological response. (3) It should be possible to potentiate the hormone (i.e. increase the magnitude of the physiological response) by administering the hormone together with theophylline or other phosphodiesterase inhibitors. The hormone and the phosphodiesterase inhibitor should act synergistically. (4) It should be possible to mimic the physiological effect of the hormone by the addition of exogenous cyclic AMP.

With the subsequent discovery of cyclic AMP-dependent protein kinase and the demonstration that cyclic AMP biological activity is propagated through this protein kinase, an added criteria was revealed: (5) the hormone-dependent rise of cyclic AMP is followed by activation of the cyclic AMP-dependent protein kinase. On a qualitative level, most of 'Sutherland's guidelines' for the second messenger role of cyclic AMP in the ACTH-induced process of steroidogenesis were applicable [36, 43, 44, 76–81, reviewed in 82]. There were, however, two critical deficiencies. One, the physiological concentrations of ACTH, 5 μU (7×10^{-12} M) to 10 μU (1.4×10^{-11} M), stimulated steroidogenesis but caused no elevation in the cyclic AMP level [83–85]. And two, these physiological concentrations of ACTH did not stimulate cyclic AMP-dependent protein kinase activity either [83–85]. Several laboratories, including this investigator's, attempted to address these issues. The model systems of isolated rat adrenal fasciculata cells and the rat adrenocortical carcinoma were developed and used for the studies of this investigator's laboratory [43, 86]. Both these cell types had no phosphodiesterase activity, and the fasciculata cells were highly sensitive to the physiological levels of ACTH, responding to as little as 1 μU of ACTH in the synthesis of corticosterone [43]. The physiological concentrations of the hormone did not raise the level of cyclic AMP in the fasciculata cells although they stimulated steroidogenesis [83]. A significant increase in the cyclic AMP level occurred only at the supra-physiological (10 μU and higher) concentrations of ACTH [83]. An added important observation was that beyond the maximal steroidogenic concentrations of ACTH there was a rise in cyclic AMP levels and yet no further elevation of steroidogenesis [83, 84]. These findings clearly indicated that cyclic AMP was not the physiological second messenger of ACTH; it was a second messenger only for the supra-physiological and pharmacological concentrations of the hormone.

In contrast to cyclic AMP, except for 'Sutherland's guideline 1', cyclic GMP met all the remaining 'Sutherland's criteria' for being a second messenger of ACTH: (a) under physiological concentrations of the hormone, an excellent temporal correlation existed between cyclic GMP formation [83, 87, 88], phosphorylation, and corticosterone synthesis [84, 85, 89]; (b) exogenous cyclic GMP stimulated the protein kinase activity followed by the steroidogenic activity [84]; and (c) exogenously supplied cyclic GMP mimicked the exogenously supplied ACTH in stimulating the direct transformation of cholesterol to corticosterone [74]. Consistent with the earlier findings [42, 90, 91], these studies found that Ca^{2+} was an obligatory factor in the process of steroidogenesis, but it was unable to promote steroidogenesis by itself [92]. The conclusions of these studies were that cyclic GMP and Ca^{2+} were the two complementary physiological second messengers of ACTH. Cyclic AMP was ACTH second messenger only at its supra-physiological and pharmacological concentrations. (The perceived role of Ca^{2+} in 1970's as a potential messenger in hormonal action has been presented in an excellent monograph by Howard Rasmussen, a leading investigator in the field of Ca^{2+} signaling [93]).

Because the isolated fasciculata cells used in these studies contained no phosphodiesterase activity [94], the above findings also indicated that the ACTH-dependent cyclic GMP formation reflected the presence of the ACTH-dependent membrane guanylate cyclase in the fasciculata cells of the adrenal cortex. Also, because at that time neither the presence of clear-cut particulate hormone-dependent guanylate cyclase in any tissue, nor the presence of cyclic GMP-dependent protein kinase in the adrenocortical cells was known, it was obvious that, despite a strong indirect evidence favoring mediation of the ACTH-induced steroidogenesis by cyclic GMP, the proposition [83, 92] would remain speculative until direct evidences for the existence of a specific ACTH-dependent particulate guanylate cyclase and cyclic GMP-dependent protein kinase were demonstrated in the rat adrenal cortex. In addition, the proposition was challenged by the results of two studies. One study found that dibutyryl cyclic GMP possessed far less steroidogenic activity than its parent compound, cyclic GMP [95]. Because the earlier studies had shown that dibutyryl cyclic AMP is threefold more potent than cyclic AMP in its steroidogenic activity [43, 96], it was argued that dibutyryl cyclic GMP should have also been more potent [95]. But because the opposite was true, it was concluded that cyclic GMP is not the mediator of steroidogenesis. The criticism resulting from the second study was based on the observation that in preincubated adrenocortical cells, sodium nitroprusside and ascorbic acid caused an increment of cyclic GMP levels without stimulating steroidogenesis, indicating dissociation between cyclic GMP formation and the production of corticosterone [97]. Both of these criticisms were satisfactorily answered by the subsequent investigations which showed that dibutyryl cyclic GMP does not stimulate [98], and nitroprusside and ascorbate directly inhibit cyclic GMP-dependent protein kinase [63, 64]. This indicates that the lack of steroidogenic activity of both these agents can be explained at the cyclic GMP-dependent protein kinase level. Furthermore, the direct analysis of the membrane guanylate cyclase showed that most (80%) of its presence in the adrenal cortex is in the membrane-bound form and only minor (20%) in the soluble [99]. Notably, the membrane form is not stimulated by sodium nitroprusside and ascorbic acid [63, 64, 99–102]. However, these agents stimulate the soluble guanylate cyclase form [99]. Thus, in the study of Laychock and Hardman [97] sodium nitroprusside and ascorbic acid must have stimulated a soluble form of guanylate cyclase, which is minor in the adrenal gland and is not linked with the process of steroidogenesis.

Provided with the clues that the membrane guanylate cyclase was hormone-dependent, that it did not respond to the nonspecific signals of nitric oxide- and other free radical-generating compounds, and that cyclic GMP-dependent protein kinase activity was present in the adrenal cortex, the studies of this investigator's group were channeled towards the direct characterization of these two components of the cyclic GMP pathway in the adrenal gland. The cyclic GMP-dependent protein kinase was purified to homogeneity and characterized from this tissue [98] and its specificity for the cyclic GMP analogues was assessed [98]. The ordered potency of these analogues was: 8-bromo cyclic GMP > cyclic GMP > N-butyryl cyclic GMP > N,O′-dibutyryl cyclic GMP and dibutyryl cyclic GMP. It was noted that dibutyryl cyclic GMP was 10,000-fold less active while N-butyryl cyclic GMP was 100-fold less active than cyclic GMP in activation of the cyclic GMP-dependent protein kinase. These findings provided a biochemical reasoning behind dibutyryl cyclic GMP possessing less steroidogenic activity than cyclic GMP, and the study also linked the cyclic GMP-dependent protein kinase with the ACTH-dependent process of steroidogenesis [98].

Detailed analysis of the plasma membrane fraction derived from the adrenal cortex showed that this fraction contained an ACTH-specific membrane guanylate cyclase [99–101]. Many biochemical features of this cyclase were distinct from its counter part, soluble guanylate cyclase (Table 1 reproduced from ref. [99]). Most importantly, the membrane guanylate cyclase was not stimulated by the nitric oxide-and the free-radical generating agents. In contrast, the soluble form responded to the nitric oxide-generating and several other agents [102]. An intriguing feature of the membrane guanylate cyclase was that it responded only to the sub-physiological to physiological concentrations of the hormone [100, 101]. The response-profile of the cell-free cyclase was similar to the intact cell-cyclase [83]. Ca^{2+} was obligatory for the hormone-dependent cyclase activation [92]. The low concentrations of the hormone did not stimulate adenylate cyclase. With almost identical features, in parallel studies, the presence of ACTH-dependent membrane guanylate cyclase was also demonstrated in the rat adrenocortical carcinoma 494 cells [101]. Subsequent studies showed that particulate guanylate cyclase activity could indeed be stimulated by ACTH in cultured neurons [103] and by ANF in rat tissues [60, 104]. Thus, for the first time, the existence of a hormonally-dependent membrane guanylate cyclase in several different cell types, that was clearly functionally and kinetically distinct from the soluble form was established. The stage was now set to purify the cyclase to homogeneity and study its functional and physical characteristics.

First purified and functionally characterized vertebrate membrane guanylate cyclase was the atrial natriuretic factor receptor guanylate cyclase [105–107]

Central to the establishment of the concept favoring the second messenger role of cyclic GMP in receptor-mediated signaling was the complete purification of a hormonally dependent membrane guanylate cyclase. A plasma membrane guanylate cyclase was purified from the rat adrenocortical carcinoma and its absolute purity was established by the following criteria: (1) Protein staining: sodium dodecyl sulfate-polyacrylamide gel electrophoresis of the protein showed a single Coomassie blue- and silver-stained band; (2) Isoelectric focusing: native protein and its iodinated form yielded a symmetrical protein peak, which superimposed over the guanylate cyclase activity peak; (3) Western blot analysis: a monospecific polyclonal antibody was raised against the native protein. The high specificity of the antibody was established based on Western analyses of membrane fractions

Table 1. Summary of the properties of particulate and soluble guanylate cyclase of rat adrenocortical tissue [99]

Conditions	Particulate	Soluble
Abundance	80%	20%
ACTH	Stimulation	No effect
Sodium nitroprusside	No effect	Stimulation
Sodium azide	No effect	Stimulation
Tufstin	No effect	Stimulation
Cd^{2+}	Inhibition at high concentrations (EC_{50} ~ 400 μM)	Inhibition at low concentrations (EC_{50} ~ 2 μM)
Dithiothreitol	No effect	Strong stimulation
N-Ethylamide	Inhibition	Stimulation at low and inhibition at high concentrations

where it yielded a single, immunoreactive band, which co-migrated with that of the purified protein band [106]. This highly specific antibody was used to probe the features of the guanylate cyclase. The antibody blocked almost all of the guanylate cyclase activity of the homogeneous protein [106]. Thus, the protein was pure and it was a guanylate cyclase. Further analysis of the homogeneous protein showed that it bound ANF in a stoichiometric ratio, indicating that it was an ANF receptor. In this manner the discovery of a plasma membrane guanylate cyclase that was a hormone receptor for ANF was made [106]. The guanylate cyclase was subsequently named atrial natriuretic factor receptor guanylate cyclase, ANF-RGC.

ANF-RGC activity was independent of the nitric oxide-generating agents – catalase, hemin, dithiothreiotol, arachidonic acid, and tufstin [102] – which had been reported to stimulate the soluble form of the guanylate cyclase. The molecular mass of ANF-RGC was almost twice that of the soluble form [106] and the ANF-RGC antibody did not cross-react with the soluble guanylate cyclase. These evidences demonstrated that the plasma membrane-bound form of the guanylate cyclase was structurally and functionally distinct from the soluble form.

By several accounts the discovery of ANF-RGC is historic. (1) It settled the ongoing debate, which questioned the independent existence of a peptide hormone-dependent guanylate cyclase transduction system in mammalian systems. (2) It established cyclic GMP as a bonafide hormonal second messenger. (3) It showed that guanylate cyclase exists in two distinct forms, the soluble and the membrane-bound. The soluble form is a transducer of the nitric oxide signals, which are generated inside the cell; the membrane-bound form is a transducer of the peptide hormone signals, which arise outside the cell. (4) The most important portion of the discovery was that the membrane guanylate cyclase form represented a hormone receptor family. The original report disclosing the discovery noted, 'coexistence of the ANF receptor and guanylate cyclase activities on a single polypeptide chain indicates that the mechanism of trans-membrane signal transduction involving mediation by second messenger, cyclic GMP, is different from the well-established adenylate cyclase system. In hormone-dependent adenylate cyclase there is an assemblage of individual components – receptor, GTP binding protein, and catalytic moiety – for signal transduction. In contrast, the presence of dual activities – receptor binding and enzymic – on a single polypeptide chain indicates that this transmembrane protein contains both the information for signal recognition and its translation into a second messenger' [106]. Thus, the discovery defined a new paradigm of cellular signaling in which cyclic GMP was a second messenger of the peptide hormone.

The two attributes of ANF-RGC, being a surface receptor as well as a guanylate cyclase, were pictorially shown in a model [63, 64]. The model depicts the cyclase as a single transmembrane protein. The ANF-receptor domain resides on the outside plasma membrane portion of the protein and the catalytic domain protrudes into the cytosol. ANF binds to the receptor domain and activates the catalytic domain (Fig. 1).

From a historical perspective, the tentative conclusion that ANF binding and guanylate cyclase activity may reside on a single protein was also independently reported in another study where the cyclase was purified from the rat lung. The conclusion was not decisive, however. The lung cyclase was only about ~ 95% pure [108] and the possibility existed that the 5% contaminant contained the separate ligand binding or the cyclase activity. There were two other serious problems with the conclusion. (1) In contrast to the 1:1 stoichiometry, the lung enzyme bound only 14.5% of ANF at the noted theoretical value; (2) the enzyme was stimulated by hemin [108] although in actuality it is not. (For a commentary on this issue, see ref. [107]).

Soon thereafter, two other laboratories reported on the purification of ANF-RGC from the adrenal cortex, consolidating the concept that this membrane guanylate cyclase is, indeed, both a hormone receptor and a catalyst [109, 110].

Using the monospecific polyclonal antibody, the presence of ANF-RGC in the rat adrenal gland, the rat and mouse testes [111], in the rat neuronal cells located in the ventral horn region of the spinal cord, and in the cerebellum Purkinje cell somata and Purkinje cell dendrites was demonstrated [63]. With the same antibody probe, ANF-RGC was purified to homogeneity and its functional linkage with ANF was established in the rat and mouse testes [111], and in the renal glomerulus cells [112]. These studies showed ubiquity of ANF-RGC in the mammalian tissues, and the testes studies suggested its linkage with the process of steroidogenesis.

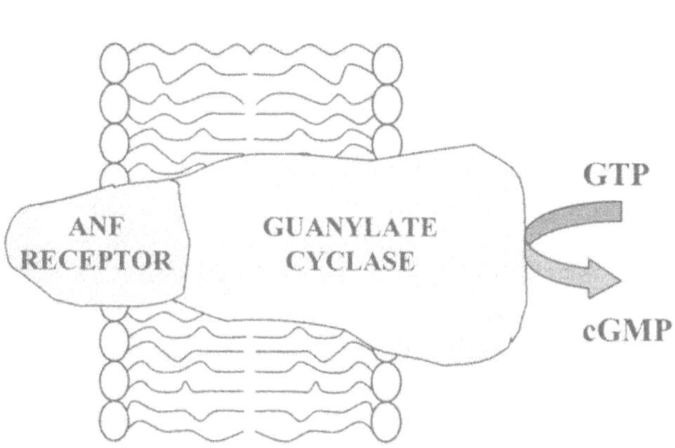

Fig. 1. A hypothetical topographical model of an ANF-dependent membrane guanylate cyclase. This model is a redrawn version of the ANF-dependent membrane guanylate cyclase model proposed in 1988 [63, 64]. The membrane guanylate cyclase was envisioned to be a transmembrane protein whose ANF receptor domain protrudes outside the plasma membrane and the catalytic domain inside the cell.

ANF was discovered by de Bold's group [113]. It regulates sodium excretion, water balance and blood pressure [114–117]. Discovery of ANF-RGC provided the identity of the ANF receptor through which ANF works. The presence of ANF-RGC in the glandular tissues of the adrenal gland and testis raised the possibility of the involvement of ANF-generated cyclic GMP in the metabolic processes of steroidogenesis in these endocrine tissues. In accord with this thought process, several studies showed that ANF raises the levels of cyclic GMP and down-regulates the aldosterone formation in the glomerulosa cells of the adrenal gland [118–121] and the progesterone production in the murine tumor cells [122]. In contrast, several other laboratories showed that ANF stimulates the process of steroidogenesis with respect to corticosterone and testosterone in rat: in the adrenal gland [123], its fasciculata cells [124], in the interstitial [125] and Leydig cells [126, 127] of testis. This suggested that the ANF-RGC-generated cyclic GMP was the second messenger in both the inhibition and the stimulation of steroidogenesis. Taken together: (1) There is an ANF-dependent stimulation of the cyclic GMP formation in the particulate fractions of these tissues; (2) ANF-RGC is present in these particulate fractions; (3) 8-bromo cyclic GMP stimulates steroidogenesis in testis and in the isolated fasciculata cells of the rat adrenal cortex; (4) in the latter cells, cyclic GMP stimulates cyclic GMP-dependent protein kinase and corticosterone production [84]; (5) cyclic GMP directly stimulates the transformation of cholesterol to corticosterone [74]. These ANF-dependent steroidogenic features, according to the 'Sutherland's guidelines' make cyclic GMP the second messenger of ANF in the process of steroidogenesis, a proposal put forth about three decades ago based on the studies with the fasciculata cells of the adrenal cortex [83].

Molecular cloning of ANF-RGC and two other peptide receptor guanylate cyclases established a surface receptor subfamily of membrane guanylate cyclases

Almost 3 years after the discovery of ANF-RGC, ANF-RGC and a very closely structurally related membrane guanylate cyclase, type C natriuretic peptide receptor guanylate cyclase (CNP-RGC), were cloned first in a joint effort [128, 129], and then reported separately by the two laboratories [130, 131]. Thereafter, cloning of the mouse analogue of ANF-RGC was reported [132]. Direct purification of the cloned form by the ANF-RGC antibody raised against the original native enzyme showed that the cloned ANF-RGC represented the native ANF-RGC [133]. Subsequently, another mRNA of a functionally defined form of mammalian guanylate cyclase (Sta-RGC) was cloned from the intestinal mucosa [134–137]. This cyclase is a receptor for the bacterial enterotoxin and for the endogenous peptide hormones: guanylin and uroguanylin [138–141, Vaanderager, this volume].

Molecular cloning of the rod outer segment membrane guanylate cyclase (ROS-GC) established a guanylate cyclase subfamily that was distinct from the surface receptor subfamily and transduced the Ca^{2+} signals arising within the sensory neurons

ROS-GC is a pivotal enzyme for operation of the phototransduction machinery (reviewed in ref. [66], Koch *et al.*, this volume). Prior to 1993, the molecular identity of the wild type ROS-GC was not known, and there was significant confusion on its biochemical properties. Four independent laboratories reported purification of the enzyme. The biochemical characteristics of the enzyme reported by three groups were quite similar [142–144]: The molecular mass of the enzyme was 110–120 kDa and the cyclase activity was unresponsive to ANF and Ca^{2+}. The fourth group reported the membrane enzyme to be nitric oxide-sensitive, and its molecular mass was 67 kDa [62, 145]. Based on the sequence of four fragments of the wild-type ROS-GC [144], purified from ROS where the process of phototransduction takes place, its cDNA was cloned from the bovine retina [146]. The cloned enzyme was unresponsive to ANF and CNP and showed only minimal, 27–30%, sequence identity with the three earlier known peptide hormone receptor guanylate cyclases – ANF-RGC, CNP-RGC and Sta-RGC [146]. Reconstitution studies showed that ROS-GC was Ca^{2+} modulated with a pattern that mimicked the native conditions of phototransduction (Koch *et al.*, this volume). This established the identity of wild-type ROS-GC, which was a critical component of the molecular machinery of phototransduction. The discovery that this cyclase was modulated by intracellular Ca^{2+} spikes represented another major advancement in the field. It revealed the existence of a second membrane guanylate cyclase subfamily that was distinct from the first subfamily. The new subfamily was meant to transduce the Ca^{2+} signals arising inside the sensory neurons of the retina instead of the transduction of the peptide signals arising outside the cells.

The disclosure of the proper identity of ROS-GC had an important historical impact. In 1992 Shyjan *et al.* reported the molecular cloning of a membrane guanylate cyclase from the human retina [147] and named it retGC. Because retGC via in situ hybridization analysis was detected in the inner segments and outer nuclear layers of the monkey's retina and the cyclase was different from ANF-RGC and CNP-RGC, the authors suggested that it may be a part of the phototransduction machinery [147]. Cloning and identification of ROS-GC proved that retGC was not ROS-GC [144, 146]. There were very significant structural differences between them. In 1995, the structure of retGC was revised (Lowe DG, accession number M92432) to show its identity with the structure of ROS-GC, which was published in the early part of 1994 [146]. Thus, ROS-GC structure helped to establish retGC as the human counter part of bovine ROS-GC.

The present evidence indicates that ROS-GC exists in two forms: the original form, which has been renamed ROS-GC1 and the other, named ROS-GC2 [148, 149]. The second form also appears to be a component of the phototransduction machinery. The relative importance of the two forms in the operation of the rod or cone outer segment phototransduction process is not known at this time (Koch *et al.*, this volume). Independent studies from three groups have revealed the molecular identity of the almost entire [150] and partial ROS-GC1 gene [151, 152]. This knowledge, together with the discovery of mutations in ROS-GC1 linked with retinal disorders has made it possible to establish the critical role of ROS-GC1 in phototransduction. Furthermore, certain phototransduction-related retinal disorders have been explained in precise biochemical terms [153–158; Duda and Koch, this volume]. To date ROS-GC2 has not been linked with any retinal disorder.

A third plasma membrane guanylate cyclase, GC-D, has been cloned from the olfactory sensory neurons of the rat [159]. GC-D has a closer structural identity with the two forms of ROS-GC than the peptide hormone receptor guanylate cyclases [159]. This, plus the fact that its residence is in the sensory neurons, makes it a potential member of the ROS-GC subfamily. However, it is not known if the Ca^{2+} spikes regulate this cyclase.

Ca^{2+}-sensitive modulation of ROS-GC is mediated by Ca^{2+}-binding proteins

Ca^{2+} signaling of ROS-GC is through Ca^{2+} binding proteins, named GCAPs. Long before the molecular identity of the first Ca^{2+}-modulated membrane guanylate cyclase, ROS-GC1, was known, the studies of Koch and Stryer provided the first evidence that GCAPs stimulate a membrane guanylate cyclase in ROS in a Ca^{2+}-dependent fashion [160]. Subsequently, two Ca^{2+}-binding proteins were purified and cloned from the retina [161–163]. They have been termed GCAP1 and GCAP2, and have been linked with phototransduction (Koch *et al.*, this volume). Subsequently, the collaborative studies of Sitaramayya's and Sharma's groups discovered and cloned a protein from the retina, which, in contrast to the conventional inhibition of ROS-GC, stimulated it. It was named Ca^{2+}-dependent guanylate cyclase activating protein, CD-GCAP [164, 165; Duda and Koch, this volume]. These studies described a new paradigm of Ca^{2+} signaling where ROS-GC is both stimulated and inhibited by Ca^{2+} spikes generated in the sensory neurons. The present knowledge of the field indicates that ROS-GC is a bimodal Ca^{2+} switch. It senses Ca^{2+} through GCAPs or CD-GCAPs. Ca^{2+} spikes inhibit GCAPs-modulated ROS-GC activity and they stimulate CD-GCAPs-modulated. There are three known GCAPs – GCAP1, GCAP2 and GCAP3 – and two CD-GCAPs: CD-GCAP and neurocalcin [161–167]. CD-GCAP was subsequently identified as S100β [165]. All these proteins are present in the retina, however, more is known about the function of GCAPs. To date evidence shows that GCAP1 and GCAP2 are involved in phototransduction (Koch *et al.*, this volume). There is, however, no evidence on the involvement of GCAP3 in phototransduction [166]. The present phototransduction model shows that ROS-GC in its native state is bound to GCAP/s (Koch *et al.*, this volume). Fluctuation levels of Ca^{2+} change the conformation of GCAP/s, which, in turn, regulate the cyclase through specified modules in ROS-GC.

Retinal neurons are not the sole domain of Ca^{2+}-modulated ROS-GC signal transduction system

With the discovery of Ca^{2+}-modulated ROS-GC signal transduction system, belief arose that this signaling system is unique to the rods and cones outer segments and it is present there as an exclusive phototransduction machinery component. Initial immunocytochemical studies suggested that this belief might not be correct. The antigenic staining of ROS-GC, besides in ROS, was observed in the inner segments and the retinal synaptic region [168, 169], which are not the sites of phototransduction. Recent studies with the pineal gland and the olfactory bulb neurons indicate that ROS-GC transduction system is present in these neuronal tissues [170, 171]. There are, however, variations on its appearance, which might define its cell specificity. In the rod and cone outer segments, ROS-GC exists in two forms, ROS-GC1 and ROS-GC2. It is accompanied by its two Ca^{2+} modulators: GCAP1 and GCAP2 [161–163]. As yet, there is no firm evidence that rod or cone outer segments contain either S100β or neurocalcin. Biochemical studies have shown that these two agents, upon binding Ca^{2+}, stimulate ROS-GC activity, which is a reversal of the scenario in phototransduction. In phototransduction, Ca^{2+} spikes through GCAPs inhibit ROS-GC; in contrast, S100β and neurocalcin stimulate ROS-GC. The physiological consequences of reversing the cyclase operation would mean, that, in contrast to phototransduction, Ca^{2+} spikes would accelerate the production of cyclic GMP when ROS-GC is co-present with S100β or neurocalcin. On the assumption that the physiological principles of phototransduction also apply to the ROS-GC stimulatory machinery, Ca^{2+} signals transmitted through S100β, or neurocalcin, to ROS-GC will depolarize the neuronal membranes. With this background the findings in the pineal gland and the olfactory neurons are briefly outlined below.

In the pineal gland, the pinealocytes, and not the glial cells, contain ROS-GC [170]. ROS-GC is present in all the pinealocytes and exists there as ROS-GC1 and not ROS-GC2. In majority of the pinealocytes, ROS-GC1 co-exists with

GCAP1; and, in minority, it co-exists with S100β. GCAP2 is missing in the pinealocytes. These results demonstrate that: (1) Ca^{2+} sensitive ROS-GC signaling system is present in the pineal gland; (2) it is specific for the pinealocytes; and (3) ROS-GC1, and not ROS-GC2, is the transducer of Ca^{2+} signals. Thus, photoreceptor cells are not the exclusive domains of the ROS-GC transduction system.

The selective presence of GCAP1 or S100β with ROS-GC1 in the pinealocytes suggests that there are two functional types of pinealocytes. The ones containing GCAP1 act like in phototransduction, i.e. Ca^{2+} spikes inhibit ROS-GC, causing reduction of cyclic GMP, closure of the hypothetical cyclic GMP-gated channels and resulting in hyperpolarization of the plasma membrane. Those containing S100β function in a reverse fashion. There, Ca^{2+} spikes stimulate ROS-GC, cause a rise in the level of cyclic GMP, open the hypothetical cyclic GMP-gated channels, and result in depolarization of the plasma membrane.

The pineal gland is light regulated. It is possible that it exhibits its light-dependence through above ROS-GC transduction modes.

Studies with the olfactory bulb neurons showed that a functional, negatively Ca^{2+}-modulated ROS-GC transduction system exists in these neurons [171]. This system mimics the phototransduction system. However, in contrast, in the olfactory bulb neurons ROS-GC exists in only one form, that of ROS-GC1, and its modulator also exists in one form, that of GCAP1. *In situ* and immunohistochemical studies show that ROS-GC1 and GCAP1 co-localize in mitral cell neurons of the bulb [171]. Thus, ROS-GC1 and its Ca^{2+}-dependent modulator, GCAP1, are synthesized and expressed in the mitral neurons; therein, they are the sole components of the Ca^{2+}-dependent ROS-GC. Thus, Ca^{2+} via GCAP1 solely modulates the mitral cell ROS-GC1 transduction system.

These illustrations, with three different neuronal systems, clearly indicate that Ca^{2+}-dependent ROS-GC transduction system is not restricted to photic modulation, and has the makings of a unified theme by which Ca^{2+} signaling controls the sensory and processing activities of neurons.

ROS-GC transduction system internets with a G-protein coupled seven transmembrane receptor signal

A new finding that a signal generated through $\alpha_{2D/A}$-AR activates ROS-GC signifies that the seven transmembrane receptor signal is internetted with the ROS-GC transduction system [172]. It took more than 25 years of investigative work to establish the nature of the biochemical linkage between the two systems. Its history is briefly outlined below.

In 1980, this investigator's laboratory made two observations [173, 174]: (1) The rat adrenocortical carcinoma 494 cell contains an epinephrine-dependent membrane guanylate cyclase transduction system, which is not present in the normal adrenal cortex; (2) epinephrine signals the cyclase activation via an α-adrenergic receptor [173–175]. This meant that the transformation of a normal to a neoplastic cell accompanies a parallel ectopic development of the epinephrine-dependent guanylate cyclase signal transduction system [86]. At that time, the heterogeneity of α-AR and guanylate cyclase was not defined. The biochemical linkage between the signaling receptor and the transduction component remained baffling. Subsequently, the receptor was purified, characterized [176–178], cloned [179, 180], and shown to belong to a new pharmacologically defined α_{2D}-AR subtype [180–182]. It has now been established that the α_{2D}-AR receptor subtype is a species variant of the pharmacologically distinct α_{2A}-AR [180, 182–184] and therefore, is referred to as $\alpha_{2D/A}$-AR [185]. Thus, the original carcinoma receptor subtype linked to the membrane guanylate cyclase was $\alpha_{2D/A}$-AR. Identity of the membrane guanylate cyclase remained elusive, however. Mystery surrounding its identity was disclosed after the discovery of ROS-GC1. Like ROS-GC1, the carcinoma membrane guanylate cyclase was Ca^{2+}-dependent and not a member of the surface receptor subfamily. Investigations with the pineal gland showed that the gland contains an epinephrine-dependent membrane guanylate cyclase transduction system with complete identity to the carcinoma cell transduction system [172]. Further analysis of the gland showed that this transduction system is exclusively present in the pinealocytes, where Ca^{2+} drives the transduction component, ROS-GC, via its modulator S100β. These studies finally resulted in the discovery of a membrane guanylate cyclase transduction system that is internetted with seven transmembrane receptor signaling system. Presence of this system may be confined to the pinealocytes and other sensory neurons.

All membrane guanylate cyclases exhibit a common topography and a composition of modular blocks

Molecular cloning studies indicate that all membrane guanylate cyclases have almost similar topography with varying indented domains. They all are single membrane-spanning proteins. A helical domain – transmembrane (TM) – divides the protein into roughly two equal regions: N-terminal, the extracellular; and C-terminal, the intracellular. In the peptide hormone receptor subfamily, the receptor domain resides in the extracellular region. To date no receptor domain in this region in the ROS-GC subfamily has been found. The intracellular region of all membrane guanylate cyclases is composed of many modular blocks [186]. The catalytic module resides at its C-terminus. Other modular domains are configured to exhibit a tight control over the catalytic domain. In general, the placement order of these domains is: juxtamembrane (JMD), a kinase homology (KHD), a dimerization

(DD) and a catalytic. In the ROS-GC subfamily there is an extra C-terminal extended domain (CTE). A pivotal ATP-regulated domain (ARM) is present in ANF-RGC and CNP-RGC, its beginning portion overlaps with JMD and overextends into KHD. ARM domain is missing in the ROS-GC subfamily.

Modular domains of the membrane guanylate cyclases define their signal transduction course

The two subfamilies of membrane guanylate cyclases respond to two different types of signals: the natriuretic factor receptor guanylate cyclases (NF-RGCs) respond to an extracellular hormone signal, while the ROS-GCs are modulated by fluctuations in $[Ca^{2+}]_i$. In the case of NF-RGCs, in addition to the hormone, binding of ATP is essential for the manifestation of hormonal response. Hence, two events – binding of the hormone and ATP – determine the NF-RGC signal transduction. In the case of ROS-GC, which is a bimodal Ca^{2+} switch, multiple proteins mediate regulation by Ca^{2+}, both at depleted and at elevated $[Ca^{2+}]_i$. Therefore, both in NF-RGCs and in ROS-GC, signal transduction necessarily involves the interaction of multiple modular domains. A major advance in the field of membrane guanylate cyclase signaling over the last two decades is the identification of some of these modular domains and the elucidation of their individual roles in signal transduction.

NF-RGC signal initiation occurs upon binding of hormone to its hormone-binding module

The first study showing that the site of signal initiation in ANF-RGC is at its extracellular domain was that of Duda *et al.* [187]. In that study, the first use of the point mutation technology in the functional analysis of a membrane guanylate cyclase was made. A variant form of ANF-RGC, GCα, was cloned [187]. It had two changes in its residues, Gln^{338}His and Leu^{364}Pro. The cyclase did not bind and was not stimulated by ANF. Restoration of the residues to their original forms restored the mutant's ANF binding activity and its dependence on ANF [187]. These results indicated that the ligand binding domain resided in the extracellular region of ANF-RGC and, one, or both of the residues – Gln^{338}His, Leu^{364}Pro – were critical in the ligand binding activity. Analysis of the two residues showed that only Leu^{364} is critical in ANF binding activity [187]. In a similar approach it was shown that the Glu^{332} residue located in the extracellular region of CNP-RGC is critical in the CNP binding to CNP-RGC [188]. These studies were then extended to show that the folding patterns of the binding domains of ANF-RGC and CNP-RGC are identical, or very similar [189]. The counter part of CNP-RGC-Glu^{332} residue is the Gln^{338} residue of ANF-RGC, and that of ANF-RGC-Leu^{364} residue in CNP-RGC is Val^{358}. Conversion of the ANF-RGC residue Gln^{338} to Glu resulted in a change from none to significant CNP signal transduction activity. Similarly, a conversion of Val^{358} to Leu generated significant ANF signal transduction activity in CNP-RGC [189]. These acquired signal transduction activities in the cyclases were in addition to their natural signal transduction activities. These studies revealed the identity of the residues that are critical in defining the ligand binding domains of ANF-RGC and CNP-RGC, showed that these domains reside in the predicted extracellular domains of the cyclases, and demonstrated that the folding patterns of these domains in ANF-RGC and CNP-RGC are very similar.

The above conclusion of the ligand-binding domain being in the extracellular domain has now been supported by a very recent crystallization study [190; Misono, this volume]. Resolution of the extracellular region structure of ANF-RGC shows the ligand-binding domain in this region of the cyclase and demonstrates that Leu^{364} is not a direct ligand-binding, but a critical hydrophobic site essential for the ANF binding activity of the cyclase.

Transduction machinery of NF-RGCs is ATP-regulated

Early studies showed that ATP accelerates ANF-dependent ANF-RGC activity [191–193]. Subsequently, two groups independently demonstrated that ATP is obligatory for ANF-dependent ANF-RGC transduction activity [194, 195]. These studies showed that neither ANF, nor ATP alone, is able to stimulate ANF-RGC. Only when the two are together, are they able to stimulate the cyclase. Because the nonhydrolyzable analogs, ATPγS and AMP-PNP, mimicked ATP, it was suggested that ATP acts directly by allosteric regulation of the cyclase [194, 195]. A study with recombinant CNP-RGC showed that ATP is also obligatory for the transduction activity of CNP-RGC [196]. It was thereby established that the transduction activities of both ANF-RGC and CNP-RGC are regulated by ATP.

The first study to demonstrate that the ATP-regulated domain resides in the intracellular region of the cyclase (KHD), between the juxtamembrane and the predicted C-terminus catalytic domain, was that of Chinkers and Garbers [193]. The study showed that the cyclase activity of the KHD^- mutant of ANF-RGC was independent of ANF/ATP, yet bound ANF, indicating that KHD represented the ATP-regulated site in ANF-RGC. With this study, the authors proposed a principal ANF signal transduction model, which to this date remains a standard mark in many laboratories [197, 198]. This model, however, does not pass rigorous experimental scrutiny [199]. It is briefly discussed below.

The authors showed that KHD^- mutant of ANF-RGC contains intrinsic saturated guanylate cyclase activity, is independent of ANF, and its stimulation by ATP is markedly reduced [193]. They formulated a model and stated 'A model for signal transduction is suggested in which binding of ANP to the extracellular domain of its receptor initiates a conformational change in the protein kinase-like domain, resulting in derepression of guanylate cyclase activity' [193]. The central idea behind the model is that KHD in native ANF-RGC suppresses the catalytic module activity; ANF functions by relieving this suppression. There are important inherent inconsistencies between the authors' data and its interpretation (Tables 1 and 2 in ref. [193]). The major inconsistency is in the expression results obtained with the ANF-RGC-KHD^- mutant and ANF-RGC. According to the model, these guanylate cyclase values should be comparable, i.e. the basal guanylate cyclase value of the KHD^- mutant should be equal to that of ANF-RGC with ANF in intact cells, and with ANF and ATP in the isolated membrane fractions. But they are not: KHD^- mutant activity is about 2-fold lower than ANF-RGC with ANF in the intact cells (Table 1 in ref. [193]). Surprisingly, the guanylate cyclase values are reversed when the activities are measured in the membrane fractions. There, KHD^- mutant guanylate cyclase activity is over 7-fold higher than ANF-RGC. Clearly, these values obtained with the intact cells and their membrane fractions are contradictory; and none of the values, by itself, shows that KHD^- mutant attains saturation in guanylate cyclase activity. Most significantly, there is no control to show the true saturation activity of ANF-RGC. It is, therefore, not possible to assess if KHD^- mutant is at its true saturation level. Therefore, the study's data does not support the proposed model [193].

Finally, the above model is not supported by a direct study, which investigated the role of KHD in ATP-dependent transduction step of ANF-RGC [200]. Partial or complete deletion of the kinase-homology domain does not cause any elevation in the basal activity of the guanylate cyclase. On the contrary, the deletions cause a 5–10 fold suppression in the basal activity [200]. Thus, ANF signaling does not involve overcoming the KHD suppression of the guanylate cyclase activity, a central theme of the model [193].

Through a detailed program of site-directed and deletion mutagenesis/expression studies, the transduction mechanism involving ATP regulation of NF-RGCs has been investigated (reviewed in ref. [201]). Direct binding experiments show that the wild-type ANF-RGC (Fig. 2A) and its recombinant form (Fig. 2B) bind ATP [133], supporting the earlier interpretation that ATP acts directly. ANF-RGC in its intracellular region contains a Glycine-rich cluster (Grc) sequence (Gly^{503}-X-Gly-X-X-X-Gly^{509}) [187, 200]. This sequence begins 40 aa after the transmembrane domain of the cyclase and is a modified form of the nucleotide-binding consensus sequence of protein kinases [202, 203]. This sequence was, therefore, targeted as a potential site to be critical in ATP

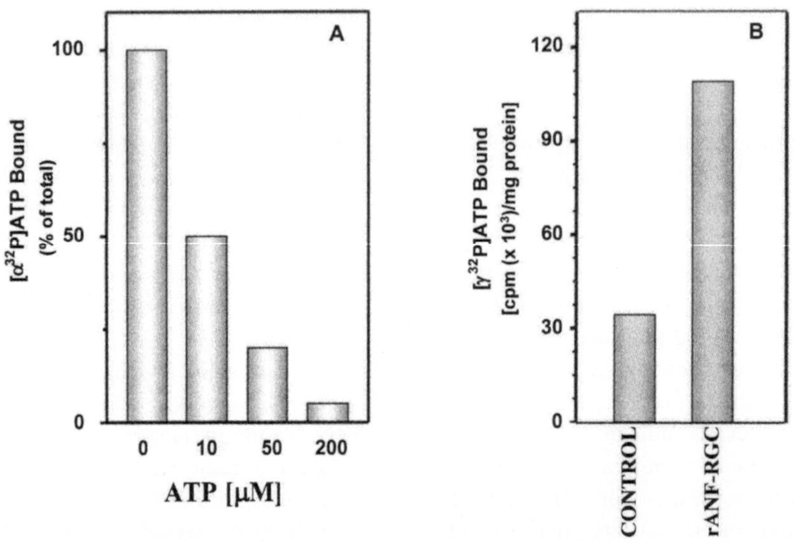

Fig. 2. ATP binding to the wild-type (A) and recombinant ANF-RGC (B). (A) ANF-RGC from rat testis was purified by immunoaffinity chromatography. The purified protein was applied to nitrocellulose membrane. The blot was incubated with [α-^{32}P]ATP (1 μCi/ml) in the absence or presence of indicated concentrations of non-radioactive ATP, washed, air-dried, and autoradiographed [194]. The quantitative values presented are derived from the intensity of the spot on the autoradiogram depicted in Fig. 3B of reference [194]. (B) COS cells were transfected with ANF-RGC cDNA or pSVL expression vector (control); their particulate fractions were prepared and used for ATP binding experiments. Membranes were incubated in a total volume of 100 μl with [γ-^{32}P]ATP (specific activity 4,500 Ci/mmol) in the presence of 4 mM Mg^{2+}, at room temperature for 5 min. Membranes were filtered through GF/C filters and washed 4 times with ice-cold phosphate buffered saline, pH 7.5. Non-specific binding was measured in the presence of 10^{-3} M ATP. Specific binding was calculated by subtracting the non-specific radioactivity from the total radioactivity bound to the filters [133]. These binding values are presented here graphically. In the original contribution they were depicted in a table [133, Table 1].

binding and ANF signaling. Two constructs, one with the entire kinase homology domain deleted, Grc$^-$K$^-$ (aa 506–677), and the other with the kinase-homology domain deleted, but the Grc sequence retained, (Grc$^+$K$^-$) (aa 555–762) were constructed and studied for the mutation consequences on the ATP binding [200] (Fig. 3). Grc$^-$K$^-$ had almost no ATP binding activity and no ATP-dependent ANF signaling. In contrast, Grc$^+$K$^-$ had both ATP binding and ATP-dependent signaling activities [200]. The findings established that the Grc sequence motif is critical in ATP binding and ANF signaling of the cyclase. For this reason, the motif was named ATP regulatory module, ARM [200]. The sequence of the motif is: Gly^{503}-Arg-Gly^{505}-Ser^{506}-Asn-Tyr-Gly^{509}.

The counterpart of the ANF-RGC ARM sequence motif in CNP-RGC is Leu^{497}-Arg-Gly^{499}-Ser-Ser-Tyr-Gly^{503}. To determine the core domain that defines the ATP-dependent signal transduction step of both ANF-RGC and CNP-RGC, the ANF-RGC-ARM sequence was changed to that of the CNP-RGC-ARM sequence [204]. The resulting mutant behaved like a natural ANF-RGC in its response to ANF/ATP signaling [204]. The study established that the respective ARM sequence motifs of ANF-RGC and CNP-RGC are interchangeable, functionally identical, and are critical in the ATP-dependent transduction activities of both ANF-RGC and CNP-RGC. These findings, together with the earlier one, showing that the entire kinase homology domains of ANF-RGC and CNP-RGC are also interchangeable and functionally identical [205], proved that the transduction mechanisms of these two guanylate cyclases are completely, or nearly, identical.

In the next studies, the functional role of individual glycine residues in the ARM sequence motifs of ANF-RGC and CNP-RGC was evaluated through point mutation studies [206]. In the initial study through Gly^{505}Val and Ser^{506}Asn substitutions ARM sequence of the ANF-RGC was disrupted [200]. The disruption changed Grc-oriented ARM architecture from Gly^{503}-X-Gly^{505}-X-X-X-Gly^{509} to Gly^{503}-X-X-X-X-X-Gly^{509}. The change eliminated most of the ATP binding and ATP-dependent ANF signaling of the cyclase, indicating that the middle Gly^{505} in the ARM sequence motif is essential in both the formation of the ATP binding pocket and signal transduction [206]. A similar approach showed that the identically placed Gly^{499} in the CNP-RGC-ARM sequence motif, Gly^{499}-X-X-X-Gly^{503}, is critical in the ATP binding and signal transduction activities of CNP-RGC [206]. Thus, the core ARM sequence critical in the formation of functional ATP binding pocket in both ANF-RGC and CNP-RGC is common, and it is represented by the Gly-X-X-X-Gly motif. Further analysis of the Gly motif showed that the C-terminal Gly in the motif has no functional role [206]. The studies proved that single ARM residue, Gly^{505} in ANF-RGC and the correspond-

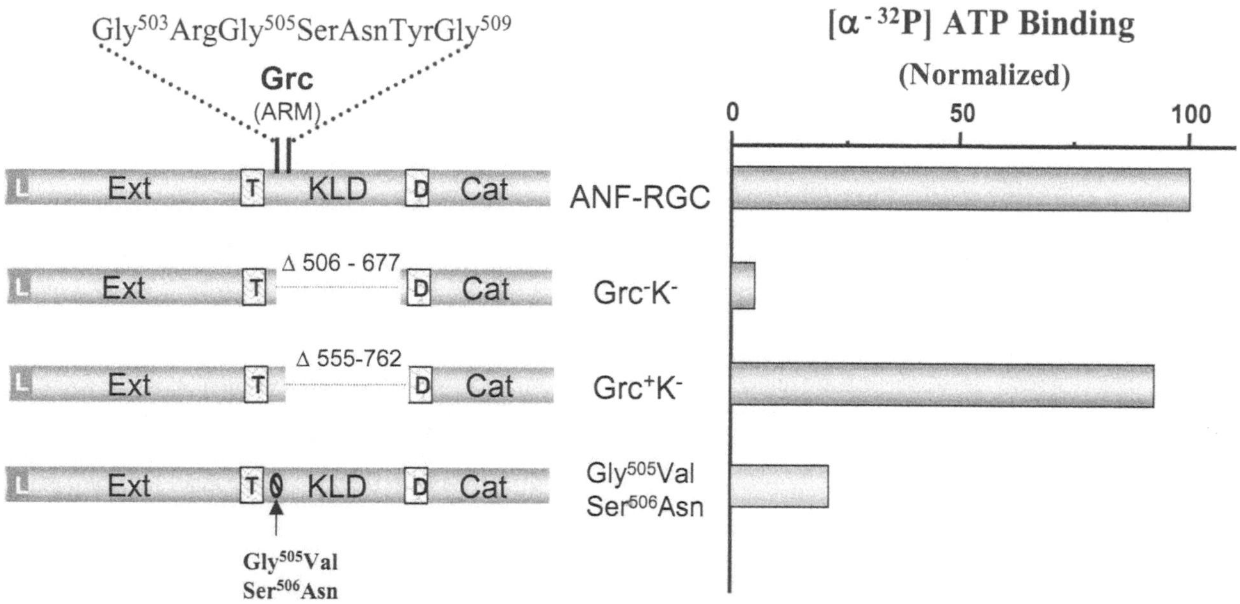

Fig. 3. The effect of the Grc (ARM) sequence of ANF-RGC on ATP binding. Three ANF-RGC mutants were prepared: (1) with the entire kinase homology domain deleted (Grc$^-$K$^-$); (2) with the kinase homology domain deleted, but the Grc sequence retained (Grc$^+$K$^-$); and (3) with disrupted Grc sequence (Gly^{505}Val,Ser^{506}Asn). These are schematically represented in the left panel. ANF-RGC and the three mutants were individually expressed in COS cells. Particulate fractions of the transfected cells were examined for ATP binding as described in the legend to Fig. 2B. The bars (right panel) represent the ATP binding values depicted in Table 2 of ref. [200]. The specific ATP binding of ANF-RGC was taken as 100, and the binding exhibited by the mutants was calculated accordingly.

ing Gly^{499} in CNP-RGC, is critical in formation of the ATP binding pocket and also in signal transduction activity of the respective cyclase.

Acquisition of the knowledge that the ARM-Gly^{505} residue in ANF-RGC and the ARM-Gly^{499} residue in CNP-RGC are critical in ATP-dependent signal transduction step offered the possibility of explaining an earlier observed ATP-dependent biochemical phenomenon. Kinetic studies had indicated that ATP inhibits ANF binding activity of ANF-RGC [207]. ATP also converted the ANF receptor in ANF-RGC from high to low affinity [208]. The issue was: does the respective ARM residue, Gly^{505} in ANF-RGC and Gly^{499} in CNP-RGC, mediate this phenomenon? The answer is yes. In each case there is about a 5-fold reduction in the affinity of the respective ligand for its receptor [209, 210]. Thus, the individual residues act as ATP bimodal switches in regulating both ligand binding and signal transduction activities of ANF-RGC and CNP-RGC, the Gly residue representing a critical site to turn the hormonal signal 'on' and 'off'.

These studies indicated that there is a very tight control over the transmission of the signal generated at the receptor level to the catalytic site where the production of cyclic GMP occurs. This suggested that NF-RGCs are modular proteins, and led to the proposal that 'Plasma membrane guanylate cyclase is a multimodule transduction system' [211]. In this 'multimodule transduction' concept, ANF signal transduction was envisioned to occur with the following sequential steps: (1) A signal is initiated by the binding of the hormone to the ligand-binding module, which is located at the extracellular region of the cyclase. Leu^{364} of ANF-RGC and Glu^{332} of CNP-RGC are critical residues for this step. (2) There is a transmembrane migration of the binding signal. (3) ARM resides just adjacent to the transmembrane domain. ATP binds to this module and causes an allosteric change in two modules located at opposite ends of the cyclase. One module is the catalytic, and the other is ligand binding. A single defined Gly residue of the ARM regulates the activities of both of these modules. (4) The catalytic module is activated, generating the accelerated production of cyclic GMP. (5) The receptor module is inactivated, resulting in the dissociation of the hormone (Fig. 4).

Three dimensional atomic model of the ARM domain shows configuration of its ATP-binding pocket and supports the concept that ATP binding to the pocket is the primary step in ANF signaling of ANF-RGC

Recently, a three-dimensional atomic model for the ARM of ANF-RGC has been created through molecular modeling based on the structures of the tyrosine kinase domains of insulin receptor (PDBId: 1IRK) and haematopoietic cell kinase (PDBId: 1AD5). The model has been experimentally validated [199; reviewed in ref. 201]. Important features of the model are that it defines the precise geometry of the ARM domain, its ATP-binding pocket, its ATP-induced changes in the pocket and global hydrophobicity changes in the ARM domain. The model supports the vital role of Gly^{505} residue in the sequence motif Gly^{505}-X-X-X-Gly^{509} for the ATP transduction event in ANF-RGC. Finally, it demonstrates that allosteric regulation by ATP is the primary mechanistic step in the ANF-dependent signaling of ANF-RGC and not the currently proposed phosphorylation step [198, 212–214]. These features of the model are briefly described below and are depicted in Fig. 5.

The ARM domain stretches from amino acid residue 496–771 of the ANF-RGC. The domain consists of two interconnected lobes: the smaller, N-terminal, and the larger, C-terminal (Fig. 5A). The smaller lobe consists of 91 residues, aa 496–586. It is composed of five β strands, four of which are arranged in a twisted β-sheet, and an α helix. The glycine rich motif, Gly-X-Gly^{505}-X-X-X-Gly, which plays a pivotal role in the ATP-dependent signaling of ANF (*vide supra*), spans the tip of the β1 strand and part of the connecting loop between β1 and β2 strands. The larger lobe consists of aa 587–771 (Figs 5A and 5B). It is predominantly helical, composed of eight α-helices and two β strands. The ATP-binding pocket is located at the interface of these two lobes.

The glycine-rich motif together with the following residues – Leu^{511}, Thr^{513}, Lys^{535}, Glu^{551}, Thr^{580}, Cys^{583}, Ser^{587}, Val^{635}, $Lysl^{630}$, Ser^{632}, Asn^{633}, and Asp^{646} – constitute the ATP-binding pocket of the ARM domain (Fig. 5B). The segment covering residues Gly^{503} to Leu^{511} includes the previously identified core ARM sequence motif Gly^{505}-X-X-X-Gly^{509}. This motif has no direct chemical interaction with ATP. It constitutes the floor of the ATP binding pocket and produces a design that is conducive to fit the ATP molecule. The other residues of the binding pocket directly interact with ATP: Leu^{511}, Thr^{513}, Thr^{580}, Cys^{583}, Ser^{587} and Val^{635} fold to adopt the ribose ring; Asp^{590}, via hydrogen bonding, interacts with the 2′- and 3′-hydroxyl groups of ATP; Lys^{535}, Lys^{630}, Ser^{632} and Asn^{633} surround the triphosphate moiety of ATP. This creates a positively charged environment, which accommodates the negatively charged environment created by the triphosphate moiety; Asp^{646} interacts with the triphosphate group of ATP through the formation of a coordinate bond with the metal ion, Mg^{2+}. The ARM residue Glu^{551} does not directly interact with ATP, it assists to stabilize the steric arrangement of the pocket.

ATP binding to its pocket induces three significant changes in the ARM configuration: (1) It moves and rotates the floor of the pocket comprising β1, β2 strands, and the loop between the strands. The movement is of 3–4 Å; the rotation is by 15° (Fig. 5C, arrow 1). Gly^{505} is a critical PIVOT for both the movement and the rotation; (2) It moves by 2–7 Å, but does not rotate, the β4 and β5 strands, and the loop (Fig. 5C, ar-

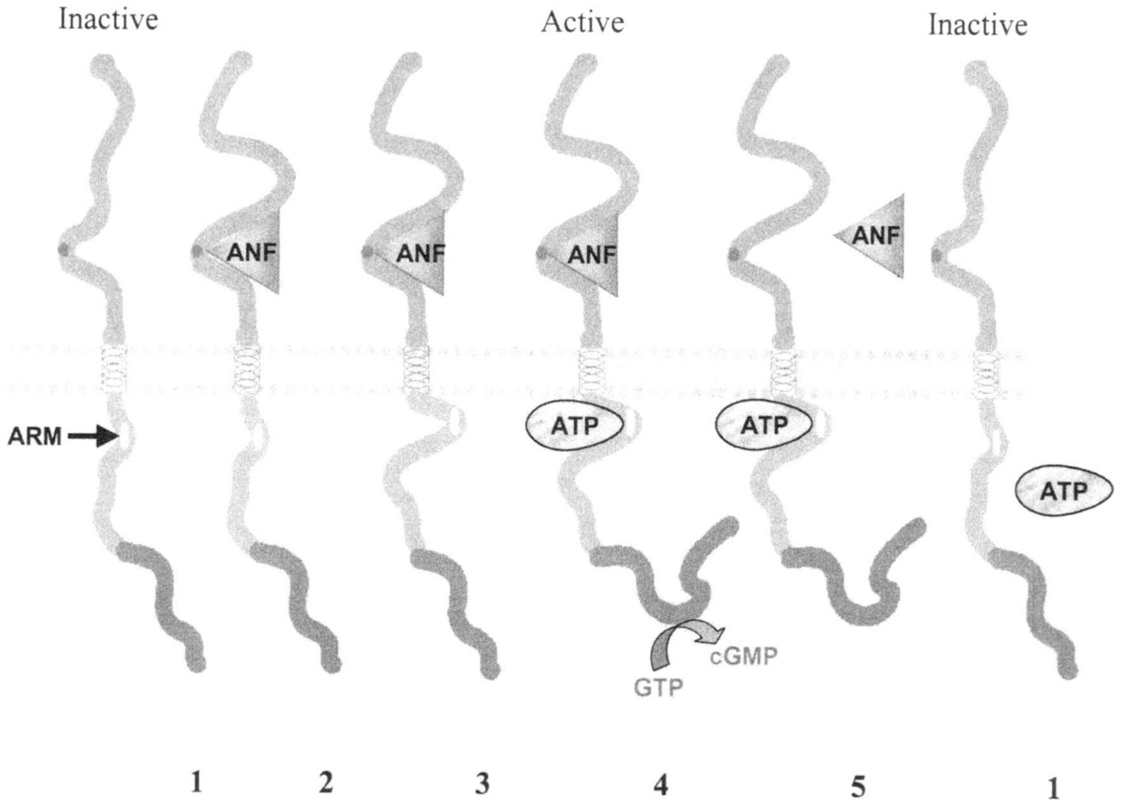

Fig. 4. Multimodule transduction concept of ANF-RGC. In the ANF-RGC (or CNP-RGC) 'multimodule transduction' concept, the following steps are envisioned to occur: (1) in the inactive form the ANF binding site in the extracellular domain and ARM in the intracellular domain are unoccupied; (2) ANF (or CNP) binding to the ligand binding module initiates the signal; (3) conformational changes induced by the binding signal are transduced from the extracellular module to the intracellular portion of the cyclase; (4) this caused ATP binding to the ARM located in the intracellular portion of the cyclase between the transmembrane and catalytic domains; as a result the catalytic module is activated; (5) the receptor module is transformed from a high affinity to a low affinity state and the hormone dissociates. Eventually, ANF-RGC (or CNP-RGC) regains the inactive state and ATP is dissociated. This multimodule signal transduction concept was envisioned and first published in 1994 [211] and then based on further evidence [209, 210] was refined and the model was depicted in [65].

row 2); (3) It moves the αEF helix by 2–5 Å (Fig. 5C, arrow 3). The model predicts that this movement exposes the conserved motif D which, is then involved in the hydrophobic interaction between the helix and the catalytic domain. This results in the ATP-dependent signaling of the cyclase.

The ARM model not only defines ARM's stereochemistry, it also predicts important ATP-dependent surface changes in the ARM domain. In two distinct regions there is a striking change from their lipophobic to lipophilic characteristics. One region encompasses the β1 and β2 strands; the second, comprises the β8 strand and helices EF and F (Fig. 5D). The second change occurs in the vicinity of the conserved motif D [199, 201], which is predicted to interact with the cyclase catalytic domain. Thus, the model explains, at a molecular level, how the ATP-dependent changes within the ARM domain result in activation of the cyclase. The model has been supported by the point mutation/expression studies (*vide infra*).

Consistent with earlier studies [133], AMP-PNP, a nonhydrolyzable analog of ATP, stimulates ANF-dependent ANF-RGC activity in a dose-dependent manner [199]. The maximal cyclase stimulation is about 10-fold over its basal value, and EC_{50} for AMP-PNP is 0.32 mM. Similar to AMP-PNP, ATP and its nonhydrolyzable analog, ATPγS, also stimulate ANF-RGC, but there are changes in the achievable saturation level of the cyclase. It is about 25% higher with ATP and about 45% with ATPγS. These results are consistent with the interpretation that ATP regulation of ANF-RGC is at least a two-step transduction phenomenon [199]. First, the principal step is allosteric. In this step, there is activation of the cyclase. However, the activation is partial, about 55% of its total saturation value. In the second step/s the cyclase achieves full activation. Recent studies have shown that phosphorylation plays an important role in NF-RGCs' activation [198, 212]. To determine if phosphorylation affected the allosteric step, the effect of ATP, AMP-PNP and ATPγS on ANF-dependent signaling of ANF-RGC was assessed in the presence of microcystin, a phosphatase inhibitor [199]. Microcystin showed no significant effect on the basal or on the ATP-, AMP-PNP- or ATPγS-dependent transduction ac-

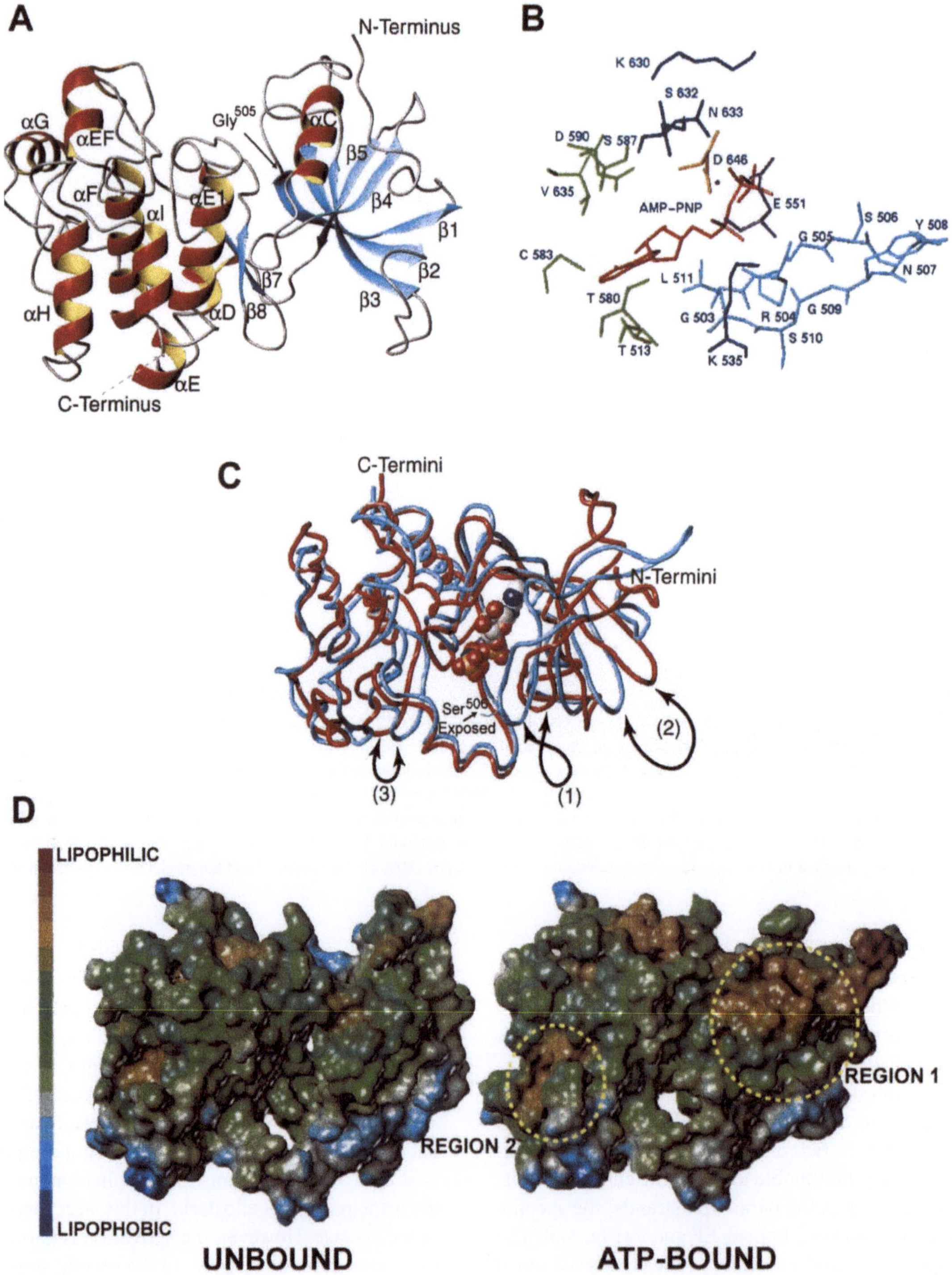
A
N-Terminus
αG
αEF
Gly505
αC
β5
αF
αI
αE1
β4
β1
β7
β2
αH
αD
β8
β3
αE
C-Terminus
B
K 630
S 632
N 633
D 590
S 587
D 646
V 635
AMP–PNP
E 551
S 506
Y 508
G 505
C 583
N 507
L 511
T 580
G 509
G 503
R 504
S 510
T 513
K 535
C
C-Termini
N-Termini
Ser506
Exposed
(2)
(3)
(1)
D
LIPOPHILIC
REGION 1
REGION 2
LIPOPHOBIC
UNBOUND
ATP-BOUND

tivity of the cyclase. Thus, the ATP-dependent allosteric step is not affected by phosphorylation of the cyclase and is the principal signal transduction step involved in ANF-RGC activation.

Based on the predicted unique chemical feature of the residue bestowing a corresponding functional element to the ATP binding pocket, four residues – Gly^{505}, Lys^{535}, Glu^{551}, and Asp^{646} – were chosen to test the functional integrity of the ATP binding pocket [199]. None of the introduced mutations affected basal activity of ANF-RGC, but they almost completely eliminated the ANF signaling. Thus, consistent with the early studies ARM domain has no role in suppressing the native catalytic activity of ANF-RGC [200]. However, it is critical in ANF signaling of ANF-RGC and the four residues, according to the model, participate in the control of the cyclase.

A special feature of the model is the mechanism by which Gly^{505} controls the activity of the ARM domain. There is 55% reduction with Gly^{505}Ala mutation [199]. Other three mutations involving Lys^{535}Ala, Glu^{551}Ala and Asp^{646}Ala almost totally incapacitate ATP-dependent transduction activity of ANF-RGC [199]. Thus, the model predicted-role of each residue is attested by experimentation.

The findings on the pivotal role of Gly^{505} in ATP-dependent transduction step deserve special comment. The model predicts that the ATP-dependent change in the orientation of Gly^{505} controls Ser^{506} orientation. It brings Ser^{506} from the hidden to the exposed state. Are Gly^{505} and Ser^{506} functionally linked? Experimental results indicate that they are. The mutation Gly^{505}Ala reduces the ATP-dependent transduction activity of ANF-RGC by more than 55%; Ser^{506}Ala reduces it by 64%; and Gly^{505}Ala, Ser^{506}Ala reduces it by about 75%. Thus, Gly^{505} is critical for the ATP-dependent PIVOT of the cyclase; the PIVOT, in turn, controls Ser^{506}. The model shows that Ser^{506} has two roles: One, it is essential for the structural integrity of the ATP binding pocket; two, in its exposed state it is possibly vulnerable to phosphorylation to bring the cyclase to full activation state. Remarkably, Gly^{505} controls both of these Ser^{506} roles.

In recent investigations, the role of the ANF-RGC and CNP-RGC ARMs in the signal transduction mechanism has been expanded to include their phosphorylation and dephosphorylation steps [198, 212–214]. More than a decade ago it was shown that ANF-RGC exists in the phosphorylated state, which is influenced by protein kinases [215, 216]. More recently, independent studies by two groups suggest that phosphorylation of NF-RGCs is an absolute requirement for their activation [198, 212–214]. The mutation studies show that the phosphorylated residues involved in activation of the cyclases reside within the first 132 aa residues next to the juxtamembrane domain of both ANF-RGC and CNP-RGC [212]. Screening of the 132 aa residue segment localized a domain, termed 'ATP binding domain' that is the residence of the six putative phosphorylation sites [212]. These 'ATP binding domain' motifs in ANF-RGC and CNP-RGC are reproduced below and the six residues identified by the investigators to be critical in the signaling of the cyclases are in bold and underlined: **Ser**^{497}-X-X-**Thr**^{500}-X-**Ser**^{502}-Gly^{503}-X-Gly^{505}-**Ser**^{506}-X-X-Gly^{509}-**Ser**^{510}-X-X-**Thr**^{513} for ANF-RGC and **Ser**^{491}-X-X-**Thr**^{494}-X-**Ser**^{496}-X-X-Gly^{499}-**Ser**^{500}-X-X-Gly^{503}-**Ser**^{504}-X-X-**Thr**^{507} for CNP-RGC. The authors conclude that the phosphorylation of the ATP binding domain is a critical event in the regulation of the NF-RGCs [198, 212–214]. With this conclusion, the authors have proposed a model in which ATP has a dual function in the regulation of ANF-RGC [198]. In the first step of the model there is phosphorylation of the cyclase and in the next step, allosteric regulation. This ordered sequence of the transduction steps is not supported by the three-dimensional ARM model of ANF-RGC, and will be discussed below.

The ATP binding domain motif found to be critical in the ATP-dependent transduction event by the studies of Potter and Hunter [212–214] is in accord with the three-dimensional ARM model [199]. It resides within the ARM domain and the motif contains the pivotal Gly^{505} and Ser^{506} residues of the ARM model. Consistent with the studies of Potter and Hunter [212], the ARM model shows a pivotal role of Ser^{506}. However, in contrast to the studies of Potter and Hunter, the ARM model shows that Ser^{506} does not play an independent role in the cyclase activation. This role is subordinate to Gly^{505} residue, which in the first step controls the orientation of Ser^{506}. It is only then possible that the Ser residue may play a fur-

←

Fig. 5. Model of ANF-RGC ARM domain [199]. (A) Ribbon diagram of the ARM domain. The α helices are shown in red and the β strands are in cyan. The N- and C-termini are indicated by arrows. (B) ATP binding pocket. Amino acid residues that participate in binding of the nucleotide (AMP-PNP) are color-coded based on the component of AMP-PNP with which they interact: β1 and β2 strands, which make the floor of the ATP binding pocket are shown in cyan; amino acid residues surrounding the ribose ring are in green; residues surrounding the triphosphate moiety are in blue; residue coordinating Mg^{2+} is in gold; the residue that stabilizes the pocket is in magenta; AMP-PNP is in red and Mg^{2+} is a red sphere. (C) Superposition of the AMP-PNP-free and bound forms of the ARM domain. The backbone of the ARM domain is shown as a tube model. The nucleotide-free form is depicted in red and the bound form is in cyan. The N- and C-termini of both forms are marked. AMP-PNP molecule is represented as a space-filling model in atom type colors: carbon is gray, oxygen is red, phosphorus is orange and nitrogen is blue. Arrows indicate three major movements observed upon nucleotide binding: (1) strands β1 and β2 and the connecting loop, (2) strands β4 and β5 with the connecting loop, and (3) the αEF helix. Movement of Ser^{506} side-chain from buried to exposed position upon binding of ATP is shown. (D) Lipophobic and lipophilic potentials at the surface of the ANF-RGC ARM domain in the unbound and ATP-bound forms. The effect of ATP binding on the distribution of the lipophobic and lipophilic potentials of the surface of the ARM domain was analyzed using SYBYL program. The two regions exhibiting striking changes in potential are circled. The color ramp to the left indicates the range from the most lipophobic (blue) to the most lipophilic (brown).

ther role in the activation of the cyclase. Thus, the model of Potter and Hunter [212–214] and that of Foster and Garbers [198] differs from that of Duda *et al.* [199] in the sequence of the transduction steps. The first two groups propose that phosphorylation of the cyclase is the first and the principal signaling step of the cyclases. In contrast, Duda *et al.* [199] show that allosteric regulation of the cyclase is the first and the principal step. In the allosteric step the cyclase is partially stimulated, about 55%, and in the next step/s it is stimulated to the full extent. It is envisioned that phosphorylation of Ser^{506} plays a role in the full activation of the cyclase. It is also possible that the other five residues recognized by Hunter's group, via phosphorylation, may also be involved in the full cyclase activation.

The above discussion shows that the transduction machinery of ANF-RGC and CNP-RGC is ATP-regulated. It operates by almost identical signal transduction steps and it is modular in nature. The modules are the sites where signal processing occurs and then is transferred to the next module for further processing. ATP is obligatory for the signal transduction machinery. ATP acts directly, by binding to its pocket in the ARM. ARM resides next to the trans-membrane module covering the aa segment 496–771. It contains a critical Gly^{505}-Ser^{506}-X-X-Gly motif in ANF-RGC and Gly^{499}-Ser^{500}-X-X-Gly in CNP-RGC. This motif is not involved in the direct binding of ATP, but is pivotal in the ATP-binding and ATP-dependent transduction activities of the cyclase. Upon binding to its pocket, ATP signals the hormone-dependent activation of the cyclase. In addition to controlling this signaling activity, the ARM module shows that ATP also induces important surface changes in the ARM domain. One change explains how the ATP-dependent ARM movement controls the activity of the catalytic module of the cyclase.

Modulation of ROS-GC by GCAPs occurs through the cyclase distinct modules

The first study to show that the modes of Ca^{2+} modulation of ROS-GC by GCAPs are different was that of Krishnan *et al.* [217]. The reconstitution of the recombinant ROS-GC1 with the recombinant forms of GCAPs showed that the principal GCAP1-modulated domain in the cyclase resided in the region, located between the transmembrane and the catalytic domains of the cyclase [218]. In contrast, the GCAP2 modulated region resided at the C-terminus of the cyclase [217]. Two earlier studies had shown that GCAP-modulated ROS-GC1 domain resides in the intracellular region [218, 219]. Duda *et al.*, used GCAP1 in their study, and localized the GCAP1-modulated domain between aa 447 and 730 of ROS-GC1 [218]. Laura *et al.* used GCAP-2 in their study and did not elaborate on the specificity of the GCAP-modulated domain in ROS-GC1 [219]. Later comprehensive studies of Krishnan *et al.* showed that compared to GCAP1, GCAP2 was one-order less specific modulator of ROS-GC1 [217]. The present evidence shows that GCAP1 does not influence ROS-GC2 activity [166, 220]. These findings indicate that GCAP1 is the primary modulator of ROS-GC1 and GCAP2 is the modulator of ROS-GC2.

GCAP1-modulated domain in ROS-GC1

Modulation of ROS-GC1 by GCAP1 has received significant attention because of the established role of GCAP1 and ROS-GC1 in phototransduction (reviewed in refs [66, 220–222]). Three attempts, by independent groups, to identify the ROS-GC1 site of interaction with GCAP1 resulted in three different conclusions. In the report by Lange *et al.* [223], the GCAP1-modulated ROS-GC1 domain has been mapped and it consists of two regions: Met^{445}-Leu^{456} and Leu^{503}-Ile^{522}. Peptides corresponding to these regions inhibited GCAP1-dependent ROS-GC1 activity both at 1 nM Ca^{2+} and at 24 µM Ca^{2+}. There was, however, a subtle difference. While region 1 (Met^{445}-Leu^{456}) was a more effective inhibitor at 1 nM Ca^{2+}, region 2 (Leu^{503}-Ile^{522}) was equally effective at both Ca^{2+} concentrations [223]. Therefore, it was concluded that the former region represents the transducer motif of ROS-GC1 and is necessary for the activation of the cyclase; the latter region represents a main regulatory site for GCAP1 interaction [223]. These conclusions were confirmed by two independent approaches: (1) deletion mutants of ROS-GC1: deletion of either region abolished GCAP1-dependent ROS-GC1 activity [223]; (2) direct binding studies: when a peptide library corresponding to the ROS-GC1 intracellular domain was scanned for ability to bind to the immobilized GCAP1 through surface plasmon resonance, the peptide corresponding to region 2 (Leu^{503}-Ile^{522}) exhibited binding (Lange *et al.*, unpublished results). The conclusions of Lange *et al.* [223] are further supported by the genetic studies identifying that the F514S mutation of ROS-GC1 is linked with Leber's Congenital Amaurosis (LCA1) [224], by the biochemical studies demonstrating that this mutant does not respond to GCAP1 [153] and the fact that this residue is located within the identified interaction site [223]. Thus, four independent lines of evidence: (1) peptide competition studies; (2) direct binding studies; (3) deletion mutation studies; and (4) etiology of LCA1 are consistent with the GCAP1-modulated domains identified by Lange *et al.* [223].

Another report concludes that the GCAP1 modulated region in ROS-GC1 resides in its catalytic domain, between Gly^{968} and Gly^{985} [225]. This study was based primarily on the ability of this peptide to bind GCAPs and its ability to inhibit GCAP-dependent stimulation of ROS-GC1. The peptide, however, is more effective with GCAP2 than GCAP1. It appears, then, that this motif represents the GCAP2 binding domain in ROS-GC1.

A recent, third, report has identified multiple regions on ROS-GC1 and GCAP1 that may serve as contact points between these proteins [226]. This study is based on cross-linking of cysteine residues on the two proteins that are in proximity followed by protease digestion and analysis of the resultant peptides by mass spectrometry. Thus, it necessitates that Cys residues in the two proteins be in proximity and that they should be at a distance appropriate for the arm of the cross-linking agent. In contrast to two of these studies that identify sequences within the kinase-homology domain of ROS-GC1 [223, 226], but in agreement with previous findings [218, 227, 228], the report by Sokal *et al.* [225], concludes that the critical region resides in the catalytic domain. It, therefore, appears that the study by Lange *et al.* [223] has identified the sequences on ROS-GC1 critical for its biological regulation by GCAP1, while the other two studies [225, 226] have identified general contact points between ROS-GC1 and GCAPs. The possibility that multiple contact points mediate GCAP1-ROS-GC1 interaction has already been proposed [223].

As noted earlier, in certain pinealocytes ROS-GC1 is co-localized with S100β [170]. It, therefore, appears that in those neurons Ca^{2+} signals the activation of ROS-GC machinery through S100β. The biochemical studies show that S100β regulates ROS-GC activity through a specified domain; this domain is located at the C-terminus region of the cyclase and is distinct from the GCAP1, GCAP2 and the neurocalcin-modulated domain of the cyclase.

Guanylate cyclase homodimers constitute the functionally active units

Single transmembrane receptor proteins, such as the receptors for growth factors, cytokines or insulin undergo ligand induced dimerization or oligomerization in order to achieve full activity [229, 230]. It is also known that dimerization of the catalytic subunits is necessary for the activity of adenylate and soluble guanylate cyclases [231–233]. Thus, it was initially hypothesized that membrane guanylate cyclases undergo similar dimerization and that high affinity ANF binding is associated with receptor dimers in ANF-RGC [110].

All known membrane guanylate cyclases contain in their intracellular portions a region that is predicted to function as a dimerization domain [234]. Experiments employing bacterial expression of truncated mutants of ANF-RGC, consisting of only the catalytic domain or this domain together with the predicted α-helical dimerization domain have shown that this α-helical region is obligatory for the cyclase activity [235, 236]. Subsequently, it was shown that natriuretic peptide receptor guanylate cyclases exist as dimers independent of ligand binding [237]. The intracellular contact of two guanylate cyclase molecules through their dimerization domains is, however, sufficient only for the basal cyclase activity. Very recent studies have shown that upon ANF binding disulfide bonds are formed between extracellular domains of two ANF-RGC molecules that allow transduction of a conformational change from the extracellular domain to the cyclase catalytic domain resulting in its maximal activation [238, 239]. It is reasonable to predict that CNP-RGC would also exhibit a similar dimerization phenomenon.

The necessity of dimerization for the cyclase activity has also been shown for the intracellularly regulated membrane guanylate cyclases of the ROS-GC subfamily. Although in this case the extracellular domain has no role in either basal or fully stimulated cyclase activity [218], the Ca^{2+}-regulated ROS-GC modulators, the GCAPs, S100β or neurocalcin, facilitate close contact between two monomeric cyclase molecules resulting in full catalytic activity [240]. This is achieved through the dimerization of the Ca^{2+}-free forms of GCAPs [241]. The existence of S100β in a dimeric form has been shown previously (reviewed in ref. [242]; Rambotti *et al.*, this volume). Thus, the present day consensus is that: (1) cyclic GMP synthesis occurs at the membrane guanylate cyclase dimer interface; (2) the membrane guanylate cyclases exist as homodimers in their basal (unstimulated) state; (3) binding of an extracellular ligand or an intracellular Ca^{2+}-regulated modulator brings catalytic domains of both homodimer subunits into proximity necessary for catalysis to occur with maximal efficiency.

The physiological importance of the dimerization in membrane guanylate cyclase functioning has been provided by the genetic studies. In a form of cone-rod dystrophy, CORD6, various mutations have been identified within the predicted dimerization domain of ROS-GC1. There are three mutation-prone positions: amino acid residues 786, 787 and 788 [243–247]. Biochemical studies with ROS-GC1 mutants in which the disease-causing mutations have been introduced showed that the CORD6 phenotype could be explained by impaired dimerization of ROS-GC1 [154–158; Duda and Koch, this issue].

In a final comment, 15 years ago the first signal transduction model of ANF-RGC was proposed [63, 64] (Fig. 1). By and large, this model has withstood intense experimental scrutiny. With the added knowledge, the recent version of this model is depicted in Fig. 6. To show its similarity, or dissimilarity, with the Ca^{2+} signal transduction model, the ROS-GC1 transduction model is shown side by side with the ANF-RGC model.

Concluding remarks and future directions

About four decades ago cyclic GMP was detected in the mammalian systems. Within this time span, spectacular strides on the mediatory role of this cyclic nucleotide in the signal transduction field have been made. Some of the cornerstone developments in the field are cited below.

The first, and foremost, is the discovery of ANF-RGC. This discovery showed the independent existence of a membrane guanylate cyclase signal transduction system. It also revealed an unusual nature of the membrane guanylate cyclase enzyme. The enzyme was also a hormone receptor. This discovery dispelled the belief that the guanylate cyclase existed only in its soluble form and that it transduced only the intracellularly-generated nitric oxide signals.

The second is the discovery of ROS-GC from the rod outer segments of the retina. This discovery linked ROS-GC with the molecular machinery of phototransduction. At a biochemical level it showed the existence of a membrane guanylate cyclase class, which transduced the Ca^{2+} signals generated inside the sensory neurons of the retina, and was not involved in the transduction of the peptide hormone signals. Thus, the discovery established that the membrane guanylate cyclase family consists of two subfamilies: the peptide hormone receptor and Ca^{2+}-modulated ROS-GC subfamily. The discovery negated the concept that all members of the membrane guanylate cyclase family are also peptide hormone receptors.

The third is the discovery that ROS-GC transduction system is not uniquely expressed in the rod outer segments of the retina. It is also expressed in other retinal neurons and in the sensory neurons of the pineal gland and the olfactory bulb. This finding broadened the ROS-GC transduction field and indicated its involvement in the light-regulated processes of the pineal gland and in the process of odor perception.

The fourth is the discovery that ROS-GC transduction system through $\alpha_{2D/A}$-AR is linked with the epinephrine/norepinephrine signaling in the pinealocytes. This finding broadens the membrane guanylate cyclase transduction field. It shows that the receptor signals arising at the seven transmembrane-spanning receptor via Ca^{2+} are in certain cases, internetted with the membrane guanylate cyclase transduction machinery.

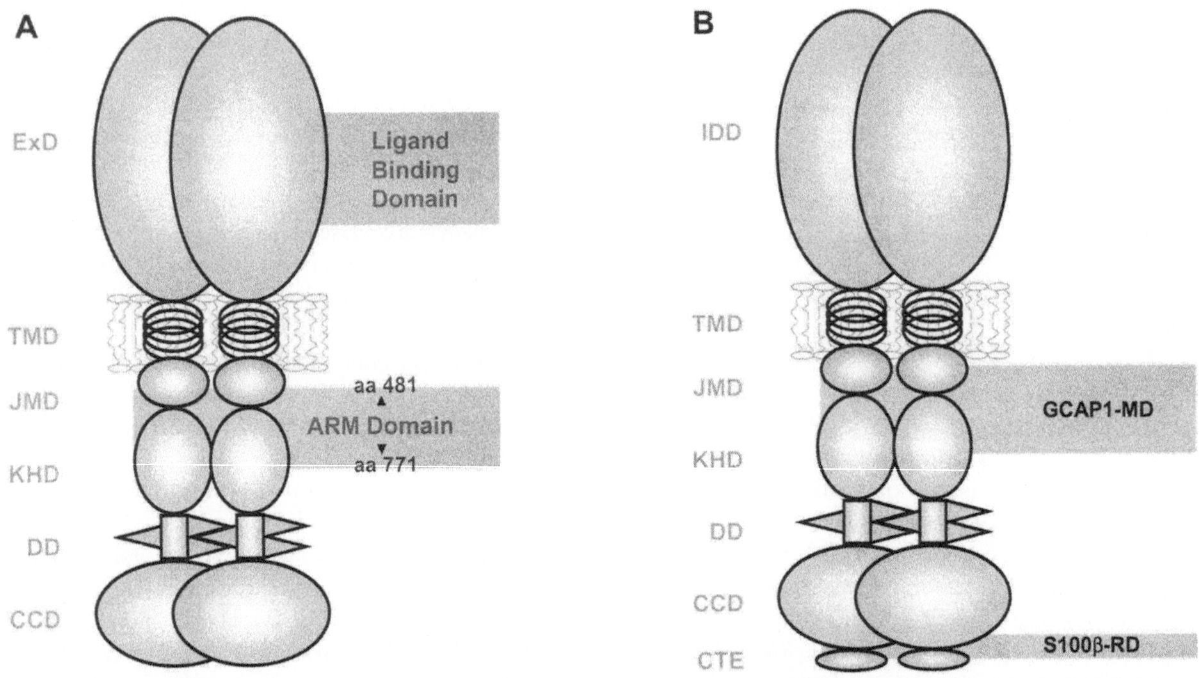

Fig. 6. Topographical model of membrane guanylate cyclase: (A) Receptor guanylate cyclase (ANF-RGC) and (B) Ca^{2+}-modulated guanylate cyclase (ROS-GC1). All membrane guanylate cyclases are single transmembrane spanning proteins. The lipophylic, helical transmembrane domain (TMD) divides the protein into two portions. The N-terminal portion in receptor guanylate cyclases (ANF-RGC, CNP-RGC and Sta-RGC) constitutes the extracellular domain (ExD); in Ca^{2+}-modulated guanylate cyclases (ROS-GC1 and ROS-GC2) this domain is termed 'intradiscal' (IDD) in photoreceptor outer segments but may be located extracellularly elsewhere. To date no ligand for this domain has been found. The intracellular region in all cyclases is comprised of juxtamembrane domain (JMD), kinase-homology domain (KHD), dimerization domain (DD) and a core catalytic domain (CCD). In Ca^{2+}-modulated guanylate cyclases an added segment is present at the C-terminus, which has been termed C-terminal extensions (CTE). Functionally, active NF-RGCs and ROS-GCs exist as dimers and it is believed that the dimerization is promoted by the hinge-like DD. Signal transduction in receptor guanylate cyclases involves binding of the ligand to the ExD and binding of ATP to the intracellular ARM domain (the amino acid residues indicated depict the boundaries of the ANF-RGC ARM domain) followed by allosteric changes and catalytic activation. A role for phosphorylation of specific serine residues within the intracellular domain has been proposed. In ROS-GCs, modulation by Ca^{2+} is mediated by a group of Ca^{2+}-binding proteins, each of which binds to a distinct domain in the intracellular region of the cyclase. The location of the GCAP1 modulary domain (GCAP1-MD) and S100b regulatory domain (S100b-RD) in ROS-GC1 are indicated.

Now that the membrane guanylate cyclase signal transduction field has been established and is shown to be very broadly linked with both the peptide hormone receptor signals and the Ca^{2+} spikes generated in the sensory neurons, many new problems and challenges have come to the forefront. A first group of the problems have to do with the membrane guanylate cyclases themselves. Have all their subtypes been discovered? Have all the regulators of each of the subtype been disclosed? What is the tissue distribution of the various cyclase subtypes and their modulators? What is their function and mechanism of regulation in these specific tissues?

A second group of problems concerns the increasing diversity of the pathways controlled by each of the cyclase. What are the molecular mechanisms involved in those controls? For instance, both ROS-GC1 and ROS-GC2 are linked with phototransduction. The activities of both cyclases appear to be modulated by both GCAP1 and GCAP2. However, in the pinealocytes, there exists only one ROS-GC, ROS-GC1, and one GCAP, GCAP-1. Similarly, in the olfactory bulb neurons, there exists only ROS-GC1 and GCAP-1. Are these the neurosensory cell-specific peculiarities of the ROS-GC transduction system? Are there other modulators of ROS-GC1 present, which make these pathways more cell-specific?

The third group of problems deals with elucidation of the precise biochemical mechanisms involved in each of the membrane guanylate cyclase regulation. All membrane guanylate cyclases are modular in nature. Each module exhibits a tight control over catalytic activity. The challenge is to define the molecular nature of these modules and how they exhibit this control. For instance, recent evidence suggests that the natriuretic factor receptor membrane guanylate cyclases, ANF-RGC and CNP-RGC, are regulated by ATP via multi-step processes, involving allosteric regulation and phosphorylation. What are the sequences of those steps and how do they operate? Similarly, ROS-GC1 and ROS-GC2 are differentially regulated by Ca^{2+}; what are their precise mechanisms of regulation?

The fourth group of problems relate with the physiological significance of each of the membrane guanylate cyclase transduction system. To date, the only physiological system with which a membrane guanylate cyclase has been firmly linked is phototransduction. Gene-knock out studies suggests that ANF-RGC is linked with salt-sensitive hypertension. The future challenge is to link each of the transduction system with a specific physiological function, which will, in turn, assist to diagnose the related pathology in molecular terms and then design a better therapy for it.

Finally, to resolve the biochemical mechanistic pathway in absolute terms, availability of the crystal structure of each of the membrane guanylate cyclase and its regulator/s is essential. To obtain these structures is a challenge of the future.

Acknowledgements

This investigator gratefully acknowledges the continuous support for the last 27 years by the numerous USPHS awards from the National Institutes of Health, National Science Foundation and the beginning award from the Damon Runyon Walter Winchell Cancer Fund. The present work is supported by NEI (EY10828). He also expresses his gratitude to all the fellows for their contributions in his laboratory; to Drs. Teresa Duda, Ari Sitaramayya and Karl-Wilhelm Koch for fruitful collaborations; to Mrs. Joan Sharma for continuous moral support; and to Ms. Sanya S. Khajuria for editing the manuscript.

References

1. Sutherland EW, Rall TW: Fractionation and characterization of a cyclic adenine ribonucleotide formed by tissue particles. J Biol Chem 232: 1077–1091, 1958
2. Sutherland EW, Rall TW: The relation of adenosine-3′,5′-phosphate and phosphorylase to the actions of catecholamines and other hormones. Pharmacol Rev 12: 265–299, 1960
3. Robison GA, Butcher RW, Sutherland EW: Adenyl cyclase as an adrenergic receptor. Ann NY Acad Sci 139: 703–723, 1967
4. Ariens EJ, Simonis AM: Aspects of molecular pharmacology. II. Theories of receptors and structure-action relationships. Farmaco Sci 21: 581–608, 1966
5. Ariens EJ: The structure-activity relationships of beta adrenergic drugs and beta adrenergic blocking drugs. Ann NY Acad Sci 139: 606–631, 1967
6. Birnbaumer L, Pohl SL, Michiel H, Krans MJ, Rodbell M: The actions of hormones on the adenyl cyclase system. Adv Biochem Psychopharmacol: 185–208, 1970
7. Rodbell M, Birnbaumer L, Pohl SL, Krans HM: Properties of the adenyl cyclase systems in liver and adipose cells: The mode of action of hormones. Acta Diabetol Lat 7(suppl 1): 9–63, 1970
8. Rodbell M, Lin MC, Salomon Y, Londos C, Harwood JP, Martin BR, Rendell M, Berman M: The role of adenine and guanine nucleotides in the activity and response of adenylate cyclase systems to hormones: Evidence for multi-site transition states. Acta Endocrinol (Copenhagen) 191(suppl): 11–37, 1974
9. Rodbell M, Lin MC, Salomon Y, Londos C, Harwood JP, Martin BR, Rendell M, Berman M: Role of adenine and guanine nucleotides in the activity and response of adenylate cyclase systems to hormones: Eevidence for multisite transition states. Adv Cyclic Nucleotide Res 5: 3–29, 1975
10. Hepler JR, Gilman AG: G proteins. Trends Biochem Sci 17: 383–387, 1992
11. Birnbaumer L, Birnbaumer M: Signal transduction by G proteins: 1994 edition. J Recept Signal Transduct Res 15: 213–252, 1995
12. Feder D, Im MJ, Klein HW, Hekman M, Holzhofer A, Dees C, Levitzki A, Helmreich EJ, Pfeuffer T: Reconstitution of beta 1-adrenoceptor-dependent adenylate cyclase from purified components. EMBO J 5: 1509–1514, 1986. Erratum in: EMBO J 5: 3408, 1986
13. Negishi M, Ito S, Yokohama H, Hayashi H, Katada T, Ui M, Hayaishi O: Functional reconstitution of prostaglandin E receptor from bovine adrenal medulla with guanine nucleotide binding proteins. J Biol Chem 263: 6893–6900, 1988

14. Bansal VS, Majerus PW: Phosphatidylinositol-derived precursors and signals. Annu Rev Cell Biol 6: 41–67, 1990
15. Harden TK: G-protein-regulated phospholipase C. Identification of component proteins. Adv Sec Mess Phosphoprot Res 26: 11–34, 1992
16. Berridge MJ, Irvine RF: Inositol phosphates and cell signalling. Nature 341: 197–205, 1989
17. Nishizuka Y: The molecular heterogeneity of protein kinase C and its implications for cellular regulation. Nature 334: 661–665, 1988
18. Strader CD, Fong TM, Tota MR, Underwood D, Dixon RA: Structure and function of G-protein-coupled receptors. Annu Rev Biochem 63: 101–132, 1994
19. Ashman DF, Lipton R, Melicow NM, Price TD: Isolation of adenosine 3′,5′-monophosphate and guanosine 3′,5′-monophosphate from rat urine. Biochem Biophys Res Commun 11: 330–334, 1963
20. Goldberg ND, Dietz SB, O'Toole AG: Cyclic guanosine 3′,5′-monophosphate in mammalian tissues and urine. J Biol Chem 244: 4458–4466, 1969
21. Ishikawa E, Ishikawa S, Davis JW, Sutherland EW: Determination of guanosine 3′,5′-monophosphate in tissues and of guanyl cyclase in rat intestine. J Biol Chem 244: 6371–6376, 1969
22. Goldberg ND, O'Dea RF, Haddox MK: Cyclic GMP. Adv Cyclic Nucleotide Res 3: 155–223, 1973
23. White AA, Aurbach GD: Detection of guanyl cyclase in mammalian tissues. Biochim Biophys Acta 191: 686–697, 1969
24. Schultz G, Bohme E, Munske K: Guanyl cyclase. Determination of enzyme activity. Life Sci 8: 1323–3213, 1969
25. Hardman JG, Sutherland EW: Guanyl cyclase, an enzyme catalyzing the formation of guanosine 3′,5′-monophosphate from guanosine triphosphate. J Biol Chem 244: 6363–6370, 1969
26. Kimura H, Murad F: Evidence for two different forms of guanylate cyclase in rat heart. J Biol Chem 249: 6910–6916, 1974
27. Kimura H, Murad F: Subcellular localization of guanylate cyclase. Life Sci 17: 837–843, 1975
28. Hardman JG, Sutherland EW: A cyclic 3′,5′-nucleotide phosphodiesterase from heart with specificity for uridine 3′,5′-phosphate. J Biol Chem 240: 3704–3705, 1965
29. Goldberg ND, Haddox MK: Cyclic GMP metabolism and involvement in biological regulation. Annu Rev Biochem 46: 823–896, 1977
30. George WJ, Polson JB, O'Toole AG, Goldberg ND: Elevation of guanosine 3′,5′-cyclic phosphate in rat heart after perfusion with acetylcholine. Proc Natl Acad Sci USA 66: 398–403, 1970
31. Goldberg ND, Haddox MK, Nicol SE, Glass DB, Sanford CH, Kuehl FA Jr, Estensen R: Biologic regulation through opposing influences of cyclic GMP and cyclic AMP: the Yin Yang hypothesis. Adv Cyclic Nucleotide Res 5: 307–330, 1975
32. Goldberg ND, Haddox MK, Hartle DK, Hadden JW: The biological role of cyclic 3′,5′-guanosine monophosphate. In: Krager and Basel (eds). Pharmacology and the Future of Man. Fifth International Congress on Pharmacology, San Francisco. Vol. 5, 1972, pp 146–169
33. Goldberg ND, Haddox MK, Dunham E, Lopez C, Hadden JW: The Yin Yang hypothesis of biological control: Opposing influences of cyclic GMP and cyclic AMP in the regulation of cell proliferation and other biological processes. In: B. Clarkson and R. Baserga (eds). The Cold Spring Harbor Symposium on the Regulation of Proliferation in Animal Cells. Cold Spring Harbor Laboratory, New York, 1974, pp 609–625
34. Goldberg ND, Haddox MK, Estensen R, White JG, Lopez C, Hadden JW: Evidence for a dualism between cyclic GMP and cyclic AMP in the regulation of cell proliferation and other cellular processes. In: W. Brown, L. Lichenstein, C. Parker (eds). Cyclic AMP, Cell Growth and the Immune Response. Springer-Verlag, New York, 1974, pp 247–262
35. Glinsmann WH, Hern EP: Inactivation of rat liver glycogen synthetase by 3′:5′-cyclic nucleotides. Biochem Biophys Res Commun 36: 931–936, 1969
36. Glinsmann WH, Hern EP, Linarelli LG, Farese RV: Similarities between effects of adenosine 3′,5′-monophosphate and guanosine 3′,5′-monophosphate on liver and adrenal. Endocrinology 85: 711–719, 1969
37. Conn HO, Karl IS, Steiner A, Kipnis DM: Studies of the mechanism of action of 3′,5′-cyclic nucleotides on hepatic glucose production. Biochem Biophys Res Commun 45: 436–443, 1971
38. Exton JH, Hardman JG, Williams TF, Sutherland EW, Park CR: Effects of guanosine 3′,5′-monophosphate on the perfused rat liver. J Biol Chem 246: 2658–2664, 1971
39. Guder W, Wieland O: The effect of cyclic nucleotides on glucose synthesis in isolated rat kidney tubules. Hoppe Seylers Z Physiol Chem 351: 291–292, 1970
40. Friedmann N, Somlyo AV, Somlyo AP: Cyclic adenosine and guanosine monophosphates and glucagon: Effect on liver membrane potentials. Science 171: 400–402, 1971
41. Somlyo AP, Somlyo AV, Friedmann N: Cyclic adenosine monophosphate, cyclic guanosine monophosphate, and glucagon: Effects on membrane potential and ion fluxes in the liver. Ann NY Acad Sci 185: 108–114, 1971
42. Sayers G, Beall RJ, Seelig S: Isolated adrenal cells: Adrenocorticotropic hormone, calcium, steroidogenesis, and cyclic adenosine monophosphate. Science 175: 1131–1133, 1972
43. Kitabchi AE, Sharma RK: Corticosteroidogenesis in isolated adrenal cells of rats. I. Effect of corticotropins and 3′,5′-cyclic nucleotides on corticosterone production. Endocrinology 88: 1109–1116, 1971
44. Mahaffee D, Ney RL: Effects of nucleotides possessing a 3′,5′-cyclic monophosphate on adrenal steroidogenesis. Metabolism 19: 1104–1108, 1970
45. Brandwein H, Lewicki J, Murad F: Production and characterization of monoclonal antibodies to soluble rat lung guanylate cyclase. Proc Natl Acad Sci USA 78: 4241–4245, 1981
46. Wallach D, Pastan I: Stimulation of membranous guanylate cyclase by concentrations of calcium that are in the physiological range. Biochem Biophys Res Commun 72: 859–865, 1976
47. Murad F, Mittal C, Arnold WP, Ichikara K, Braughler M, El-Zayat M: Properties and regulation of guanylate cyclase: Activation by azide, nitro compounds, and hydroxyl radical and effects of heme containing proteins. In: G.. Folco, R. Paolottie (eds). Molecular Biology and Pharmacology of Cyclic Nucleotides. Elsevier, Amsterdam, Netherlands,1978, pp 33–42
48. Murad F, Mittal CK, Arnold WP, Katsuki S, Kimura H: Guanylate cyclase: Activation by azide, nitro compounds, nitric oxide, and hydroxyl radical and inhibition by hemoglobin and myoglobin. Adv Cyclic Nucleotide Res 9: 145–158, 1978
49. Mittal CK, Murad F: Properties and oxidative regulation of guanylate cyclase. J Cyclic Nucleotide Res 3: 381–391, 1977
50. Arnold WP, Mittal CK, Katsuki S, Murad F: Nitric oxide activates guanylate cyclase and increases guanosine 3′:5′-cyclic monophosphate levels in various tissue preparations. Proc Natl Acad Sci USA 74: 3203–3207, 1977
51. Mittal CK, Kimura H, Murad F: Purification and properties of a protein required for sodium azide activation of guanylate cyclase. J Biol Chem 252: 4384–4390, 1977
52. Waldman SA, Lewicki JA, Brandwein HJ, Murad F: Partial purification and characterization of particulate guanylate cyclase from rat liver after solubilization with trypsin. J Cyclic Nucleotide Res 8: 359–370, 1982
53. Arnold WP, Aldred R, Murad F: Cigarette smoke activates guanylate cyclase and increases guanosine 3′,5′-monophosphate in tissues. Science 198: 934–936, 1977
54. Mittal CK, Murad F: Activation of guanylate cyclase by superoxide dismutase and hydroxyl radical: A physiological regulator of guanosine 3′,5′-monophosphate formation. Proc Natl Acad Sci USA 74: 4360–4364, 1977

55. Katsuki S, Arnold W, Mittal C, Murad F: Stimulation of guanylate cyclase by sodium nitroprusside, nitroglycerin and nitric oxide in various tissue preparations and comparison to the effects of sodium azide and hydroxylamine. J Cyclic Nucleotide Res 3: 23–35, 1977
56. Katsuki S, Arnold WP, Murad F: Effects of sodium nitroprusside, nitroglycerin, and sodium azide on levels of cyclic nucleotides and mechanical activity of various tissues. J Cyclic Nucleotide Res 3: 239–247, 1977
57. Gill GN, McCune RW: Guanosine 3′,5′-monophosphate-dependent protein kinase. Curr Top Cell Regul 15: 1–45, 1979
58. Murad F, Arnold WP, Mittal CK, Braughler JM: Properties and regulation of guanylate cyclase and some proposed functions for cyclic GMP. Adv Cyclic Nucleotide Res 11: 175–204, 1979
59. Bohme E, Grossmann G, Herz J, Mulsch A, Spies C, Schultz G: Regulation of cyclic GMP formation by soluble guanylate cyclase: Stimulation by NO-containing compounds. Adv Cyclic Nucleotide Prot Phosphoryl Res 17: 259–266, 1984
60. Waldman SA, Rapoport RM, Murad F: Atrial natriuretic factor selectively activates particulate guanylate cyclase and elevates cyclic GMP in rat tissues. J Biol Chem 259: 14332–14334, 1984
61. Waldman SA, Sinacore MS, Lewicki JA, Chang LY, Murad F: Selective activation of particulate guanylate cyclase by a specific class of porphyrins. J Biol Chem 259: 4038–4042, 1984
62. Horio Y, Murad F: Purification of guanylyl cyclase from rod outer segments. Biochim Biophys Acta 1133: 81–88, 1991
63. Sharma RK, Jaiswal RK, Duda T: Second messenger role of cyclic GMP in atrial natriuretic factor receptor mediated signal transduction: 180-kD membrane guanylate cyclase, its coupling with atrial natriuretic factor receptor and its regulation by protein kinase C. In: Biological and Molecular Aspects of Atrial Factors. Alan R. Liss, Inc., 1988, pp 77–96
64. Sharma RK, Marala RB, Paul AK: Mediatory role of cyclic GMP in receptor-mediated signal transduction: Membrane guanylate cyclase and its coupling with atrial natriuretic factor receptor. In: B.M. Brenner, J.H. Laragh (eds). Advances in Peptide research, vol. II: American Society of Hypertension Symposium Series. Raven Press, New York, 1988, pp 61–77
65. Sharma RK, Duda T, Goraczniak RM, Sitaramayya A: Membrane guanylate cyclase signal transduction system. Indian J Biochem Biophys 34: 40–49, 1997
66. Pugh EN Jr, Duda T, Sitaramayya A, Sharma RK: Photoreceptor guanylate cyclases: A review. Biosci Rep 17: 29–73, 1997
67. Rall TW, Sutherland EW: The regulatory role of adenosine-3′,5′-phosphate. Cold Spring Harbor Symposia on Quantitative Biology 26: 347–354, 1961
68. Haynes RC Jr, Berthet L: Studies on the mechanism of action of the adrenocortictropic hormone. J Biol Chem 225: 115–124, 1957
69. Haynes RC Jr: The activation of adrenal phosphorylase by the adrenocorticotropic hormone. J Biol Chem 233: 1220–1222, 1958
70. Sayers G: In: C.H. Gray (ed). Hormones in Blood. Academic Press, New York and London, 1967, pp 169–194
71. Garren LD: The mechanism of action of adrenocorticotropic hormone. Vitam Horm 26: 119–145, 1968
72. Bronsome ED Jr: Adrenal cortex. Annu Rev Physiol 30: 171–212, 1968
73. Halkerston IDK: Cyclic AMP and adrenocortical function. Adv Cyclic Nucleotide Res 6: 99–136, 1975
74. Sharma RK, Sawhney RS: Metabolic regulation of steroidogenesis in isolated adrenal cell. Investigation of the adrenocorticotropic hormone, guanosine 3′,5′-monophosphate, and adenosine 3′,5′-monophosphate control step. Biochemistry 17: 316–321, 1978
75. Robinson GA, Butcher RW, Sutherland EW: In: Cyclic AMP. Academic Press, New York, 1971
76. Haynes RC Jr, Koritz SB, Peron FG: Influence of adenosine 3′,5′-monophosphate on corticoid production by rat adrenal glands. J Biol Chem 234: 1421–1423, 1959
77. Farese RV, Linarelli LG, Glinsmann WH, Ditzion BR, Paul MI, Pauk GA: Persistence of the steroidogenic effect of adenosin-3′,5′-monophosphate in vitro: evidence for a third factor during the steroidogenic effect of ACTH. Endocrinology 85: 867–874, 1969
78. Tsang CP, Peron FG: Effects of adenosine-3′,5′-monophosphate on steroidogenesis and glycolysis in the rat adrenal gland incubated *in vitro*. Steroids 17: 453–469, 1971
79. Rivkin I, Chasin M: Nucleotide specificity of the steroidogenic response of rat adrenal cell suspensions prepared by collagenase digestion. Endocrinology 88: 664–670, 1971
80. Scarpa A, Baldassare J, Inesi G: The effect of calcium ionophores on fragmented sarcoplasmic reticulum. J Gen Physiol 60: 735–749, 1972
81. Sharma RK, Hashimoto K, Kitabchi AE: Steroidogenesis in isolated adrenal cells of rat. 3. Morphological and biochemical correlation of cholesterol and cholesterol ester content in ACTH and N6-2′-O-dibutyryl-adenosine-3′,5′-monophosphate activated adrenal cells. Endocrinology 91: 994–1003, 1972
82. Sharma RK: Abnormal Adrenocorticotropic hormone control in adrenocortical carcinoma. In: Sharma and Criss (eds). Endocrine Control of Neoplasia. Raven Press, New York, 1978, pp 13–52
83. Sharma RK, Ahmed NK, Sutliff L, Brush JS: Metabolic regulation of steroidogenesis in isolated adrenal cells of the rat. ACTH regulation of cGMP and cAMP levels and steroidogenesis. FEBS Lett 45: 107–110, 1974
84. Sharma RK, Ahmed NK, Shanker G: Metabolic regulation of steroidogenesis in isolated adrenal cells of rat. Relationship of adrenocorticotropin – adenosine 3′:5′-monophosphate – and guanosine 3′:5′-monophosphate-stimulated steroidogenesis with the activation of protein kinase. Eur J Biochem 70: 427–433, 1976
85. Perchellet JP, Shanker G, Sharma RK: Regulatory role of guanosine 3′,5′-monophosphate in adrenocorticotropin hormone-induced steroidogenesis. Science 199: 311–312, 1978
86. Sharma RK, Hashimoto K: Ultrastructural studies and metabolic regulation of isolated adrenocortical carcinoma cells of rat. Cancer Res 32: 666–674, 1972
87. Harrington CA, Fenimore DC, Farmer RW: Regulation of adrenocortical steroidogenesis by cyclic 3′-5′-guanosine monophosphate in isolated rat adrenal cells. Biochem Biophys Res Commun 85: 55–61, 1978
88. Neri G, Gambino AM, Mazzocchi G, Nussdorfer GG: Effects of chronic treatment with ACTH on the intracellular levels of cyclic-AMP and cyclic-GMP in the rat adrenal cortex. Experientia 34: 815–817, 1978
89. Sharma RK: Metabolic regulation of steroidogenesis in adrenocortical carcinoma cells of rat. Effect of adrenocorticotropin and adenosine cyclic 3′:5′-monophosphate on corticosteroidogenesis. Eur J Biochem 32: 506–512, 1973
90. Haksar A, Peron FG: The role of calcium in the steroidogenic response of rat adrenal cells to adrenocorticotropic hormone. Biochim Biophys Acta 313: 363–371, 1973
91. Bowyer F, Kitabchi AE: Dual role of calcium in steroidogenesis in the isolated adrenal cell of rat. Biochem Biophys Res Commun 57: 100–105, 1974
92. Perchellet JP, Sharma RK: Mediatory role of calcium and guanosine 3′,5′-monophosphate in adrenocorticotropin-induced steroidogenesis by adrenal cells. Science 203: 1259–1261, 1979
93. Rasmussen H: In: Calcium and Cyclic AMP as Synarchic Messengers. John Wiley & Sons Inc., New York, 1981
94. Kitabchi AE, Wilson DB, Sharma RK: Steroidogenesis in isolated adrenal cells of rat. II. Effect of caffeine on ACTH and cyclic nucleotide-induced steroidogenesis and its relation to cyclic nucleotide

phosphodiesterase (PDE). Biochem Biophys Res Commun 44: 898–904, 1971
95. Hayashi K, Sala G, Catt K, Dufau ML: Regulation of steroidogenesis by adrenocorticotropic hormone in isolated adrenal cells. J Biol Chem 254: 6678–6683, 1979
96. Sayers G, Ma RM, Giordano ND: Isolated adrenal cells: corticosterone production in response to cyclic AMP (adenosine-3′,5′-monophosphate). Proc Soc Exp Biol Med 136: 619–622, 1971
97. Laychock SG, Hardman JG: Effects of sodium nitroprusside and ascorbic acid on rat adrenocortical cell cGMP levels and steroidogenesis. J Cyclic Nucleotide Res 4: 335–344, 1978
98. Ahrens H, Paul AK, Kuroda Y, Sharma RK: Adrenocortical cyclic GMP-dependent protein kinase: purification, characterization, and modification of its activity by calmodulin, and its relationship with steroidogenesis. Arch Biochem Biophys 215: 597–609, 1982
99. Nambi P, Aiyar NV, Sharma RK: Adrenocorticotropin-dependent particulate guanylate cyclase in rat adrenal and adrenocortical carcinoma: Comparison of its properties with soluble guanylate cyclase and its relationship with AC TH-induced steroidogenesis. Arch Biochem Biophys 217: 638–646, 1982
100. Nambi P, Sharma RK: Adrenocorticotropic hormone-responsive guanylate cyclase in the particulate fraction of rat adrenal glands. Endocrinology 108: 2025–2027, 1981
101. Nambi P, Sharma RK: Demonstration of ACTH-sensitive particulate guanylate cyclase in adrenocortical carcinoma. Biochem Biophys Res Commun 100: 508–514, 1981
102. Sharma RK, Marala RB, Duda T: Purification and characterization of the 180-kDa membrane guanylate cyclase containing atrial natriuretic factor receptor from rat adrenal gland and its regulation by protein kinase C. Steroids 53: 437–460, 1989
103. Anglard P, Zwiller J, Vincendon G, Louis JC: Regulation of cyclic AMP and cyclic GMP levels by adrenocorticotropic hormone in cultured neurons. Biochem Biophys Res Commun 133: 286–292, 1985
104. Hamet P, Tremblay J, Pang SC, Garcia R, Thibault G, Gutkowska J, Cantin M, Genest J: Effect of native and synthetic atrial natriuretic factor on cyclic GMP. Biochem Biophys Res Commun 123: 515–527, 1984
105. Paul AK: Particulate guanylate cyclase from adrenocortical carcinoma 494. Purification, biochemical and immunological characterization. Doctoral thesis, University of Tennessee, 1986
106. Paul AK, Marala RB, Jaiswal RK, Sharma RK: Coexistence of guanylate cyclase and atrial natriuretic factor receptor in a 180-kD protein. Science 235: 1224–1226, 1987
107. Sharma RK: Guanylate cyclase and the adrenal natriuretic factor receptor. Response to S.A. Waldman, D.C. Leitman, J. Andresen, F. Murad. Science 240: 805–806, 1988
108. Kuno T, Andresen JW, Kamisaki Y, Waldman SA, Chang LY, Saheki S, Leitman DC, Nakane M, Murad F: Co-purification of an atrial natriuretic factor receptor and particulate guanylate cyclase from rat lung. J Biol Chem 261: 5817–5823, 1986
109. Takayanagi R, Inagami T, Snajdar RM, Imada T, Tamura M, Misono KS: Two distinct forms of receptors for atrial natriuretic factor in bovine adrenocortical cells. Purification, ligand binding, and peptide mapping. J Biol Chem 262: 12104–12113, 1987
110. Meloche S, McNicoll N, Liu B, Ong H, De Lean A: Atrial natriuretic factor R1 receptor from bovine adrenal zona glomerulosa: Purification, characterization, and modulation by amiloride. Biochemistry 27: 8151–8158, 1988
111. Marala RB, Sharma RK: Characterization of atrial-natriuretic-factor-receptor-coupled membrane guanylate cyclase from rat and mouse testes. Biochem J 251: 301–304, 1988
112. Ballermann BJ, Marala RB, Sharma RK: Characterization and regulation by protein kinase C of renal glomerular atrial natriuretic peptide receptor-coupled guanylate cyclase. Biochem Biophys Res Commun 157: 755–761, 1988
113. de Bold AJ: Atrial natriuretic factor of the rat heart. Studies on isolation and properties. Proc Soc Exp Biol Med 170: 133–138, 1982
114. Cantin M, Genest J: The heart, an endocrine gland. Ann Endocrinol (Paris) 46: 219–228, 1985
115. Schwartz D, Geller DM, Manning PT, Siegel NR, Fok KF, Smith CE, Needleman P: Ser-Leu-Arg-Arg-atriopeptin III: The major circulating form of atrial peptide. Science 229: 397–400, 1985
116. de Bold AJ: Atrial natriuretic factor: An overview. Fed Proc 45: 2081–2085, 1986
117. Atlas SA, Laragh JH: Atrial natriuretic peptide: A new factor in hormonal control of blood pressure and electrolyte homeostasis. Annu Rev Med 37: 397–414, 1986
118. Atarashi K, Mulrow PJ, Franco-Saenz R, Snajdar R, Rapp J: Inhibition of aldosterone production by an atrial extract. Science 224: 992–994, 1984
119. Chartier L, Schiffrin E, Thibault G, Garcia R: Atrial natriuretic factor inhibits the stimulation of aldosterone secretion by angiotensin II, ACTH and potassium *in vitro* and angiotensin II-induced steroidogenesis *in vivo*. Endocrinology : 2026–2028, 1984
120. De Lean A, Racz K, Gutkowska J, Nguyen TT, Cantin M, Genest J: Specific receptor-mediated inhibition by synthetic atrial natriuretic factor of hormone-stimulated steroidogenesis in cultured bovine adrenal cells. Endocrinology 115: 1636–1638, 1984
121. Kudo T, Baird A: Inhibition of aldosterone production in the adrenal glomerulosa by atrial natriuretic factor. Nature 312: 756–757, 1984
122. Pandey KN, Kovacs WJ, Inagami T: The inhibition of progesterone secretion and the regulation of cyclic nucleotides by atrial natriuretic factor in gonadotropin responsive murine Leydig tumor cells. Biochem Biophys Res Commun 133: 800–806, 1985
123. Nakamura M, Odaguchi K, Shimizu T, Nakamura Y, Okamoto M: Stimulation of corticosterone production by atrial natriuretic polypeptide in hypophysectomized rats. Eur J Pharmacol 117: 285–286, 1985
124. Jaiswal N, Paul AK, Jaiswal RK, Sharma RK: Atrial natriuretic factor regulation of cyclic GMP levels and steroidogenesis in isolated fasciculata cells of rat adrenal cortex. FEBS Lett 199: 121–124, 1986
125. Bex F, Corbin A: Atrial natriuretic factor stimulates testosterone production by mouse interstitial cells. Eur J Pharmacol 115: 125–126, 1985
126. Mukhopadhyay AK, Schumacher M, Leidenberger FA: Steroidogenic effect of atrial natriuretic factor in isolated mouse Leydig cells is mediated by cyclic GMP. Biochem J 239: 463–467, 1986
127. Pandey KN, Inagami T, Misono KS: Atrial natriuretic factor receptor on cultured Leydig tumor cells: Ligand binding and photoaffinity labeling. Biochemistry 25: 8467–8472, 1986
128. Chinkers M, Garbers DL, Chang MS, Lowe DG, Chin HM, Goeddel DV, Schulz S: A membrane form of guanylate cyclase is an atrial natriuretic peptide receptor. Nature 338: 78–83, 1989
129. Lowe DG, Chang MS, Hellmiss R, Chen E, Singh S, Garbers DL, Goeddel DV: Human atrial natriuretic peptide receptor defines a new paradigm for second messenger signal transduction. EMBO J: 1377–13784, 1989
130. Schulz S, Singh S, Bellet RA, Singh G, Tubb DJ, Chin H, Garbers DL: The primary structure of a plasma membrane guanylate cyclase demonstrates diversity within this new receptor family. Cell: 1155–1162, 1989
131. Chang MS, Lowe DG, Lewis M, Hellmiss R, Chen E, Goeddel DV: Differential activation by atrial and brain natriuretic peptides of two different receptor guanylate cyclases. Nature 341: 68–72, 1989
132. Pandey KN, Singh S: Molecular cloning and expression of murine guanylate cyclase/atrial natriuretic factor receptor cDNA. J Biol Chem 265: 12342–12348, 1990

133. Marala R, Duda T, Goraczniak RM, Sharma RK: Genetically tailored atrial natriuretic factor-dependent guanylate cyclase. Immunological and functional identity with 180 kDa membrane guanylate cyclase and ATP signaling site. FEBS Lett 296: 254–258, 1992
134. Schulz S, Green CK, Yuen PS, Garbers DL: Guanylyl cyclase is a heat-stable enterotoxin receptor. Cell 63: 941–948, 1990
135. Singh S, Singh G, Heim JM, Gerzer R: Isolation and expression of a guanylate cyclase-coupled heat stable enterotoxin receptor cDNA from a human colonic cell line. Biochem Biophys Res Commun 179: 1455–1463, 1991
136. de Sauvage FJ, Camerato TR, Goeddel DV: Primary structure and functional expression of the human receptor for *Escherichia coli* heat-stable enterotoxin. J Biol Chem 266: 17912–17918, 1991
137. Wada A, Hirayama T, Kitao S, Fujisawa J, Hidaka Y, Shimonishi Y: Pig intestinal membrane-bound receptor (guanylyl cyclase) for heat-stable enterotoxin: cDNA cloning, functional expression, and characterization. Microbiol Immunol 38: 535–541, 1994
138. Currie MG, Fok KF, Kato J, Moore RJ, Hamra FK, Duffin KL, Smith CE: Guanylin: An endogenous activator of intestinal guanylate cyclase. Proc Natl Acad Sci USA 89: 947–951, 1992
139. Wiegand RC, Kato J, Currie MG: Rat guanylin cDNA: Characterization of the precursor of an endogenous activator of intestinal guanylate cyclase. Biochem Biophys Res Commun 185: 812–817, 1992
140. Hamra FK, Forte LR, Eber SL, Pidhorodeckyj NV, Krause WJ, Freeman RH, Chin DT, Tompkins JA, Fok KF, Smith CE, Duffin KL, Siegel NR, Currie MG: Uroguanylin: Structure and activity of a second endogenous peptide that stimulates intestinal guanylate cyclase. Proc Natl Acad Sci USA: 10464–10468, 1993
141. Khare S, Wilson D, Wali RK, Tien XY, Bissonnette M, Niedziela SM, Bolt MJ, Sitrin MD, Brasitus TA: Guanylin activates rat colonic particulate guanylate cyclase. Biochem Biophys Res Commun 203: 1432–1437, 1994
142. Koch KW: Purification and identification of photoreceptor guanylate cyclase. J Biol Chem 266: 8634–8637, 1991
143. Hayashi F, Yamazaki A: Polymorphism in purified guanylate cyclase from vertebrate rod photoreceptors. Proc Natl Acad Sci USA 88: 4746–4750, 1991
144. Margulis A, Goraczniak RM, Duda T, Sharma RK, Sitaramayya A: Structural and biochemical identity of retinal rod outer segment membrane guanylate cyclase. Biochem Biophys Res Commun 194: 855–861, 1993
145. Horio Y, Murad F: Solubilization of guanylyl cyclase from bovine rod outer segments and effects of lowering Ca^{2+} and nitro compounds. J Biol Chem 266: 3411–3415, 1991
146. Goraczniak RM, Duda T, Sitaramayya A, Sharma RK: Structural and functional characterization of the rod outer segment membrane guanylate cyclase. Biochem J 302: 455–461, 1994
147. Shyjan AW, de Sauvage FJ, Gillett NA, Goeddel DV, Lowe DG: Molecular cloning of a retina-specific membrane guanylyl cyclase. Neuron 9: 727–737, 1992
148. Lowe DG, Dizhoor AM, Liu K, Gu Q, Spencer M, Laura R, Lu L, Hurley JB: Cloning and expression of a second photoreceptor-specific membrane retina guanylyl cyclase (RetGC), RetGC-2. Proc Natl Acad Sci USA 92: 5535–5539, 1995
149. Goraczniak RM, Duda T, Sharma RK: Calcium modulated signaling site in type 2 rod outer segment membrane guanylate cyclase (ROS-GC2). Biochem Biophys Res Commun 245: 447–453, 1998
150. Duda T, Venkataraman V, Krishnan A, Sharma RK: Rod outer segment membrane guanylate cyclase type 1 (ROS-GC1) gene: Structure, organization and regulation by phorbol ester, a protein kinase C activator. Mol Cell Biochem 189: 63–70, 1998
151. Yang RB, Fulle HJ, Garbers DL: Chromosomal localization and genomic organization of genes encoding guanylyl cyclase receptors expressed in olfactory sensory neurons and retina. Genomics 31: 367–372, 1996
152. Johnston JP, Farhangfar F, Aparicio JG, Nam SH, Applebury ML: The bovine guanylate cyclase GC-E gene and 5′ flanking region. Gene 193: 219–227, 1997
153. Duda T, Venkataraman V, Goraczniak R, Lange C, Koch KW, Sharma RK: Functional consequences of a rod outer segment membrane guanylate cyclase (ROS-GC1) gene mutation linked with Leber's congenital amaurosis. Biochemistry 38: 509–515, 1999
154. Duda T, Krishnan A, Venkataraman V, Lange C, Koch KW, Sharma RK: Mutations in the rod outer segment membrane guanylate cyclase in a cone-rod dystrophy cause defects in calcium signaling. Biochemistry 38: 13912–13919, 1999
155. Duda T, Venkataraman V, Jankowska A, Lange C, Koch KW, Sharma RK: Impairment of the rod outer segment membrane guanylate cyclase dimerization in a cone-rod dystrophy results in defective calcium signaling. Biochemistry 39: 12522–12533, 2000
156. Tucker CL, Woodcock SC, Kelsell RE, Ramamurthy V, Hunt DM, Hurley JB: Biochemical analysis of a dimerization domain mutation in RetGC-1 associated with dominant cone-rod dystrophy. Proc Natl Acad Sci USA 96: 9039–9044, 1999
157. Wilkie SE, Newbold RJ, Deery E, Walker CE, Stinton I, Ramamurthy V, Hurley JB, Bhattacharya SS, Warren MJ, Hunt DM: Functional characterization of missense mutations at codon 838 in retinal guanylate cyclase correlates with disease severity in patients with autosomal dominant cone-rod dystrophy. Hum Mol Genet 9: 3065–3073, 2000
158. Ramamurthy V, Tucker C, Wilkie SE, Daggett V, Hunt DM, Hurley JB: Interactions within the coiled-coil domain of RetGC-1 guanylyl cyclase are optimized for regulation rather than for high affinity. J Biol Chem 276: 26218–26229, 2001
159. Fulle HJ, Vassar R, Foster DC, Yang RB, Axel R, Garbers DL: A receptor guanylyl cyclase expressed specifically in olfactory sensory neurons. Proc Natl Acad Sci USA 92: 3571–3575, 1995
160. Koch KW, Stryer L: Highly cooperative feedback control of retinal rod guanylate cyclase by calcium ions. Nature 334: 64–66, 1988
161. Palczewski K, Subbaraya I, Gorczyca WA, Helekar BS, Ruiz CC, Ohguro H, Huang J, Zhao X, Crabb JW, Johnson RS, Walsh KA, Gray-Keller MP, Detwiler PB, Baehr W: Molecular cloning and characterization of retinal photoreceptor guanylyl cyclase-activating protein. Neuron 13: 395–404, 1994
162. Dizhoor AM, Olshevskaya EV, Henzel WJ, Wong SC, Stults JT, Ankoudinova I, Hurley JB: Cloning, sequencing, and expression of a 24-kDa Ca(2+)-binding protein activating photoreceptor guanylyl cyclase. J Biol Chem 270: 25200–25206, 1995
163. Frins S, Bonigk W, Muller F, Kellner R, Koch KW: Functional characterization of a guanylyl cyclase-activating protein from vertebrate rods. Cloning, heterologous expression, and localization. J Biol Chem 271: 8022–8027, 1996
164. Pozdnyakov N, Yoshida A, Cooper NG, Margulis A, Duda T, Sharma RK, Sitaramayya A: A novel calcium-dependent activator of retinal rod outer segment membrane guanylate cyclase. Biochemistry 34: 14279–14283, 1995
165. Pozdnyakov N, Goraczniak R, Margulis A, Duda T, Sharma RK, Yoshida A, Sitaramayya A: Structural and functional characterization of retinal calcium-dependent guanylate cyclase activator protein (CD-GCAP): Identity with S100beta protein. Biochemistry 36: 14159–14166, 1997
166. Haeseleer F, Sokal I, Li N, Pettenati M, Rao N, Bronson D, Wechter R, Baehr W, Palczewski K: Molecular characterization of a third member of the guanylyl cyclase-activating protein subfamily. J Biol Chem 274: 6526–6535, 1999
167. Kumar VD, Vijay-Kumar S, Krishnan A, Duda T, Sharma RK: A second calcium regulator of rod outer segment membrane guanylate

cyclase, ROS-GC1: Neurocalcin. Biochemistry 38: 12614–12620, 1999
168. Liu X, Seno K, Nishizawa Y, Hayashi F, Yamazaki A, Matsumoto H, Wakabayashi T, Usukura J: Ultrastructural localization of retinal guanylate cyclase in human and monkey retinas. Exp Eye Res 59: 761–768, 1994
169. Cooper N, Liu L, Yoshida A, Pozdnyakov N, Margulis A, Sitaramayya A: The bovine rod outer segment guanylate cyclase, ROS-GC, is present in both outer segment and synaptic layers of the retina. J Mol Neurosci 6: 211–222, 1995
170. Venkataraman V, Nagele R, Duda T, Sharma RK: Rod outer segment membrane guanylate cyclase type 1-linked stimulatory and inhibitory calcium signaling systems in the pineal gland: Biochemical, molecular, and immunohistochemical evidence. Biochemistry 39: 6042–6052, 2000
171. Duda T, Venkataraman V, Krishnan A, Nagele RG, Sharma RK: Negatively calcium-modulated membrane guanylate cyclase signaling system in the rat olfactory bulb. Biochemistry 40: 465446-465462, 2001
172. Venkataraman V, Duda T, Sharma RK: The alpha$_{2D/A}$-adrenergic receptor-linked membrane guanylate cyclase: A new signal transduction system in the pineal gland. FEBS Lett 427: 69–73, 1998
173. Perchellet JP, Sharma RK: Ectopic alpha-adrenergic mediated accumulation of guanosine 3′,5′-monophosphate in isolated adrenocortical carcinoma cells. Endocrinology 106: 1589–1593, 1980
174. Shanker G, Sharma RK: Characterization of ectopic alpha-adrenergic binding receptors of adrenocortical carcinoma cells. Endocrinology 106: 1594–1598, 1980
175. Sharma RK: Two operational modes of transmembrane migration of cyclic GMP signaling pathway. In: K.N. Prasad, F.L. Meyskens Jr (eds). Nutrients and Cancer Prevention. The Humana Press, 1990, pp 3–18
176. Jaiswal RK, Sharma RK: Purification and biochemical characterization of alpha 2-adrenergic receptor from the rat adrenocortical carcinoma. Biochem Biophys Res Commun 130: 58–64, 1985
177. Nambi P, Aiyar NV, Sharma RK: Solubilization of epinephrine-specific alpha-adrenergic receptors from adrenocortical carcinoma. FEBS Lett 140: 98–102, 1982
178. Nambi P, Aiyar NV, Sharma RK: Identification and characterization of ectopic alpha2-adrenergic receptors in adrenocortical carcinoma membranes. J Nutr Growth Cancer 1: 77–84, 1983
179. Chalberg SC, Duda T, Rhine JA, Sharma RK: Molecular cloning, sequencing and expression of an alpha2-adrenergic receptor complementary DNA from rat brain. Mol Cell Biochem 97: 161–172, 1990
180. Lanier SM, Downing S, Duzic E, Homcy CJ: Isolation of rat genomic clones encoding subtypes of the alpha2-adrenergic receptor. Identification of a unique receptor subtype. J Biol Chem 266: 10470–10478, 1991
181. O'Rourke MF, Iversen LJ, Lomasney JW, Bylund DB: Species orthologs of the alpha-2A adrenergic receptor: The pharmacological properties of the bovine and rat receptors differ from the human and porcine receptors. J Pharmacol Exp Ther 271: 735–740, 1994
182. Wypijewski K, Duda T, Sharma RK: Structural, genetic and pharmacological identity of the rat alpha2-adrenergic receptor subtype cA2-47 and its molecular characterization in rat adrenal, adrenocortical carcinoma and bovine retina. Mol Cell Biochem 144: 181–190, 1995
183. Kobilka BK, Matsui H, Kobilka TS, Yang-Feng TL, Francke U, Caron MG, Lefkowitz RJ, Regan JW: Cloning, sequencing, and expression of the gene coding for the human platelet alpha2-adrenergic receptor. Science 238: 650–656, 1987
184. Regan JW, Nakata H, DeMarinis RM, Caron MG, Lefkowitz RJ: Purification and characterization of the human platelet alpha2-adrenergic receptor. J Biol Chem 261: 3894–3900, 1986
185. Bylund DB, Eikenberg DC, Hieble JP, Langer SZ, Lefkowitz RJ, Minneman KP, Molinoff PB, Ruffolo RR Jr, Trendelenburg U: International Union of Pharmacology nomenclature of adrenoceptors. Pharmacol Rev 46: 121–136, 1994
186. Sharma RK, Duda T: Plasma membrane guanylate cyclase. A multimodule transduction system. Adv Exp Med Biol 407: 271–279, 1997
187. Duda T, Goraczniak RM, Sharma RK: Site-directed mutational analysis of a membrane guanylate cyclase cDNA reveals the atrial natriuretic factor signaling site. Proc Natl Acad Sci USA 88: 7882–7886, 1991
188. Duda T, Goraczniak RM, Sharma RK: Glutamic acid-332 residue of the type C natriuretic peptide receptor guanylate cyclase is important for signaling. Biochemistry 33: 7430–7433, 1994
189. Duda T, Goraczniak RM, Sharma RK: Single amino acid residue-linked signaling shifts in the transduction activities of atrial and type C natriuretic factor receptor guanylate cyclases. Biochem Biophys Res Commun 212: 1046–1053, 1995
190. van den Akker F, Zhang X, Miyagi M, Huo X, Misono KS, Yee VC: Structure of the dimerized hormone-binding domain of a guanylyl-cyclase-coupled receptor. Nature 406: 101–104, 2000
191. Kurose H, Inagami T, Ui M: Participation of adenosine 5′-triphosphate in the activation of membrane-bound guanylate cyclase by the atrial natriuretic factor. FEBS Lett 219: 375–379, 1987
192. Chang CH, Kohse KP, Chang B, Hirata M, Jiang B, Douglas JE, Murad F: Characterization of ATP-stimulated guanylate cyclase activation in rat lung membranes. Biochim Biophys Acta 1052: 159–165, 1990
193. Chinkers M, Garbers DL: The protein kinase domain of the ANP receptor is required for signaling. Science 245: 1392–1394, 1989
194. Marala RB, Sitaramayya A, Sharma RK: Dual regulation of atrial natriuretic factor-dependent guanylate cyclase activity by ATP. FEBS Lett 281: 73–76, 1991
195. Chinkers M, Singh S, Garbers DL: Adenine nucleotides are required for activation of rat atrial natriuretic peptide receptor/guanylyl cyclase expressed in a baculovirus system. J Biol Chem 266: 4088–4093, 1991
196. Duda T, Goraczniak RM, Sitaramayya A, Sharma RK: Cloning and expression of an ATP-regulated human retina C-type natriuretic factor receptor guanylate cyclase. Biochemistry 32: 1391–1395, 1993
197. Lowe DG: The guanylate cylase-B receptor. Natriuretic peptides in health and disease. Series: Contemp Endocrinol 5: 35–50, 1997
198. Foster DC, Garbers DL: Dual role for adenine nucleotides in the regulation of the atrial natriuretic peptide receptor, guanylyl cyclase-A. Biol Chem 273: 16311–16318, 1998
199. Duda T, Yadav P, Jankowska A, Venkataraman V, Sharma RK: Three dimensional atomic model and experimental validation for the ATP-Regulated Module (ARM) of the atrial natriuretic factor receptor guanylate cyclase. Mol Cell Biochem. 214: 7–14, 2000. Reprinted in: *ibid* 217: 165–172, 2001
200. Goraczniak RM, Duda T, Sharma RK: A structural motif that defines the ATP-regulatory module of guanylate cyclase in atrial natriuretic factor signalling. Biochem J 282: 533–537, 1992
201. Sharma RK, Yadav P, Duda T: Allosteric regulatory step and configuration of the ATP-binding pocket in atrial natriuretic factor receptor guanylate cyclase transduction mechanism. Can J Physiol Pramacol 79: 682–691, 2001
202. Wierenga RK, Hol WG: Predicted nucleotide-binding properties of p21 protein and its cancer-associated variant. Nature 302: 842–844, 1983
203. Hanks SK, Quinn AM, Hunter T: The protein kinase family: conserved features and deduced phylogeny of the catalytic domains. Science 241: 42–52, 1988
204. Duda T, Goraczniak RM, Sharma RK: Core sequence of ATP regulatory module in receptor guanylate cyclases. FEBS Lett 315: 143–148, 1993

205. Koller KJ, de Sauvage FJ, Lowe DG, Goeddel DV: Conservation of the kinase-like regulatory domain is essential for activation of the natriuretic peptide receptor guanylyl cyclases. Mol Cell Biol 12: 2581–2590, 1992
206. Duda T, Goraczniak RM, Sharma RK: The glycine residue of ATP regulatory module in receptor guanylate cyclases that is essential in natriuretic factor signaling. FEBS Lett 335: 309–314, 1993
207. Larose L, McNicoll N, Ong H, De Lean A: Allosteric modulation by ATP of the bovine adrenal natriuretic factor R1 receptor functions. Biochemistry 30: 8990–8995, 1991
208. Jewett JR, Koller KJ, Goeddel DV, Lowe DG: Hormonal induction of low affinity receptor guanylyl cyclase. EMBO J 12: 769–777, 1993
209. Duda T, Sharma RK: ATP bimodal switch that regulates the ligand binding and signal transduction activities of the atrial natriuretic factor receptor guanylate cyclase. Biochem Biophys Res Commun 209: 286–292, 1995
210. Duda T, Sharma RK: ATP modulation of the ligand binding and signal transduction activities of the type C natriuretic peptide receptor guanylate cyclase. Mol Cell Biochem 152: 179–183, 1995
211. Sharma RK, Duda T, Sitaramayya A: Plasma membrane guanylate cyclase is a multimodule transduction system. Amino Acids 7: 117–127, 1994
212. Potter LR, Hunter T: Identification and characterization of the major phosphorylation sites of the B-type natriuretic peptide receptor. J Biol Chem 273: 15533–15539, 1998
213. Potter LR, Hunter T: A constitutively 'phosphorylated' guanylyl cyclase-linked atrial natriuretic peptide receptor mutant is resistant to desensitization. Mol Biol Cell 10: 1811–1820, 1999
214. Potter LR, Hunter T: Identification and characterization of the phosphorylation sites of the guanylyl cyclase-linked natriuretic peptide receptors A and B. Methods 19: 506–520, 1999
215. Duda T, Sharma RK: Regulation of guanylate cyclase activity by atrial natriuretic factor and protein kinase C. Mol Cell Biochem 93: 179–184, 1990
216. Larose L, Rondeau JJ, Ong H, De Lean A: Phosphorylation of atrial natriuretic factor R1 receptor by serine/threonine protein kinases: Evidences for receptor regulation. Mol Cell Biochem 115: 203–211, 1992
217. Krishnan A, Goraczniak RM, Duda T, Sharma RK: Third calcium-modulated rod outer segment membrane guanylate cyclase transduction mechanism. Mol Cell Biochem 178: 251–259, 1998
218. Duda T, Goraczniak R, Surgucheva I, Rudnicka-Nawrot M, Gorczyca WA, Palczewski K, Sitaramayya A, Baehr W, Sharma RK: Calcium modulation of bovine photoreceptor guanylate cyclase. Biochemistry 35: 8478–8482, 1996
219. Laura RP, Dizhoor AM, Hurley JB: The membrane guanylyl cyclase, retinal guanylyl cyclase-1, is activated through its intracellular domain. J Biol Chem 271: 11646–11651, 1996
220. Sokal I, Li N, Verlinde CL, Haeseleer F, Baehr W, Palczewski K: Ca(2+)-binding proteins in the retina: From discovery to etiology of human disease. Biochim Biophys Acta 1498: 233–251, 2000
221. Downes SM, Holder GE, Fitzke FW, Payne AM, Warren MJ, Bhattacharya SS, Bird AC: Autosomal dominant cone and cone-rod dystrophy with mutations in the guanylate cyclase activator 1A gene-encoding guanylate cyclase activating protein-1. Arch Ophthalmol 119: 96–105, 2001
222. Dizhoor AM: Regulation of cGMP synthesis in photoreceptors: Role in signal transduction and congenital diseases of the retina. Cell Signal 12: 711–719, 2000
223. Lange C, Duda T, Beyermann M, Sharma RK, Koch KW: Regions in vertebrate photoreceptor guanylyl cyclase ROS-GC1 involved in Ca(2+)-dependent regulation by guanylyl cyclase-activating protein GCAP-1. FEBS Lett 460: 27–31, 1999
224. Perrault I, Rozet JM, Calvas P, Gerber S, Camuzat A, Dollfus H, Chatelin S, Souied E, Ghazi I, Leowski C, Bonnemaison M, Le Paslier D, Frezal J, Dufier JL, Pittler S, Munnich A, Kaplan J: Retinal-specific guanylate cyclase gene mutations in Leber's congenital amaurosis. Nat Genet 14: 461–464, 1996
225. Sokal I, Haeseleer F, Arendt A, Adman ET, Hargrave PA, Palczewski K: Identification of a guanylyl cyclase-activating protein-binding site within the catalytic domain of retinal guanylyl cyclase 1. Biochemistry 38: 1387–1393, 1999
226. Krylov DM, Hurley JB: Identification of proximate regions in a complex of retinal guanylyl cyclase 1 and guanylyl cyclase activating protein-1 by a novel mass spectrometry based method. J Biol Chem 2001
227. Tucker CL, Laura RP, Hurley JB: Domain-specific stabilization of photoreceptor membrane guanylyl cyclase by adenine nucleotides and guanylyl cyclase activating proteins (GCAPs) Biochemistry 36: 11995–12000, 1997
228. Laura RP, Hurley JB: The kinase homology domain of retinal guanylyl cyclases 1 and 2 specifies the affinity and cooperativity of interaction with guanylyl cyclase activating protein-2. Biochemistry 37: 11264–11271, 1998
229. McInnes C, Sykes BD: Growth factor receptors: structure, mechanism, and drug discovery. Biopolymers 43: 339–366, 1997
230. Heldin CH: Dimerization of cell surface receptors in signal transduction. Cell 80: 213–223, 1995
231. Tesmer JJ, Sunahara RK, Gilman AG, Sprang SR: Crystal structure of the catalytic domains of adenylyl cyclase in a complex with Gsalpha. GTPgammaS. Science 278: 1907–1916, 1997
232. Giuili G, Scholl U, Bulle F, Guellaen G: Molecular cloning of the cDNAs coding for the two subunits of soluble guanylyl cyclase from human brain. FEBS Lett 304: 83–88, 1992
233. Harteneck C, Koesling D, Soling A, Schultz G, Bohme F: Expression of soluble guanylyl cyclase. Catalytic activity requires two enzyme subunits. FEBS Lett 272: 221–223, 1990
234. Chinkers M, Wilson E: Ligand-independent oligomerization of natriuretic peptide receptors. Identification of heteromeric receptors and a dominant negative mutant. J Biol Chem 267: 18589–18597, 1992
235. Thorpe DS, Niu S, Morkin E: Overexprression of dimeric guanylyl cyclase cores of an atrial natriuretic peptide receptor. Biochem Biophys Res Commun 180: 538–544, 1991
236. Wilson E, Chinkers M: Identification of sequences mediating guanylyl cyclase dimerization. Biochemistry 34: 4696–4701, 1995
237. Lowe DG: Human natriuretic peptide receptor-A guanylyl cyclase is self-associated prior to hormone binding. Biochemistry 31: 10421–10425, 1992
238. Labrecque J, McNicoll N, Marquis M, DeLean A: A disulfide-bridged mutant of natriuretic peptide receptor-A displays constitutive activity. J Biol Chem 274: 9752–9759, 1999
239. Labrecque J, Deschenes J, McNicoll N, DeLean A: Agonistic induction of a covalent dimer in a mutant of natriuretic peptide receptor-A documents a juxtamembrane interaction that accompanies receptor activation. J Biol Chem 276: 8064–8072, 2001
240. Yu H, Olshevskaya E, Duda T, Seno K, Hayashi F, Sharma RK, Dizhoor AM, Yamazaki A: Activation of retinal guanylyl cyclase-1 by Ca^{2+}-binding proteins involves its dimerization. J Biol Chem 274: 15547–15555, 1999
241. Olshevskaya E, Ermilow AE, Dizhoor AM: Dimerization of guanylyl cyclase-activating protein and a mechanism of photoreceptor guanylyl cyclase activation. J Biol Chem 274: 25583–25587, 1999
242. Donato R: S100: A multigenic family of calcium-modulated proteins of the EF-hand type with intracellular and extracellular functional roles. Int J Biochem Cell Biol 33: 637–668, 2001

243. Kelsell RE, Gregory-Evans K, Payne AM, Perrault I, Kaplan J, Yang R-B, Garbers DL, Bird AC, Moore AT, Hunt DM: Mutations in the retinal guanylate cyclase (RETGC-1) gene in dominant cone-rod dystrophy. Hum Mol Genet 7: 1179–1184, 1998
244. Perrault I, Rozet JM, Gerber S, Kelsell RE, Souied E, Cabout A, Hunt H, Munnich A, Kaplan J: A ret-GC1 mutation in autosomal dominant cone-rod dystrophy. Am J Hum Genet 63: 651–654, 1998
245. van Ghelue M, Eriksen HL, Ponjavic V, Fagerheim T, Andreasson S, Forsman-Semb K, Sandgren O, Holmgren G, Tranebjaerg L. Autosomal dominant cone-rod dystrophy due to a missense mutation (R838C) in the guanylate cyclase 2D gene (GUCY2D) with preserved rod function in one branch of the family. Ophthalmic Genet 21: 197–209, 2000
246. Gregory-Evans K, Kelsell RE, Gregory-Evans CY, Downes SM, Fitzke FW, Holder GE, Simunovic M, Mollon JD, Taylor R, Hunt DM, Bird AC, Moore AT: Autosomal dominant cone-rod retinal dystrophy (CORD6) from heterozygous mutation of GUCY2D, which encodes retinal guanylate cyclase. Ophthalmology 107: 55–61, 2000
247. Weigell-Weber M, Fokstuen S, Torok B, Niemeyer G, Schinzel A, Hergersberg M: Codons 837 and 838 in the retinal guanylate cyclase gene on chromosome 17p: Hot spots for mutations in autosomal dominant cone-rod dystrophy? Arch Ophthalmol 118: 300, 2000

Molecular and Cellular Biochemistry **230**: 31–47, 2002.

Biochemistry and physiology of the natriuretic peptide receptor guanylyl cyclases

Johanne Tremblay, Richard Desjardins, David Hum, Jolanta Gutkowska and Pavel Hamet

Research Centre, Centre hospitalier de l'Université de Montréal, CHUM, Montreal, Quebec, Canada

Abstract

Guanylyl cyclases (GC) exist as soluble and particulate, membrane-associated enzymes which catalyse the conversion of GTP to cGMP, an intracellular signalling molecule. Several membrane forms of the enzyme have been identified up to now. Some of them serve as receptors for the natriuretic peptides, a family of peptides which includes atrial natriuretic peptide (ANP), brain natriuretic peptide (BNP) and C-type natriuretic peptide (CNP), three peptides known to play important roles in renal and cardiovascular physiology. These are transmembrane proteins composed of a single transmembrane domain, a variable extracellular natriuretic peptide-binding domain, and a more conserved intracellular kinase homology domain (KHD) and catalytic domain. GC-A, the receptor for ANP and BNP, also named natriuretic peptide receptor-A or -1 (NPR-A or NPR-1), has been studied widely. Its mode of activation by peptide ligands and mechanisms of regulation serve as prototypes for understanding the function of other particulate GC. Activation of this enzyme by its ligand is a complex process requiring oligomerization, ligand binding, KHD phosphorylation and ATP binding. Gene knockout and genetic segregation studies have provided strong evidence for the importance of GC-A in the regulation of blood pressure and heart and renal functions. GC-B is the main receptor for CNP, the latter having a more paracrine role at the vascular and venous levels. The structure and regulation of GC-B is similar to that of GC-A. This chapter reviews the structure and roles of GC-A and GC-B in blood pressure regulation and cardiac and renal pathophysiology. (Mol Cell Biochem **230**: 31–47, 2002)

Key words: guanylyl cyclases, natriuretic peptides, hypertension, cardiomyopathy, genetics

Introduction

There are 2 main forms of guanylyl cyclase (GC), a soluble and a particulate form [1–5]. In the 1970s, most studies focused on the soluble form, and particulate GC attracted less attention because no physiological activators were found for this enzyme.

While research on sea urchin sperm suggested receptor-mediated regulation of the enzyme by peptide ligands [6], it was mainly with the discovery of the first natriuretic peptide (NP) by Adolfo de Bold in 1981 [7] that a great deal of interest was stimulated on the particulate form of GC. The first reports that atrial natriuretic peptide (ANP) simulates cGMP production in several tissues [8, 9] and that the peptide activates the particulate form of GC [10–12] appeared in the mid 1980s. Subsequent co-purification of ANP binding and GC catalytic activity [13–15] ascertained the initial suggestion that the 2 entities were part of the same molecule [16]. This field of research then rapidly expanded with parallel molecular cloning of several members of the NP family on the one hand and of the particulate GC isoenzyme family on the other hand.

In mammals, there are 7 membrane forms of GC (GC-A to G) identified up to now. GC-A and GC-B are receptors for the NPs. GC-C, the heat-stable enterotoxin receptor, is expressed primarily in the small intestine [17, 18]. GC-D is expressed in olfactory sensory neurons of the rat [19], while GC-E, present in the eye and pineal gland, is believed to be retinal GC, and GC-F is formed exclusively in the eye [20–22]. The latter 3 cyclases are still orphan receptors. The ex-

Address for offprints: J. Tremblay, Laboratory of Cellular Biology of Hypertension, Centre de recherche, CHUM – Hôtel-Dieu, 3850 St. Urbain St., Montréal, Québec H2W 1T7, Canada (E-mail: johanne.tremblay@umontreal.ca)

tracellular domain of GC-G is very homologous to that of natriuretic peptide receptor (NPR) cyclases, suggesting similar peptide ligands [23].

The NP family and their receptors

NPs are members of a family of vasoactive hormones that play an important role in the regulation of blood pressure and cardiovascular homeostasis. ANP, the first peptide identified, is a cardiac peptide of 28 amino acids that acts on the vasculature, the kidneys, the adrenals and the nervous system via specific membrane receptors possessing intrinsic cGMP-synthesizing activity [24]. ANP also shows antimitogenic activity in several cell types, including mesangial cells, vascular smooth muscle cells (VSMC), endothelial cells and cardiac fibroblasts [25–28]. These physiological functions of ANP oppose those of the renin-angiotensin system (RAS). ANP is a member of the NP family, which includes brain natriuretic peptide (BNP, 32aa) [29, 30] and C-type natriuretic peptide (CNP, 23aa) [31, 32]. A fourth NP, DNP, has recently been found to have potent natriuretic activity in humans [33].

ANP and BNP have similar functions; both are synthesized in the heart and secreted in the circulation, while CNP is more a paracrine factor synthesized in the central nervous system and the endothelium of the blood vessels. The 3 peptides are encoded by different genes. All NP family members contain a disulfide bond between 2 cystein residues, producing a ring structure that is essential for receptor recognition and biological function [34]. They are synthesized as precursors that are cleaved from the C-terminus end of the precursor molecule. In the kidney, ANP prohormone cleavage results in a 33 amino acid peptide called urodilatin which has autocrine/paracrine activity [35].

The NPR family includes 3 major receptor subtypes. GC-A (also called NPR-A or NPR-1) is a transmembrane protein with an extracellular binding site for ANP and BNP, a single short transmembrane domain separating the protein into 2 halves, an intracellular domain containing a regulatory kinase homology domain, and a catalytic GC domain at its C-terminal end [36–38] (Fig. 1). GC-B has a topology similar to that of GC-A but binds CNP preferentially [38–40]. There is also a truncated form of the receptor known as NPR-C which appears to have a clearance function [41]. This receptor binds all 3 peptides at its extracellular ligand binding site with similar affinity. Since its intracellular domain is very short, it lacks GC activity, but some studies have shown that it can inhibit adenylyl cyclase and stimulate phospholipase C [42, 43].

This chapter will focus on the physiology and biochemistry of the NPRs GC-A, GC-B and NPR-C. As NPs have a major impact on cardiovascular homeostasis, we will concentrate this review on cardiac, renal and vascular regulation as well as the effects of NPRs in health and disease.

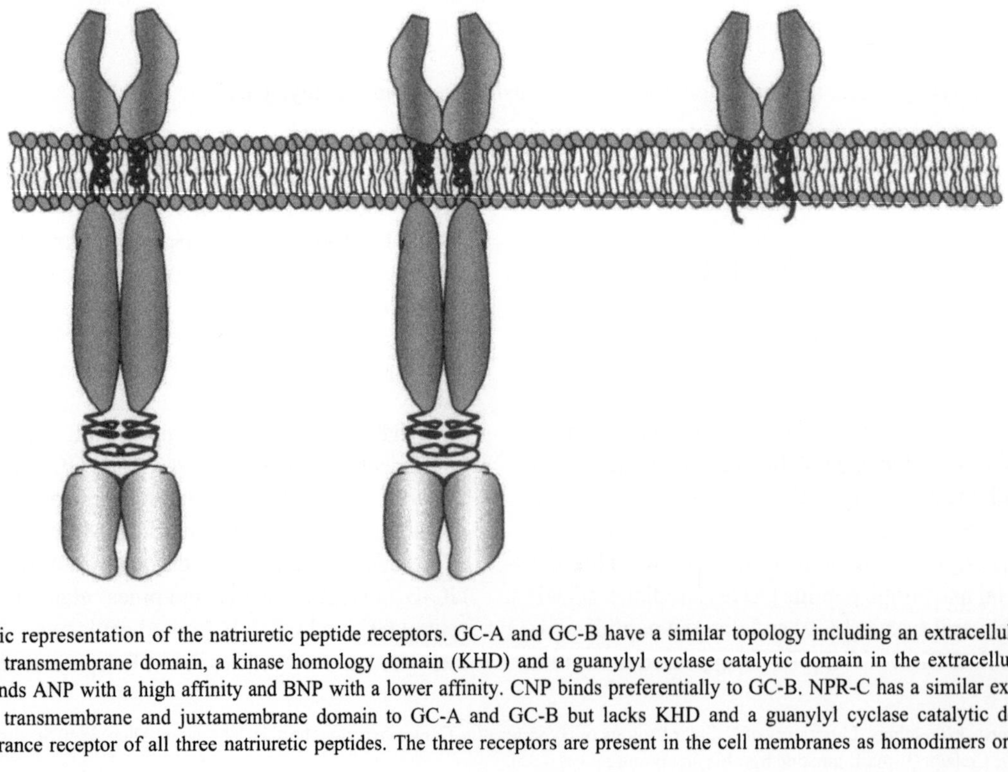

Fig. 1. Schematic representation of the natriuretic peptide receptors. GC-A and GC-B have a similar topology including an extracellular ligand binding domain, a single transmembrane domain, a kinase homology domain (KHD) and a guanylyl cyclase catalytic domain in the extracellular portion of the protein. GC-A binds ANP with a high affinity and BNP with a lower affinity. CNP binds preferentially to GC-B. NPR-C has a similar extracellular ligand-binding domain, transmembrane and juxtamembrane domain to GC-A and GC-B but lacks KHD and a guanylyl cyclase catalytic domain. NPR-C is defined as a clearance receptor of all three natriuretic peptides. The three receptors are present in the cell membranes as homodimers or homooligomers.

GC-A is the main and possibly the only receptor for ANP [44]. It is the one that has been studied the most and will thus be discussed in more detail here.

GC-A and GC-B genes

GC-A and GC-B genes have similar structures with lengths of 17.5 Kb for GC-A and 16.5 Kb for GC-B. GC-A gene is composed of 22 exons [45]. The regulation of GC-A gene expression is still not well defined. GC-A has a TATA-less promoter that contains an inverted CCAAT box at positions –141 to –136, and at least 3 putative Sp1-binding sites between –350 and –50 from the transcription site [45]. Functional promoter activity and gel retardation assays have determined that the Sp1 sites are essential for the activity of the promoter [46]. The CCAAT box is also important for GC-A gene transcription. Recently, nuclear protein NF-Y (nuclear factor Y) has been identified as binding to the CCAAT box. NF-Y and Sp1 appear to act synergistically on the transcription of GC-A. Mutations of the CCAAT box and all Sp1 sites completely abolish the promoter activity of GC-A [47]. On the other hand, ANP has been shown to be able to regulate the transcription of its own receptor [48]. Studies in VSMC have revealed that ANP stimulation can reduce the mRNA levels of GC-A in a dose- and time-dependent fashion. This decrease seems to be related to diminution of gene transcription [48, 49].

Maturation of GC-A and GC-B proteins

GC-A and GC-B are membrane enzymes that contain, in addition to their peptide-binding domain and GC catalytic domains, a kinase homology domain (KHD) that plays an essential role in maturation, activity and activation of these proteins (Fig. 1). After transcription, the GC-A protein is maturated. Heat shock proteins (HSPs) are known to act as chaperons, allowing correct folding of the proteins. Recently, Kumar *et al.* [50] have shown that HSP90 and P50 (cdc37) are associated with the KHD of GC-A, suggesting that GC-A is regulated by a complex containing HSP90, and maturated in a way similar to protein kinases [50]. Furthermore, during maturation, all receptors containing GC, with the exception of GC-F, undergo a step of glycosylation of residues located in the extracellular portion at the N-terminal end of GC [51]. Glycosylation of the extracellular domain appears to be responsible for the heterogeneity and specificity of the different GC receptors. It plays a role in the binding of ligands to the peptide-binding domain, replacement of glycosylation sites by alanine [52] and removal of carbohydrate residues with endoglycosidase [53], abolishing the binding of ligands and further activation of the cyclases.

Structure and activation of GC-A and GC-B

In the basal state, in the absence of ligands, GC-A and GC-B are present in the cell membrane in the form of homodimers or homotetramers, in a highly phosphorylated state, and catalytic GC activity is low [53–57]. The binding of a NP to a high affinity site formed by a homodimer induces a conformational change of the KHD leading to the binding of adenosine triphosphate (ATP), an essential step for transmembrane signal transduction [58]. The KHD has been proposed to normally repress catalytic activity [51]. This model, however, is not supported by the studies of Goraczniak *et al.* [20] who reported that KHD-deleted mutants showed a 5–10-fold suppression of basal cyclase activity. Both groups agree however that ATP is necessary for maximal activation of the receptor, and that loss of the KHD leads to a constitutively active cyclase in the absence of NP binding. In fact, Potter and Hunter [59] have identified phosphorylation sites rich in serine and threonine within the KHD of GC-A. These sites also appear to be involved in the homologous desensitization of GC-A via dephosphorylation [60], which will be discussed further in the following section.

The interaction of ATP with the KHD increases the V_{max} of cGMP without affecting affinity for the substrate [61–63]. Purification of GC-A and its overexpression in Sf9 cells showed that the enzyme cannot be activated by ANP in the absence of ATP [64–66]. These results suggest that ATP is an essential prerequisite for receptor effector coupling of this family of receptors. Deletions of the KHD of GC-A and GC-B revealed that this domain represses the catalytic site of GC in the absence of ligand [51, 67]. Thus, transfection of cells with cDNA coding for GC-A or GC-B deleted of the KHD leads to constitutive cGMP production independently of ligand binding or of the presence of ATP. The ATP-binding motif has been defined [20, 68, 69]. A sequence rich in glycine (GXGXXG) within the KHD of GC-A and GC-B serves as an anchor for the terminal phosphate of ATP [70]. A GC-A mutant deleted of this glycine-rich domain by site-directed mutagenesis became unresponsive to ANP or ATP [69]. These investigations thus suggest that activation of the catalytic domain by ATP is mediated by the KHD. An activation model has been proposed for particulate GC regulation by their ligands based on studies of GC-A [24]. In the basal state, particulate GC exists in an homo-oligomerized complex, each monomer being phosphorylated on key residues in the KHD. Binding of the ligand to its receptor leads to transduction of an activatory signal through the plasma membrane, allowing the KHD to bind ATP. ATP binding to the KHD results in derepression of the catalytic domain of GC. Two catalytic domains then interact to form an active site capable of GC activity. The requirement for preformed receptor oligomers in the absence of ligand may reflect the general requirement of the cyclases for 2 catalytic subunits to convert nucleotide triphosphates to cyclic nucleotides. A hinge region between the KHD and the catalytic domain is essential for dimerization of the catalytic domain [71, 72] (Fig. 1). Cystein residues in the extracellular domain appear to mediate ligand-independent oligomerization of the receptors [56].

In addition to its action on receptor effector coupling, ATP also allosterically regulates the affinity of GC-A and GC-B for its ligands. Thus, GC-A undergoes a ligand-induced and time-dependent shift from a higher to a lower affinity state [65, 73–75]. ATP potentiates this shift in affinity that is associated with nucleotide effects on activation. The shift from high to low affinity results in interaction of ligand-receptor with the catalytic site. Amiloride, an antagonist of ATP binding to the kinase catalytic domain, blocks the effects of ATP on ligand binding and catalytic activation [75, 76]. It appears that amiloride can 'freeze' GC-A in a high affinity state unresponsive to ANP. The allosteric regulation of GC-A and GC-B by adenine nucleotides coordinately modulates ligand receptor interaction and receptor effector coupling [64, 77, 78]. Again, the allosteric activation of catalytic activity and receptor binding by adenine nucleotides appear to be mediated by the KHD. Thus, mutants of GC-A and GC-B lacking the KHD are unresponsive to ATP and ANP [64, 67, 75]. Non-hydrolyzable analogs of ATP can substitute for ATP in their allosteric regulation [61, 62, 64] of GC-A and GC-B. However, the superior effect of ATP and ATP-γS as compared to non-hydrolyzable analogs of ATP suggest that both, allosteric activation and phosphorylation of cyclases, are involved in the activation of GC by adenosine nucleotides [61, 63, 79]. Dephosphorylation of the KHD, resulting in dissociation of ATP and ligand, leads to termination of the activatory signal and subsequent homologous desensitization.

Desensitization and downregulation of NPR

In very early research on NPR regulation, it became apparent that these receptors can be regulated similarly to other receptor types. This has been observed *in vitro* in membrane preparations, in cultured cells, in whole tissues and *in vivo* studies in animals.

As discussed previously, a phosphorylated state of the receptors is required for the acquisition of sensitivity towards NPs [59, 60]. Although ATP binding to the KHD has been reported, this domain lacks kinase catalytic activity. Only retinal particulate GC appears to possess intrinsic kinase activity. Activation of GC by the NPs leads to their dephosphorylation, resulting in subsequent homologous desensitization [80–84]. This dephosphorylation occurs very early and is associated with reduced activation of GC-A on subsequent stimulation by the ligand. In membrane preparations, GC-A can be dephosphorylated by the addition of protein phosphatase [80], and this is blocked by a phosphatase inhibitor [60, 80], suggesting that phosphorylation/dephosphorylation is an important mechanism of NPR regulation. Early desensitization does not appear to be dependent on cGMP as the addition of analogs of cGMP, 8-Bromo cGMP or dibutyryl-cGMP, did not induce early desensitization [81]. Recently, Potter and Hunter [83, 84] identified phosphorylation sites rich in serine and threonine in the KHD of GC-A and GC-B. These sites would be implicated in the desensitization of these enzymes.

Multiple factors, including growth factors, vasoconstrictive hormones such as vasopressin, angiotensin II and endothelin, counteract the action of NPs and reduce the increases in cGMP levels they cause. These hormones are able to activate protein kinase C (PKC), and it has been shown that in stable transfectants of NIH 3T3 fibroblasts with GC-A, addition of the PKC agonist PMA produces a drastic decrease in GC-A activity and stimulation by ANP [81]. Heterologous desensitization has been correlated with the state of GC-A dephosphorylation. In the same study, it was also found that the phosphorylation sites following PKC activation are different from those observed after stimulation by ANP. The authors concluded that the types of desensitization, homologous vs heterologous, are distinct events mediated by the dephosphorylation of different residues through PKC-dependent and independent pathways, respectively.

Other studies have shown that PKC can negatively modulate the activity of NPRs [85–87]. However, one of them suggested that, in contrast to results obtained by Potter and Garbers [81], desensitization can be potentially correlated with the state of GC phosphorylation [87]. Desensitization by dephosphorylation is a phenomenon that can be observed very early after receptor activation by its ligand. It is not known if this dephosphorylation is reversible, and the protein kinases and protein phosphatases involved in this process have not been identified yet. With a yeast 2-hybrid system, Chinkers [88] isolated a protein of 55 kDa that binds specifically to the KHD of particulate GC. This protein, containing a putative domain of serine/threonine phosphatase, could be involved in the desensitization process.

It should be mentioned here that the desensitization of NPRs via dephosphorylation is apparently NPR-specific. Thus, most membrane receptors, including G protein-coupled receptors, are desensitized via a mechanism involving phosphorylation rather than dephosphorylation.

Downregulation of NPRs has also been demonstrated *in vitro* in VSMC [89–94]. This downregulation has been associated with increased internalization of the receptors and degradation of the ligand receptor complex via a receptor-mediated endocytosis mechanism [95]. A portion of the receptor is recycled towards the plasma membrane, and *de novo* receptor synthesis is required for reconstitution of the receptor population [95, 96]. This has been observed also for GC-B and NPR-C in PC12 cells [97]. Downregulation of these receptors by receptor-mediated endocytosis is thus another mechanism that participates in regulation of the biological response to NPs. Again, specific sites in the KHD and catalytic domains of GC-A appear to play a role in the endocytosis and intracellular sequestration of these receptors [98].

We have discussed up to now the results of studies indicating that the regulation of NPR activity by their ligands, and particularly of GC-A, occurs at the post-transcriptional level. This negative regulation is observed very early after stimulation with ANP. In addition, several studies have reported that GC activity and receptor number are reduced in cultured cells stimulated for a prolonged period of time with ANP [48, 49, 89–93, 99, 100]. Some of these investigations have recorded time- and concentration-dependent reductions of GC activity, of the number of ligand-binding domains as well as of GC-A mRNA levels. Transforming growth factor-β_1 can also reduce the mRNA levels and enzyme activity of GC-A and GC-B in cultured VSMC [100]. In these cells and in the inner medullary collecting duct [48, 49], the decrease in GC-A mRNA levels induced by its ligand has been correlated with reduced transcription of the GC-A gene. Thus, functional analysis of the GC-A promoter has demonstrated that a region of 285 bp (–1575 to –1290) is sensitive to the incubation of cells with ANP, BNP and 8-Bromo cGMP [48]. These results suggest that the long-term negative regulation of GC-A occurs at the level of gene transcription involving a putative cGMP-responsive element located at more than 1 Kbp upstream of the transcription start site. As the incubation of cells with 8-Bromo cGMP was shown to be as effective as NPs in reducing GC-A mRNA levels and its promoter activity, downregulation at the level of GC-A transcription appears to be dependent on intracellular increases of cGMP and not on the ligand itself [49]. These studies did not identify the sequence of the putative cGMP-responsive element (RE). In an attempt to localize this element, we have recently subcloned a 1396 bp fragment of the rat GC-A promoter in an expression vector containing the luciferase reporter gene. Data presented in Fig. 2 show that pre-incubation with ANP or 8-Bromo cGMP inhibited the promoter activity of GC-A of transiently transfected fibroblasts or aortic smooth muscle cells in culture with the plasmid containing the GC-A promoter fragment. The region containing the cGMP-RE was further narrowed down by deletion and mutational gel shift experiments [101]. However, desensitization of GC-A after prolonged exposure to NP has not been observed in all studies. Thus, some investigations have shown a decrease in the number of receptors and mRNA levels of NPR-C with no de-

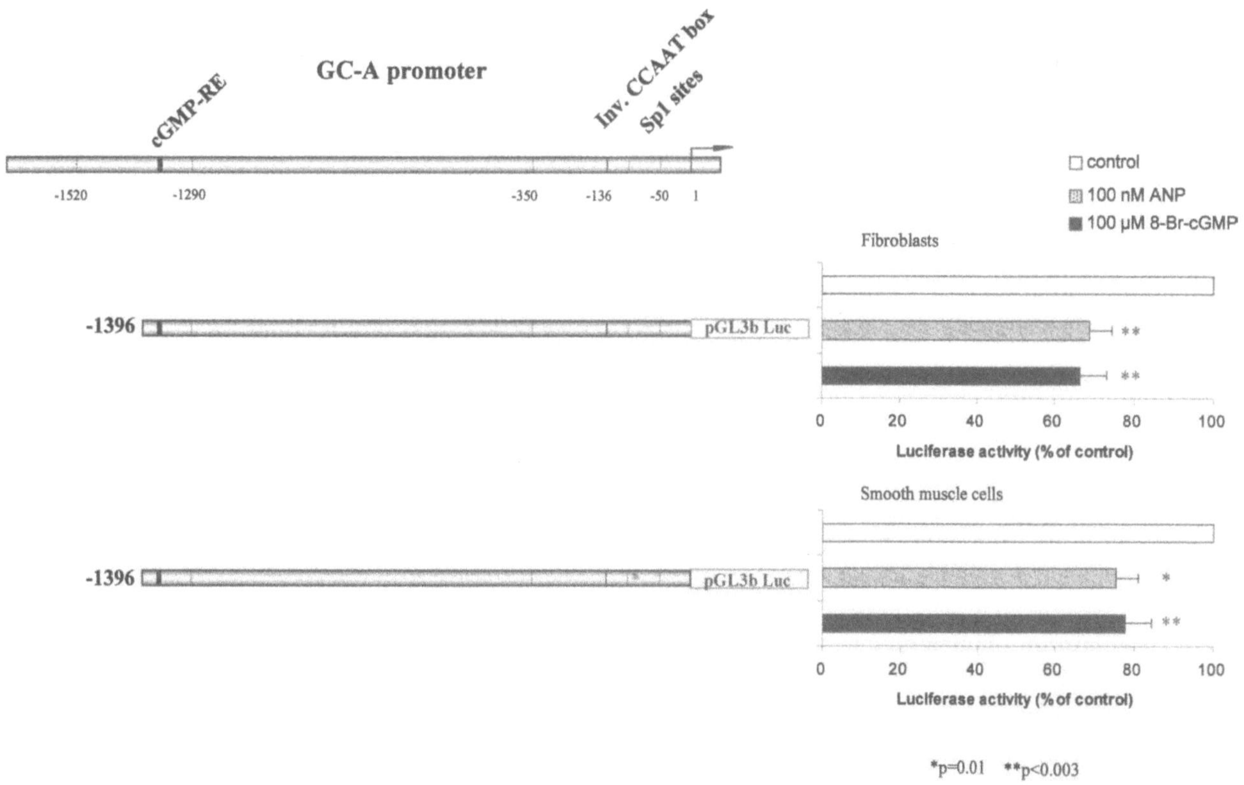

Fig. 2. Luciferase activity of GC-A promoter fragment inserted into an expression vector containing luciferase reporter gene and transiently transfected into cultured NIH 3T3 fibroblasts and aortic smooth muscle pre-incubated 8 h or not with ANP or 8-Bromo cGMP. (Data from [101].)

sensitization of guanylyl cyclase receptor in cultured cells previously treated with ANP or CNP [102, 103]. Downregulation of this receptor would not be a direct effect of the binding of NPs to NPR-C but rather of GC-B activation [102–104]. In fact, intracellular increases in cGMP levels mediate this effect as exposure of cells for several hours to 8-Bromo cGMP or sodium nitroprusside has been found to diminish the number of receptors and mRNA levels of NPR-C [102–104]. On the other hand, another study has shown coordinated downregulation of NPR-C and GC-B in VSMC in culture [93]. The apparent lack of downregulation of GC-A could thus be a consequence of a more drastic downregulation of NPR-C that would decrease NP clearance and result in greater GC stimulation. The length of pre-incubation and the relative proportion of individual NPRs in cells could be important factors modulating the downregulation.

In summary, NPR-GC activation is a complex process requiring oligomerization, ligand binding, KHD phosphorylation and adenine nucleotide binding. It serves as a model system to understanding other members of the GC family.

Effect of NPs on volume control and blood pressure

ANP and BNP secretion is stimulated by atrial stretch and pressure. Peptides released into the circulation reach their targets, which include the kidneys, adrenals, vascular beds and heart. While BNP has biological effects similar to those of ANP, its circulating levels are much lower under normal conditions. They increase to similar values in pathological states such as myocardial hypertrophy and congestive heart failure (CHF). Urodilatin, which is synthesized in the kidneys, has been shown to have the same effects as ANP when administered to experimental models. On the other hand, CNP, which has been localized in endothelial cells and in proximal, distal and medullary collecting duct tubule cells, lacks the renal effects of ANP, even when injected at high doses in the renal artery, and is therefore considered to be more a cardiovascular-selective peptide [105].

The main hemodynamic actions of ANP (1) directly and indirectly target the kidney to alter renal hemodynamics, and to increase fluid and electrolyte excretion, (2) functionally antagonize the renin-angiotensin-aldosterone system by inhibiting the synthesis and release of renin and aldosterone, and by opposing all the known effects of angiotensin, (3) produce vasodilatation, (4) shift fluid from the intravascular to the interstitial compartments by augmenting capillary hydraulic permeability [106]. All these actions contribute to the regulation of pressure-volume homeostasis.

At the level of the kidney, ANP increases the glomerular filtration rate (GFR) mainly via an elevation of glomerular capillary hydrostatic pressure that results from efferent arteriolar constriction and afferent arteriolar dilatation. It is the only physiological agent capable of increasing the GFR in the presence of decreased blood pressure or renal blood flow. The renal effects of ANP appear to be mediated by GCs abundant in the glomeruli [12] rather than by NPR-C as they are blocked by HS-142, a GC-specific antagonist [107, 108]. In the ANP knockout mouse model, adaptation to a high-salt diet is impaired, resulting in a salt-sensitive form of hypertension [109]. These data suggest that ANP is an important factor in GFR control and that this system may play a significant role in adaptation to chronic volume expansion occurring in CHF and in the nephrotic syndrome. In these conditions, ANP's major purpose is to defend the organism against excessive volume expansion.

The renal content of transforming growth factor-beta1 (TGF-β_1) is elevated in both the acute and the chronic phases of many renal diseases [110]. Increased expression of glomerular TGF-β_1 for a prolonged period leads to irreversible glomerulopathy. Previously mentioned effects of TGF-β_1 on expression of the natriuretic peptide receptors GC-A and GC-B [100] could contribute to this deleterous renal impact of the cytokine [110]. Transgenic mice overexpressing BNP are more resistant to glomerular injury [111]. On the other hand, NPR-C knockout mice have slightly decreased blood pressure and mild diuresis that are believed to be due to the longer half-life of circulating NPs arising from their reduced clearance [112]. It was recently shown that in chronic renal failure patients, low doses of ANP, that did not lower blood pressure, promoted a marked increase in sodium excretion to the same extent observed in normal subjects [113]. This could be explained by the abnormal elevation of plasma ANP most probably owing to its reduced clearance or degradation in these patients after administration of exogenous ANP [113]. This study suggests that ANP, injected at low, non-hypotensive doses, may have therapeutic potential in subjects with chronic renal failure.

The effects of ANP on cardiovascular hemodynamics are complex. Its administration decreases blood pressure slightly but consistently in normotensive rats and humans and markedly in several models of hypertension. In normotension, ANP lowers blood pressure by reducing cardiac output while total peripheral resistance is unaffected. This effect is due to a diminution of plasma volume caused by a shift of fluid from the intravascular to the interstitial compartment [114, 115] and to a decrease of central venous pressure. In experimental models of hypertension, ANP's impact on cardiac output depends on renin status. Thus, in renin-dependent hypertension, ANP acts as an antagonist of vasoconstriction with little or no effect on cardiac output. Thus, ANP is more an antagonist of vasoconstriction than a classical vasodilator, which may explain, in part, why contradictory ANP actions have been reported in the literature. In this context, it is noteworthy that the novel combination of neutral endopeptidase

(NEP) and angiotensin-converting enzyme (ACE) produces cardiovascular effects greater than those resulting from suppression of either enzyme alone [116].

Effects of NPs on veins and arteries

Elevation of cGMP in vascular smooth muscle by such agents as ANP, nitric oxide (NO) and exogenous nitrovasodilators is known to cause relaxation [9, 117–123]. Both GC-A and GC-B are present in cultured VSMC. CNP increases cGMP levels more than ANP in culture. However, ANP is more potent than CNP in relaxing aortic rings. *In vitro*, CNP is a stronger inhibitor of VSMC proliferation [124]. This can be explained by up-regulation of GC-B in aortic SMC in culture [100]. On the other hand, CNP has effects comparable to ANP in isolated saphenous, femoral and renal veins. However, relative to ANP, CNP is a selective endothelium-independent venodilator. Thus, it may have more important control of venous return than ANP, and this may explain its greater impact in reducing cardiac filling pressure and output. In contrast, CNP is less potent than ANP in relaxing peripheral arteries, suggesting a differential localization of GC-A and GC-B between peripheral arteries and veins [105]. The vasorelaxant effects of ANP have been correlated with the activation of cGMP-dependent protein kinase G (PKG) [122]. G kinase activation is sufficient to account for decreases in intracellular calcium, myosin light chains phosphorylation and smooth muscle relaxation [125–127].

Effect of NPs on cellular growth

As mentioned previously, ANP suppresses the proliferation of many cell types, including renal mesangial cells, VSMC, endothelial cells and cardiac fibroblasts [25–28]. The peptide can also inhibit extracellular matrix production in VSMC as well as collagen synthesis in cardiac fibroblasts [128, 129]. ANP gene overexpression found in the hypertrophied heart has been suggested to be a compensatory mechanism against cardiac hypertrophy and fibrosis. ANP and BNP serve as early markers of CHF [130]. Both are augmented in CHF, and circulating BNP levels, which are normally lower than those of ANP, exceed the ANP increases in CHF.

GC-A and ANP knockout mice [109, 131] present cardiac enlargement which is partly independent of the blood pressure elevation observed in these strains. Both BNP and ANP levels are reduced in ANP knockout mice while ANP levels are up-regulated in BNP knockouts [132]. BNP knockout mice present severe cardiac fibrosis with no hypertrophy or hypertension, indicating that BNP acts mainly as an anti-fibrolytic factor within the heart. Thus, although ANP and BNP effects are mainly mediated by GC-A, disruption of one peptide by gene knockout does not have the same consequences as disruption of the other, pointing to the importance of local signalling systems. Yet, in cardiac fibroblasts, GC-B is the predominant receptor that could mediate the anti-proliferative response in fibroblasts. Genetic studies in F2 hybrids of crosses between 2 rat strains with normal heart or hypertrophic cardiomyopathy have demonstrated co-segregation of the ANP locus on rat chromosome 5 and left ventricular mass with significant inverse relationships between myocardial ANP mRNA or peptide levels and left ventricular mass [133, 134].

The growth-inhibitory effect of ANP is mediated by GC-A [135] probably through PKG-mediated suppressive actions on regulators of G-protein signalling proteins [136]. PKG can phosphorylate Raf1 on serine 43, which is adjacent to its Ras-binding domain [137]. Phosphorylation of Ser43 is thought to uncouple Ras/Raf1 kinase interaction [138]. Elevations of intracellular cGMP levels have been shown to inhibit ERK activation by EGF. ANP treatment also results in phosphorylation and activation of MAPK, but the precise mechanism and significance in its growth-inhibitory effects remain to be established. Cardiomyocyte hypertrophy does not appear to require activation of ERK and p38.

Both ANP and CNP inhibit VSMC proliferation. This may have a role in vascular remodeling in hypertension. For instance, CNP has been postulated to be a local regulator of ACE activity, and inhibition of angiotensin II formation reduces vascular hypertrophy and remodeling [139]. A major problem of angioplasty is the restenosis that occurs in a significant proportion of patients. In the rat carotid model of injury, CNP infusion at a rate of 1 µg/kg/min for 5 days resulted in a 60% reduction of neointima formation [140], suggesting its therapeutic potential in restenosis. Gene transfer of ANP by adenoviral vectors [141] has also evoked a decrease of VSMC and endothelial cell growth as well as protection against hypoxia-induced pulmonary hypertension [142]. These studies propose alternative ways for the local administration of NPs that may hold promise for the treatment of restenosis and pulmonary hypertension.

NPR in hypertension

Since many physiological systems are involved in blood pressure control, it is plausible to think that a mutation in any of the genes involved in these systems is potentially responsible for the development of hypertension. However, hypertension is a complex and heterogeneous pathology with variable impact of genetic components and strong influence of the environment [143]. As the natriuretic peptides have profound cardiovascular and renal effects, clinical studies of their regulation and effects in hypertension have received early interest [144].

It has been shown that ANP infusion for short time periods in patients with mild hypertension produces enhanced increases of plasma cGMP but normal natriuresis and diuresis [145] while in genetically hypertensive rats (SHR), normal cGMP elevation is accompanied by exaggerated diuresis and natriuresis [146]. On the other hand, prolonged infusion of low ANP doses in mildly hypertensive patients leads to exaggerated diuretic and natriuretic responses but with normal cGMP elevations [147].

The Dahl salt-sensitive rat develops hypertension when fed a high-salt diet [148, 149]. The increased renal vascular resistance and reduced salt excretion observed in these animals have been postulated to play a central role in the pathogenesis of hypertension. Dahl salt-sensitive rats are hypo-responsive to the natriuretic and diuretic effects of ANP, suggesting defects at the level of GC-A expression in this rat strain which is a hypertension model of sodium sensitivity [150, 151].

In SHR, data on ANP-binding sites are conflicting, as both increased and decreased density have been reported [152, 153]. This discrepancy can be explained by the organs studied, by differences in NPR populations, the type of hypertension, and age. For instance, increased receptor density or affinity is observed in young SHR [154, 155] while older animals manifest reduced receptor density [155, 156]. We have reported previously that a heightened response to ANP in glomeruli and in aortic smooth muscle cells of young SHR, as compared to WKY rats, is accompanied by GC-A mRNA elevation [100, 157]. Considering that ANP modulates the transcription of its main receptor gene [48], our results suggested abnormalities at this level [100, 157].

The locus on rat chromosome 2 (between D2Cebr4s8 anonymous marker and Nep (neutral endopeptidase gene) contains GC-A (Fig. 3). This locus has been shown to co-segregate positively with blood pressure in 6 different rat crosses with widely distant hypertensive and normotensive progenitors. Data from congenic strains have confirmed the existence of a blood pressure quantitative trait locus (QTL) in GC-A [158]. The BP QTL encompassing GC-A gene correlates with kidney GC-A activity in isolated glomeruli and diuretic and blood pressure responses to ANP, suggesting that lower GC-A activity contributes to the development of hypertension in the Dahl salt-sensitive rat [159]. The involvement of GC-A in hypertension is further supported by studies showing that knock-out mice lacking GC-A develop hypertension, which is not altered by salt intake [160], and that diuresis and natriuresis induced by volume expansion are completely abolished in GC-A-deficient mice, indicating that NPs act predominantly through GC-A [161].

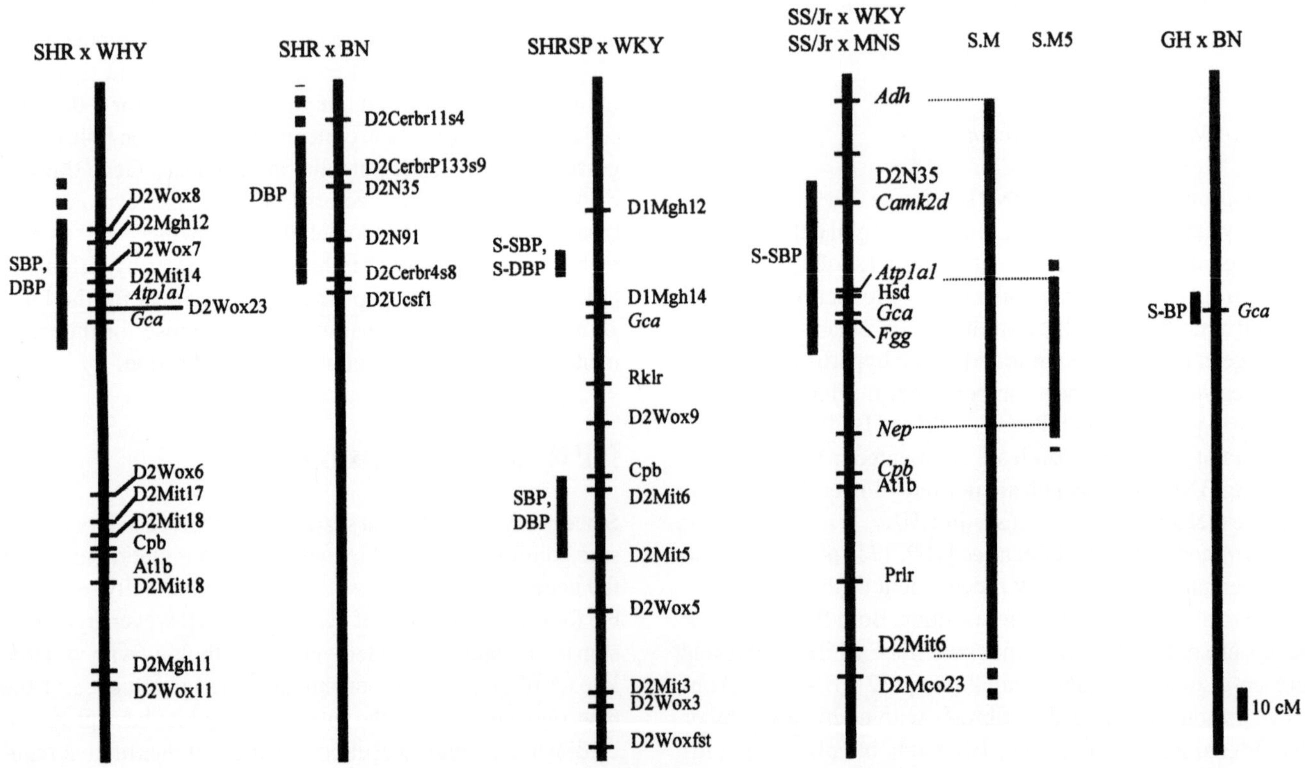

Fig. 3. Quantitative trait loci of blood pressure on rat chromosome 2 identified in co-segregation studies of F_2 hybrids and congenic strains. GC-A gene is located in a blood pressure QTL of several genetically hypertensive rat models. (Adapted from [143].)

NPR in cardiomyopathy and congestive heart failure

Cardiomyopathy is a clinical disease characterized by a complex pattern of biochemical and molecular alterations, such as ventricular enlargement, activation of vasoconstrictive hormonal systems and systolic contractile dysfunction, before the onset of hemodynamic overload and CHF developing as a consequence. CHF is the most widely recognized pathology with alteration of the NP system. Vasoconstriction and fluid and salt retention, the hallmarks of CHF syndromes, are associated with high plasma NP levels. Patients with CHF have elevated plasma ANP and BNP with these peptide levels being related to the severity of the disease [130]. Furthermore, high NP levels reflect increased ANP and BNP synthesis in ventricular myocytes observed during the development of cardiac hypertrophy [162]. This elevation of plasma NP seems to be at odds with avid water and salt retention in patients with CHF and in experimental animals. Several groups have tried to determine why salt and water retention is severe in CHF despite high levels of circulating NP. Furthermore, a blunted effect of exogenous ANP has been observed in CHF, suggesting either the hyporesponsiveness of target organs to ANP or activation of mechanisms opposing its actions.

Several mechanisms have been proposed to explain this phenomenon, such as decreased renal perfusion pressure, increased renal sympathetic activity [163] and activation of the RAS [164], both antagonizing NP activity, heightened activity of various NP-degrading enzymes, post-translational mechanisms inducing the formation of NPs devoid of biological activity, and downregulation of GC receptors. Several data suggest that stimulated sodium-retaining hormonal systems may overwhelm the diuretic and natriuretic potency of NP. However, treatment with converting-enzyme inhibitors or angiotensin II receptor antagonists significantly improves but does not restore the responses to exogenous ANP [164, 165]. Furthermore, the hypothesis suggesting that circulating ANP in CHF has lower bio-activity has to be ruled out, since blunted responses to exogenous ANP have been observed.

The question of downregulation of NPR in CHF remains controversial. The decreased density of NPR in target tissues could be a consequence of the high level of circulating hormones in CHF. However, this contention has not been substantiated. Other models of salt retention, such as dogs with CHF induced by caval ligation [166] or dogs with hepatic cirrhosis, show similar blunted responses to ANP infusion. Plasma ANP levels are not elevated in these models and may even be reduced [166].

The existence of more than one type of NPR, the models of experimental CHF and the use of different species have made it difficult to compare and directly assess renal receptors and draw conclusions.

Renal NPR

Reports on renal NPR are controversial. Most of the published data on renal receptors in CHF are based on studies in the cardiomyopathic (CMO) hamster, a paradigm of human CHF. The CMO hamster is a naturally-occurring, genetically-transmitted model in which cardiomyopathy with end stage heart failure is caused by an inherited autosomal mutation in the gene coding for 5-sarcoglycan [167]. This genetic model is characterized by progressive myocyte hypertrophy and myocyte death through apoptosis and necrosis. Plasma ANP levels are increased in CMO (Table 1). The glomerular density of NPR in CHF animals has been reported to be decreased [168, 169], increased [170] or not changed [171]. However, these NPR-C are mainly 'clearance' type devoid of GC activity.

Reduction of renal glomerular NP-binding sites has been shown in mild and moderate but not in severe heart failure in CMO hamsters, and no correlation has been observed by autoradiography between the degree of CHF and the density of ANP binding [172], indicating that the impairment of renal sodium excretion is not related to glomerular ANP receptor density at any stage of the disease [172].

The decrease in both receptor density and cGMP generation capacity in glomeruli isolated from CMO hamsters reported by Levin *et al.* [168] suggested that glomerular GC receptors are indeed downregulated, and could underlie the resistance to ANP seen in heart failure. However, unchanged renal receptors [173, 174] or increased glomerular NPR have also been reported in CMO hamsters [170]. This latter observation has been confirmed by the same group in aorta-caval (A-V) shunt rats [174]. A-V fistula is an experimental model of CHF, characterized by renal and neurohormonal changes that closely mimic those detected in patients with advanced CHF such as reduced renal blood flow, glomerular filtration rate-activated RAAS and blunted responses to ANP. The controversy in NPR evaluation could be attributed to the various experimental designs, different models of heart

Table 1. Plasma and tissue ANP content in normal and cardiomyopathic hamsters

	Normal	CMO
Plasma (pg/ml)	50.50 ± 11.1	309.40 ± 32.6*
Atria (μg/mg protein)		
Left	0.51 ± 0.08	0.24 ± 0.06*
Right	0.36 ± 0.05	0.27 ± 0.03
Ventricles (ng/mg protein)	0.37 ± 0.11	7.76 ± 0.92*
Lungs (ng/lung)	16.10 ± 1.5	64.70 ± 10.6*
Lungs (ng/mg protein)	0.44 ± 0.04	2.10 ± 0.04*

Values are means ± S.E. ANP – atrial natriuretic peptide; CMO – cardiomyopathic. *$p < 0.001$ vs. corresponding normal. (Reproduced from [189], with permission.)

failure and disease severity [175] or different species. Most of the early studies focused on the total NPR population in glomeruli. However, the majority of glomerular receptors are NPR-C type. Differential modification of the proportion of NPR-C vs. GC as well as GC-A or GC-B may lead to different physiological effects. The presence of GC-B mRNA has been shown in the rat kidney, [157, 176, 177] but the sheep kidney does not exhibit functional GC-B [178].

Thus, using various methodologies, such as competitive binding, autoradiography, cross-linking, stimulation of cGMP production and reverse transcription-polymerase chain reaction (RT-PCR), we have determined the level of transcripts of both GC-A and CC-B and their protein activity in the renal cortex and outer and inner medulla [179]. Compared to normal hamsters, in CMO animals, GC-A mRNA levels were similar in the cortex but significantly increased in the outer and inner medulla, with no change in GC-B mRNA. However, competitive binding of renal glomerular membranes revealed no change in total receptor density or the dissociation constant. Inner medullary receptors, exclusively of the GC-A type, were also not altered by cardiomyopathy, and glomerular cGMP production was similar in normal and CMO hamsters [179]. These data are consistent with those of Bianchi *et al.* [172] and Isnard *et al.* [173] but do not support the hypothesis that downregulation of renal receptors may contribute as a major factor to the blunted responses to high plasma NP levels or exogenous ANP.

Renal resistance to NP may thus be secondary to an antagonizing action, to enhanced efferent renal sympathetic nerve activity or to an intracellular defect beyond the second messenger system.

Platelet, vascular and adrenal NPR

The data on NPR in platelets are also controversial. Platelets exclusively possess NPR-C type receptors. Downregulated platelet receptors in patients with severe CHF have been reported by Schiffrin *et al.* [180]. However, Strom *et al.* [181] failed to show any differences in receptor density or affinity in CHF patients.

Interestingly, NPR in adrenal zona glomerulosa are highly insensitive to changes in plasma ANP levels [172] that are known to inhibit aldosterone secretion. Aldosterone has both myocardial and renal effects that may have profound implications for left ventricular remodeling. Aldosterone, secreted from the adrenal gland, stimulates cardiac collagen synthesis and fibroblast proliferation. The concentration and affinity of ANP binding sites in the adrenals are not changed at any stage of the disease in CMO hamsters. This could possibly explain the same aldosterone level despite high plasma renin activity in the state of severe CHF of CMO hamsters [182, 183].

In the thoracic aorta, Bianchi *et al.* [172] showed that ANP binding increased with the development of moderate to severe heart failure in CMO hamsters. B_{max} was augmented progressively, similarly to the results obtained with CMO hamsters in the subfornical organ and area postrema [184], indicating that high plasma ANP levels do not produce downregulation of NPR in these tissues. Interestingly, intraarterial ANP infusion in CHF patients evoked increases in forearm blood flow similar to those in normal subjects, indicating that the ANP functions mediated by peripheral artery receptors are preserved [185].

In contrast, the vasodilatory responses to ANP infusion showed that they are attenuated in patients with decompensated heart failure [186]. Furthermore, in both patients with severe, chronic, heart failure and experimental models of severe CHF [166, 169] high circulating ANP levels are associated with decreased cGMP levels, which suggest that downregulation of GC receptors may account for decreased vascular responsiveness to ANP in CHF.

Since ANP is a powerful inhibitor of precontracted vessels, the observed upregulation of ANP-binding sites in the aorta [172] may play a buffering role, at least in part, in intense vasoconstriction, an important feature of CHF caused by several hormonal systems that are activated in CHF [187].

Lung and heart NPR

The lungs are an important source of NP and NPR synthesis as well as a major site of NP clearance. Particulate GC activity in type I and type II pneumocytes has been shown to be reduced in early hamster cardiomyopathy, suggesting that lung NPR are implicated in heart failure [188]. A study of 200–230-day-old hamsters (advanced stage of the disease) revealed that lung NPR are downregulated. ANP content is higher in CMO hamster lungs (Table 1). Transcripts of GC-A, GC-B and NPR-C were detected in lung membranes by RT-PCR [189]. Compared to normal Syrian hamster controls, CMO hamster lung GC-A mRNA was reduced by 50%, but GC-B and NPR-C were unchanged. Downregulation of GC-A was also confirmed by decreased GC activation by ANP in the lungs of CMO animals (Fig. 4). The significance of pulmonary NPR downregulation is not clear. Physiological concentrations of ANP may protect the lungs from chemically-induced pulmonary edema [190, 191]. This anti-edematous action is thought to be a receptor-mediated mechanism involving cGMP and positive correlation has also been found between cGMP levels and the anti-edematous action of ANP [192]. Whether the downregulation of pulmonary NPR makes the lungs more vulnerable to edema formation remains to be determined.

The downregulation of pulmonary receptors in CHF could also contribute to the decreased pulmonary clearance of NP

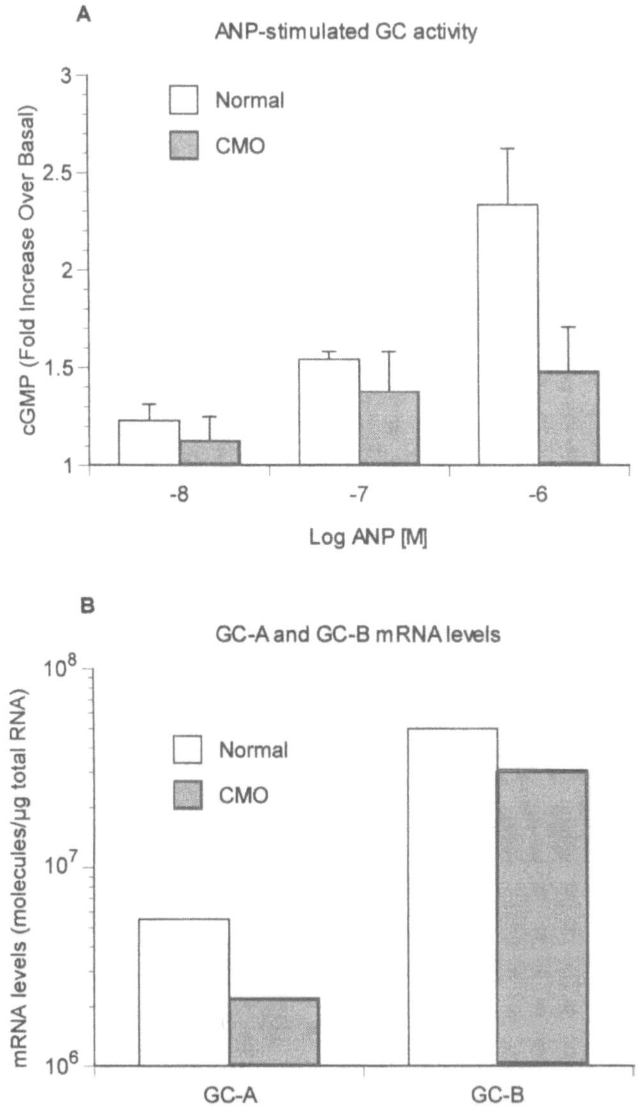

Fig. 4. (A) ANP-stimulated GC activity in lung membranes of normal and CMO hamsters. (B) GC-A and GC-B mRNA levels in hamster lungs determined by quantitative RT-PCR. (Adapted from [189].)

observed in dogs with chronic CHF [193]. Circulating ANP concentrations are regulated by 2 mechanisms: the first involves degradation by the widely-distributed neutral endopeptidase NEP 24.11; the second is through binding to receptors located in many tissues and vascular beds. Receptor-bound ANP is further internalized and degraded by lysosomal enzymes [194]. Significantly lower pulmonary clearance in CHF may also reflect increased receptor occupancy by augmented pulmonary ANP synthesis, since during development of the disease in CMO hamsters, pulmonary ANP synthesis is enhanced [195].

The genes of all 3 NPR are expressed in both ventricular myocytes and fibroblasts [196]. Rat ventricular myocytes predominantly synthesize GC-A, as determined by RT-PCR and confirmed by the generation of cGMP by ANP and BNP but not by CNP stimulation. Evidence supporting the possibility that the heart is a site of action for NPs comes from experiments with isolated myocytes. ANF administration reduces the basal level of cytosolic free calcium in guinea pig cardiomyocytes [197]. Progressive 4-chamber cardiac hypertrophy in the rat A-V model of heart failure is accompanied by the gradual disappearance of NPR-C transcripts in all heart chambers and an increase in both GC-A and GC-B. The downregulation of clearance receptors, if it is a widespread phenomenon in heart failure, could be responsible for the reduced ANP clearance [198]. The finding that fibroblasts express NPR suggests that NPs could be involved in structural remodeling of the heart [196].

Renal hyporesponsiveness in CHF is a multifactorial phenomenon. The reported data indicate that downregulation of NPR is not a predominant factor in renal hyporesponsiveness, but could be secondary to renal hemodynamics in concert with overwhelming activation of salt- and water-retaining vasoconstrictive hormonal systems. Furthermore, heart failure subjects maintain their diuresis while their natriuresis is blunted. We have previously reported that the ANP natriuretic and diuretic effects can be dissociated at the level of renal GC [199] but the relevance of this observation in heart failure and its potential therapeutic role require further investigation.

The lower renal responses in CHF are not specific to NPs since responses to furosemide have also been reduced. There is evidence that renal responsiveness to ANP and furosemide is sensitive to reduced blood pressure and decreased renal blood flow, features occurring in CHF [170, 200].

Furthermore, chronic elevation of norepinephrine and activation of circulating renin-angiotensin-aldosterone have been well documented in CHF [201, 202]. Arginine-vasopressin (AVP) may also contribute to the plasma-expanded vasoconstrictive state in CHF. Plasma AVP levels have been shown to be elevated in CHF and variably correlated with hyponatremia and osmolality.

Activation of the sympathetic and renin-angiotensin systems leads to such peripheral vasoconstriction that cannot be overcome, even by high levels of circulating NPs. In support of this contention is the fact that beneficial responses to ANP administration are partially restored after interruption of the RAS by ACE inhibitors or angiotensin II receptor antagonism [203]. Very recent data [204] show that simultaneous inhibition of NEP 24.11 and ACE increases both ANP and BNP of myocardial cell origin and CNP, considered an endothelial hormone with very beneficial effects in CHF. The simultaneous inhibition of both enzymes reduces vasoconstriction and enhances vasodilatation, thereby decreasing vascular and ventricular remodeling and improving cardiac performance.

Perspectives

The discovery of the first NP 20 years ago was followed by an explosion of research on the clinical and basic aspects of NPs as well as their receptors, the particulate GCs. Both families have rapidly expanded with the molecular cloning of new members. NPR structure-function relationships have been the subject of intensive research and serve as models for studies on the mechanisms of regulation of other GC family members. The recent development of several mouse models of gene disruption or overexpression of individual components of these systems has provided a framework for understanding the pathophysiological importance of these peptides and the signalling pathways involved in their mechanisms of action. Future challenges include, at the cellular level, the identification of downstream cytosolic and nuclear targets of NPs via cGMP or other putative signalling mechanisms, and more in-depth knowledge of their role in cellular functions (other than vasorelaxation, inhibition of hormonal secretion and sodium transport which have been the subject of early interest), such as gene transcription, cellular growth, apoptosis and differentiation. The pathophysiological significance of receptor desensitization in diseases demonstrating a blunted response to the NPs and the identification of human genetic diseases with mutations in one or the other component of this system are major focal points of research in hypertension, stroke and heart failure. It is now becoming clear that circulating NPs are markers of CHF. Their therapeutic potential for the treatment of cardiovascular diseases, including CHF, hypertrophic cardiomyopathy, stroke and hypertension, was stimulated recently with the development of novel NEP inhibitors and their combination with ACE inhibitors. A better understanding of cross-talk with other signalling systems is required. Alternative ways of modulating peptide levels, systemically or locally, are being investigated in pulmonary hypertension and restenosis. Understanding more the function of the NP/GC system has improved our knowledge of the homeostatic mechanisms of cardiovascular and renal physiopathology. While other avenues of exploration are to be contemplated, we are now at the beginning of its pharmacotherapeutic implications.

References

1. Hardman JG, Sutherland EW: Guanyl cyclase, an enzyme catalyzing the formation of guanosine 3',5'-monophosphate from guanosine triphosphate. J Biol Chem 244: 6363–6370, 1969
2. Schultz G, Bohme E, Munske K: Guanyl cyclase. Determination of enzyme activity. Life Sci 8: 1323–1332, 1969
3. Garbers DL, Hardman JG, Rudolph FB: Kinetic analysis of sea urchin sperm guanylate cyclase. Biochemistry 13: 4166–4171, 1974
4. Kimura H, Murad F: Evidence for two different forms of guanylate cyclase in rat heart. J Biol Chem 249: 6910–6916, 1974
5. Garbers DL, Parks MA, Hardman JG: Characterization of particulate and soluble guanylate cyclases from rat lung. J Biol Chem 250: 374–381, 1975
6. Ramarao CS, Garbers DL: Receptor-mediated regulation of guanylate cyclase activity in spermatozoa. J Biol Chem 260: 8390–8396, 1985
7. De Bold AJ, Borenstein HB, Veress AT, Sonnenberg H: A rapid and potent natriuretic response to intravenous injection of atrial myocardial extract in rats. Life Sci 28: 89–94, 1981
8. Hamet P, Tremblay J, Thibault G, Garcia R, Cantin M, Genest J: Effect of atrial natriuretic factor on metabolism of cGMP. Endocrinology 112: 289, 1983
9. Hamet P, Tremblay J, Pang SC, Garcia R, Thibault G, Gutkowska J, Cantin M, Genest J: Effect of native and synthetic atrial natriuretic factor on cyclic GMP. Biochem Biophys Res Commun 123: 515–527, 1984
10. Waldman SA, Rapoport RM, Murad F: Atrial natriuretic factor selectively activates particulate guanylate cyclase and elevates cyclic GMP in rat tissues. J Biol Chem 259: 14332–14334, 1984
11. Winquist RJ, Faison EP, Waldman SA, Schwartz K, Murad F, Rapoport RM: Atrial natriuretic factor elicits an endothelium-independent relaxation and activates particulate guanylate cyclase in vascular smooth muscle. Proc Natl Acad Sci USA 81: 7661–7664, 1984
12. Tremblay J, Gerzer R, Vinay P, Pang SC, Beliveau R, Hamet P: The increase of cGMP by atrial natriuretic factor correlates with the distribution of particulate guanylate cyclase. FEBS Lett 181: 17–22, 1985
13. Kuno T, Andresen JW, Kamisaki Y, Waldman SA, Chang LY, Saheki S, Leitman DC, Nakane M, Murad F: Co-purification of an atrial natriuretic factor receptor and particulate guanylate cyclase from rat lung. J Biol Chem 261: 5817–5823, 1986
14. Paul AK, Marala RB, Jaiswal RK, Sharma RK: Coexistence of guanylate cyclase and atrial natriuretic factor receptor in a 180-kD protein. Science 235: 1224–1226, 1987
15. Meloche S, McNicoll N, Liu B, Ong H, de Lean A: Atrial natriuretic factor R^1 receptor from bovine adrenal zona glomerulosa: Purification, characterization, and modulation by amiloride. Biochemistry 27: 8151–8158, 1988
16. Tremblay J, Gerzer R, Pang SC, Cantin M, Genest J, Hamet P: ANF stimulation of detergent-dispersed particulate guanylate cyclase from bovine adrenal cortex. FEBS Lett 194: 210–214, 1986
17. Schulz S, Green CK, Yuen PS, Garbers DL: Guanylyl cyclase is a heat-stable enterotoxin receptor. Cell 63: 941–948, 1990
18. Mann EA, Jump ML, Wu J, Yee E, Giannella RA: Mice lacking the guanylyl cyclase C receptor are resistant to STa-induced intestinal secretion. Biochem Biophys Res Commun 239: 463–466, 1997
19. Fulle HJ, Vassar R, Foster DC, Yang RB, Axel R, Garbers DL: A receptor guanylyl cyclase expressed specifically in olfactory sensory neurons. Proc Natl Acad Sci USA 92: 3571–3575, 1995
20. Goraczniak RM, Duda T, Sharma RK: A structural motif that defines the ATP-regulatory module of guanylate cyclase in atrial natriuretic factor signalling. Biochem J 282: 533–537, 1992
21. Margulis A, Goraczniak RM, Duda T, Sharma RK, Sitaramayya A: Structural and biochemical identity of retinal rod outer segment membrane guanylate cyclase. Biochem Biophys Res Commun 194: 855–861, 1993
22. Yang RB, Foster DC, Garbers DL, Fulle HJ: Two membrane forms of guanylyl cyclase found in the eye. Proc Natl Acad Sci USA 92: 602–606, 1995
23. Schulz S, Wedel BJ, Matthews A, Garbers DL: The cloning and expression of a new guanylyl cyclase orphan receptor. J Biol Chem 273: 1032–1037, 1998
24. Lucas KA, Pitari GM, Kazerounian S, Ruiz-Stewart I, Park J, Schulz S, Chepenik KP, Waldman SA: Guanylyl cyclases and signaling by cyclic GMP. Pharmacol Rev 52: 375–414, 2000

25. Kohno M, Ikeda M, Johchi M, Horio T, Yasunari K, Kurihara N, Takeda T: Interaction of PDGF and natriuretic peptides on mesangial cell proliferation and endothelin secretion. Am J Physiol 265: E673–E679, 1993
26. Itoh H, Pratt RE, Dzau VJ: Atrial natriuretic polypeptide inhibits hypertrophy of vascular smooth muscle cells. J Clin Invest 86: 1690–1697, 1990
27. Itoh H, Pratt RE, Ohno M, Dzau VJ: Atrial natriuretic polypeptide as a novel antigrowth factor of endothelial cells. Hypertension 19: 758–761, 1992
28. Cao L, Gardner DG: Natriuretic peptides inhibit DNA synthesis in cardiac fibroblasts. Hypertension 25: 227–234, 1995
29. Sudoh T, Kangawa K, Minamino N, Matsuo H: A new natriuretic peptide in porcine brain. Nature 332: 78–81, 1988
30. Minamino N, Kangawa K, Matsuo H: Isolation and identification of high molecular weight brain natriuretic peptide in porcine cardiac atrium. Biochem Biophys Res Commun 157: 402–409, 1988
31. Sudoh T, Minamino N, Kangawa K, Matsuo H: C-type natriuretic peptide (CNP): A new member of natriuretic peptide family identified in porcine brain. Biochem Biophys Res Commun 168: 863–870, 1990
32. Kojima M, Minamino N, Kangawa K, Matsuo H: Cloning and sequence analysis of cDNA encoding a precursor for rat C-type natriuretic peptide (CNP). FEBS Lett 276: 209–213, 1990
33. Lisy O, Jougasaki M, Heublein DM, Schirger JA, Chen HH, Wennberg PW, Burnett JC: Renal actions of synthetic dendroaspis natriuretic peptide. Kidney Int 56: 502–508, 1999
34. Inagami T, Misono KS, Fukumi H, Maki M, Tanaka I, Takayanagi R, Imada T, Grammer RT, Naruse M, Naruse K, Pandey KN, Parmentier M, Yasujima M, Abe K: Structure and physiological actions of rat atrial natriuretic factor. Hypertension 10 (suppl I): I113–I117, 1987
35. Schulz-Knappe P, Honrath U, Forssmann WG, Sonnenberg H: Endogenous natriuretic peptides: Effect on collecting duct function in rat kidney. Am J Physiol 259: F415–F418, 1990
36. Takayanagi R, Snajdar RM, Imada T, Tamura M, Pandey KN, Misono KS, Inagami T: Purification and characterization of two types of atrial natriuretic factor receptors from bovine adrenal cortex: Guanylate cyclase-linked and cyclase-free receptors. Biochem Biophys Res Commun 144: 244–250, 1987
37. Lowe DG, Chang MS, Hellmiss R, Chen E, Singh S, Garbers DL, Goeddel DV: Human atrial natriuretic peptide receptor defines a new paradigm for second messenger signal transduction. EMBO J 8: 1377–1384, 1989
38. Schulz S, Singh S, Bellet RA, Singh G, Tubb DJ, Chin H, Garbers DL: The primary structure of a plasma membrane guanylate cyclase demonstrates diversity within this new receptor family. Cell 58: 1155–1162, 1989
39. Chang MS, Lowe DG, Lewis M, Hellmiss R, Chen E, Goeddel DV: Differential activation by atrial and brain natriuretic peptides of two different receptor guanylate cyclases. Nature 341: 68–72, 1989
40. Koller KJ, Lowe DG, Bennett GL, Minamino N, Kangawa K, Matsuo H, Goeddel DV: Selective activation of the B natriuretic peptide receptor by C-type natriuretic peptide (CNP). Science 252: 120–123, 1991
41. Maack T, Suzuki M, Almeida FA, Nussenzveig D, Scarborough RM, McEnroe GA, Lewicki JA: Physiological role of silent receptors of atrial natriuretic factor. Science 238: 675–678, 1987
42. Anand-Srivastava MB, Franks DJ, Cantin M, Genest J: Atrial natriuretic factor inhibits adenylate cyclase activity. Biochem Biophys Res Commun 121: 855–862, 1984
43. Hirata M, Chang CH, Murad F: Stimulatory effects of atrial natriuretic factor on phosphoinositide hydrolysis in cultured bovine aortic smooth muscle cells. Biochim Biophys Acta 101: 346–351, 1989
44. Dubois SK, Kishimoto I, Lillis TO, Garbers DL: A genetic model defines the importance of the atrial natriuretic peptide receptor (guanylyl cyclase-A) in the regulation of kidney function. Proc Natl Acad Sci USA 97: 4369–4373, 2000
45. Yamaguchi M, Rutledge LJ, Garbers DL: The primary structure of the rat guanylyl cyclase A/atrial natriuretic peptide receptor gene. J Biol Chem 265: 20414–20420, 1990
46. Gardner DG, Schaufele F, Gardner DG: Sp1 dependence of natriuretic peptide receptor A gene transcription in rat aortic smooth muscle cells. Endocrinology 140: 1695–1701, 1999
47. Liang F, Schaufele F, Gardner DG: Functional interaction of NF-Y and Sp1 is required for type a natriuretic peptide receptor gene transcription. J Biol Chem 276: 1516–1522, 2001
48. Cao L, Wu J, Gardner DG: Atrial natriuretic peptide suppresses the transcription of its guanylyl cyclase-linked receptor. J Biol Chem 270: 24891–24897, 1995
49. Cao L, Chen SC, Humphreys MH, Gardner DG: Ligand-dependent regulation of NPR-A gene expression in inner medullary collecting duct cells. Am J Physiol 275: F119–F125, 1998
50. Kumar R, Grammatikakis N, Chinkers M: Regulation of the atrial natriuretic peptide receptor by heat shock protein 90 complexes. J Biol Chem 276: 11371–11375, 2001
51. Chinkers M, Garbers DL: The protein kinase domain of the ANP receptor is required for signaling. Science 245: 1392–1394, 1989
52. Hasegawa M, Hidaka Y, Wada A, Hirayama T, Shimonishi Y: The relevance of N-linked glycosylation to the binding of a ligand to guanylate cyclase C. Eur J Biochem 263: 338–346, 1999
53. Lowe DG, Fendly BM: Human natriuretic peptide receptor-A guanylyl cyclase. J Biol Chem 267: 21691–21697, 1992
54. Tremblay J, Huot C, Koch C, Potier M: Characterization of the functional domains of the natriuretic peptide receptor/guanylate cyclase by radiation inactivation. J Biol Chem 266: 8171–8175, 1991
55. Iwata T, Uchida-Mizuno K, Katafuchi T, Ito T, Hagiwara H, Hirose S: Bifunctional atrial natriuretic peptide receptor (type A) exists as a disulfide-linked tetramer in plasma membranes of bovine adrenal cortex. J Biochem (Tokyo) 110: 35–39, 1991
56. Chinkers M, Wilson EM: Ligand-independent oligomerization of natriuretic peptide receptors. J Biol Chem 267: 18589–18597, 1992
57. Lowe DG: Human natriuretic peptide receptor-A guanylyl cyclase is self-associated prior to hormone binding. Biochemistry 31: 10421–10425, 1992
58. Huo X, Abe T, Misono KS: Ligand binding-dependent limited proteolysis of the atrial natriuretic peptide receptor: Juxtamembrane hinge structure essential for transmembrane signal transduction. Biochemistry 38: 16941–16951, 1999
59. Potter LR, Hunter T: Identification and characterization of the phosphorylation sites of the guanylyl cyclase-linked natriuretic peptide receptors A and B. Methods 19: 506–520, 1999
60. Potter LR, Hunter T: A constitutively 'phosphorylated' guanylyl cyclase-linked atrial natriuretic peptide receptor mutant is resistant to desensitization. Mol Biol Cell 10: 1811–1820, 1999
61. Kurose H, Inagami T, Ui M: Participation of adenosine 5'-triphosphate in the activation of membrane-bound guanylate cyclase by the atrial natriuretic factor. FEBS Lett 219: 375–379, 1987
62. Chang CH, Jiang B, Douglas JG: Structural requirements of ATP for activation of basal and atrial natriuretic factor-stimulated guanylate cyclase in rat lung membranes. Eur J Pharmacol 189: 293–298, 1990
63. Gazzano H, Wu HI, Waldman SA: Adenine nucleotide regulation of particulate guanylate cyclase from rat lung. Biochim Biophys Acta 1077: 99–106, 1991
64. Chinkers M, Singh S, Garbers DL: Adenine nucleotides are required for activation of rat atrial natriuretic peptide receptor/guanylyl cyclase expressed in a baculovirus system. J Biol Chem 266: 4088–4093, 1991

65. Larose L, McNicoll N, Ong H, de Lean A: Allosteric modulation by ATP of the bovine adrenal natriuretic factor R1 receptor functions. Biochemistry 30: 8990–8995, 1991
66. Wong SK, Ma CP, Foster DC, Chen AY, Garbers DL: The guanylyl cyclase-A receptor transduces an atrial natriuretic peptide/ATP activation signal in the absence of other proteins. J Biol Chem 270: 30818–30822, 1995
67. Koller KJ, de Sauvage FJ, Lowe DG, Goeddel DV: Conservation of the kinase like regulatory domain is essential for activation of the natriuretic peptide receptor guanylyl cyclases. Mol Cell Biol 12: 2581–2590, 1992
68. Marala R, Duda T, Goraczniak RM, Sharma RK: Genetically tailored atrial natriuretic factor-dependent guanylate cyclase. Immunological and functional identity with 180 kDa membrane guanylate cyclase and ATP signaling site. FEBS Lett 296: 254–258, 1992
69. Duda T, Goraczniak RM, Sharma RK: Core sequence of ATP regulatory module in receptor guanylate cyclases. FEBS Lett 315: 143–148, 1993
70. Hanks SK, Quinn AM, Hunter T: The protein kinase family: Conserved features and deduced phylogeny of the catalytic domains. Science 241: 42–52, 1988
71. Thompson DK, Garbers DL: Dominant negative mutations of the guanylyl cyclase-A receptor. J Biol Chem 270: 425–430, 1995
72. Wilson EM, Chinkers M: Identification of sequences mediating guanylyl cyclase dimerization. Biochemistry 34: 4696–4701, 1995
73. Duda T, Sharma RK: ATP bimodal switch that regulates the ligand binding and signal transduction activities of the atrial natriuretic factor receptor guanylate cyclase. Biochem Biophys Res Commun 209: 286–292, 1995
74. Duda T, Sharma RK: ATP modulation of the ligand binding and signal transduction activities of the type C natriuretic peptide receptor guanylate cyclase. Mol Cell Biochem 152: 179–183, 1995
75. Jewett JRS, Koller KJ, Goeddel DV, Lowe DG: Hormonal induction of low affinity receptor guanylyl cyclase. EMBO J 12: 769–777, 1993
76. Rondeau JJ, McNicoll N, Gagnon J, Bouchard N, Ong H, de Lean A: Stoichiometry of the atrial natriuretic factor-R1 receptor complex in the bovine zona glomerulosa. Biochemistry 34: 2130–2136, 1995
77. Marala RB, Sitaramayya A, Sharma RK: Dual regulation of atrial natriuretic factor-dependent guanylate cyclase activity by ATP. FEBS Lett 281: 73–76, 1991
78. Duda T, Goraczniak RM, Sitaramayya A, Sharma RK: Cloning and expression of an ATP-regulated human retina C-type natriuretic factor receptor guanylate cyclase. Biochemistry 32: 1391–1395, 1993
79. Parkinson SJ, Carrithers SL, Waldman SA: Opposing adenine nucleotide-dependent pathways regulate guanylyl cyclase C in rat intestine. J Biol Chem 269: 22683–22690, 1994
80. Potter LR, Garbers DL: Dephosphorylation of the guanylyl cyclase-A receptor causes desensitization. J Biol Chem 267: 14531–14534, 1992
81. Potter LR, Garbers DL: Protein kinase C-dependent desensitization of the atrial natriuretic peptide receptor is mediated by dephosphorylation. J Biol Chem 269: 14636–14642, 1994
82. Koller KJ, Lipari MT, Goeddel DV: Proper glycosylation and phosphorylation of the type A natriuretic peptide receptor are required for hormone-stimulated guanylyl cyclase activity. J Biol Chem 268: 5997–6003, 1993
83. Potter LR, Hunter T: Identification and characterization of the major phosphorylation sites of the B-type natriuretic peptide receptor. J Biol Chem 273: 15533–15539, 1998
84. Potter LR, Hunter T: Phosphorylation of the kinase homology domain is essential for activation of the A-type natriuretic peptide receptor. Mol Cell Biol 18: 2164–2172, 1998
85. Simpson FO: Blood pressure and sodium intake. In: J.H. Laragh, B.M. Brenner (eds), Hypertension: Pathophysiology, Diagnosis and Management – Vol. I. Raven Press, New York, 1995, pp 273–282
86. Chamoux E, Breault L, Lehoux JG, Gallo-Payet N: Involvement of the angiotensin II type 2 receptor in apoptosis during human fetal adrenal gland development. J Clin Endocrinol Metab 84: 4722–4730, 1999
87. Duda T, Sharma RK: Regulation of guanylate cyclase activity by atrial natriuretic factor and protein kinase C. Mol Cell Biochem 93: 179–184, 1990
88. Chinkers M: Targeting of a distinctive protein serine phosphatase to the protein kinase-like domain of the atrial natriuretic peptide receptor. Proc Natl Acad Sci USA 91: 11075–11079, 1994
89. Hirata Y, Hirose S, Takata S, Takagi Y, Matsubara H: Down-regulation of atrial natriuretic peptide receptor and cyclic GMP response in cultured rat vascular smooth muscle cells. Eur J Pharmacol 135: 439–442, 1987
90. Neuser D, Bellemann P: Receptor binding, cGMP stimulation and receptor desensitization by atrial natriuretic peptides in cultured A10 vascular smooth muscle cells. FEBS Lett 209: 347–351, 1986
91. Roubert P, Lonchampt MO, Chabrier PE, Plas P, Goulin J, Braquet P: Down-regulation of atrial natriuretic factor receptors and correlation with cGMP stimulation in rat cultured vascular smooth muscle cells. Biochem Biophys Res Commun 148: 61–67, 1987
92. Hughes RJ, Struthers RS, Fong AM, Insel PA: Regulation of the atrial natriuretic peptide receptor on a smooth muscle cell. Am J Physiol 253: C809–C816, 1987
93. Cahill PA, Redmond EM, Keenan AK: Vascular atrial natriuretic factor receptor subtypes are not independently regulated by atrial peptides. J Biol Chem 265: 21896–21906, 1990
94. Schiffrin EL, St-Louis J, Hamet P, Garcia R: Vascular receptors for atrial natriuretic peptide in spontaneously hypertensive rats. In: B.M. Brenner, J.H. Laragh (eds), Progress in Atrial Peptide Research (American Society of Hypertension Symposium Series), Vol. 3. Raven Press, New York, 1990, pp 564–566
95. Pandey KN: Kinetic analysis of internalization, recycling and redistribution of atrial natriuretic factor-receptor complex in cultured vascular smooth-muscle cells. Ligand-dependent receptor down-regulation. Biochem J 288: 55–61, 1992
96. Hirata Y, Takata S, Takagi Y, Matsubara H, Omae T: Regulation of atrial natriuretic peptide receptors in cultured vascular smooth muscle cells of rat. Biochem Biophys Res Commun 138: 405–412, 1986
97. Rathinavelu A, Isom GE: Differential internalization and processing of atrial-natriuretic-factor B and C receptor in PC12 cells. Biochem J 276: 493–497, 1991
98. Pandey KN, Kumar R, Li M, Nguyen H: Functional domains and expression of truncated atrial natriuretic peptide receptor-A: The carboxyl-terminal regions direct the receptor internalization and sequestration in COS-7 cells. Mol Pharmacol 57: 259–267, 2000
99. Hirata Y, Tomita M, Takada S, Yoshimi H: Vascular receptor binding activities and cyclic GMP responses by synthetic human and rat atrial natriuretic peptides (ANP) and receptor down-regulation by ANP. Biochem Biophys Res Commun 128: 538–546, 1985
100. Fujio N, Gossard F, Bayard F, Tremblay J: Regulation of natriuretic peptide receptor A and B expression by transforming growth factor-β_1 in cultured aortic smooth muscle cells. Hypertension 23: 908–913, 1994
101. Hum D, Desjardins R, del Rocio Sanchez Garcia M, Devost D, Gossard F, Hamet P, Tremblay J: Characterization and localization of a cGMP response element in the guanylyl cyclase A (GC-A) promoter. 2002 (in press)
102. Kato J, Lanier Smith KL, Currie MG: Cyclic GMP down-regulates atrial natriuretic peptide receptors on cultured vascular endothelial cells. J Biol Chem 266: 14681–14685, 1991
103. Zhang LM, Tao H, Newman WH: Regulation of atrial natriuretic peptide receptors in vascular smooth muscle cells: Role of cGMP. Am J Physiol 264: H1753–H1759, 1993

104. Kishimoto I, Nakao K, Suga SI, Hosoda K, Yoshimasa T, Itoh H, Imura H: Downregulation of C-receptor by natriuretic peptides via ANP-B receptor in vascular smooth muscle cells. Am J Physiol 265: H1373–H1379, 1993
105. Chen HH, Burnett JC Jr: C-type natriuretic peptide: The endothelial component of the natriuretic peptide system. J Cardiovasc Pharmacol 32 (suppl 3): S22–S28, 1998
106. Maack T: Role of atrial natriuretic factor in volume control. Kidney Int 49: 1732–1737, 1996
107. Zhang PL, Mackenzie HS, Troy JL, Brenner BM: Effects of natriuretic peptide receptor inhibition on remnant kidney function in rats. Kidney Int 46: 414–420, 1994
108. Sakamoto K, Kikkawa R, Haneda M, Shigeta Y: Prevention of glomerular hyperfiltration in rats with streptozotocin-induced diabetes by an atrial natriuretic peptide receptor antagonist. Diabetologia 38: 536–542, 1995
109. John SW, Krege JH, Oliver PM, Hagaman JR, Hodgin JB, Pang SC, Flynn TG, Smithies O: Genetic decreases in atrial natriuretic peptide and salt-sensitive hypertension. Science 267: 679–681, 1995
110. Krag S, Osterby R, Chai Q, Nielsen CB, Hermans C, Wogensen L: TGF-beta1-induced glomerular disorder is associated with impaired concentrating ability mimicking primary glomerular disease with renal failure in man. Lab Invest 80: 1855–1868, 2000
111. Kasahara M, Mukoyama M, Sugawara A, Makino H, Suganami T, Ogawa Y, Nakagawa M, Yahata K, Goto M, Ishibashi R, Tamura N, Tanaka I, Nakao K: Ameliorated glomerular injury in mice overexpressing brain natriuretic peptide with renal ablation. J Am Soc Nephrol 11: 1691–1701, 2000
112. Matsukawa N, Grzesik WJ, Takahashi N, Pandey KN, Pang S, Yamauchi M, Smithies O: The natriuretic peptide clearance receptor locally modulates the physiological effects of the natriuretic peptide system. Proc Natl Acad Sci USA 96: 7403–7408, 1999
113. Conte G, Bellizzi V, Cianciaruso B, Minutolo R, Fuiano G, De Nicola L: Physiologic role and diuretic efficacy of atrial natriuretic peptide in health and chronic renal disease. Kidney Int Suppl 59: S28–S32, 1997
114. Maack T, Marion DN, Camargo MJF, Kleinert HD, Laragh JH, Vaughan ED, Atlas SA: Effects of auriculin (ANF) on blood pressure, renal function, and the renin-aldosterone system in dogs. Am J Med 77: 1069–1075, 1984
115. Almeida FA, Suzuki M, Maack T: Atrial natriuretic factor increases hematocrit and decreases plasma volume in nephrectomized rats. Life Sci 39: 1193–1199, 1986
116. Trippodo NC, Robl JA, Asaad MM, Fox M, Panchal BC, Schaeffer TR: Effects of omapatrilat in low, normal, and high renin experimental hypertension. Am J Hypertens 11: 363–372, 1998
117. Schultz KD, Schultz K, Schultz G: Sodium nitroprusside and other smooth muscle-relaxants increase cyclic GMP levels in rat ductus deferens. Nature 265: 750–751, 1977
118. Katsuki S, Murad F: Regulation of adenosine cyclic 3',5'-monophosphate and guanosine cyclic 3',5'-monophosphate levels and contractility in bovine tracheal smooth muscle. Mol Pharmacol 13: 330–341, 1977
119. Katsuki S, Arnold WP, Murad F: Effects of sodium nitroprusside, nitroglycerin, and sodium azide on levels of cyclic nucleotides and mechanical-activity of various tissues. J Cyclic Nucleotide Res 3: 239–247, 1977
120. Katsuki S, Arnold W, Mittal C, Murad F: Stimulation of guanylate cyclase by sodium nitroprusside, nitroglycerin and nitric oxide in various tissue preparations and comparison to the effects of sodium azide and hydroxylamine. J Cyclic Nucleotide Res 3: 23–35, 1977
121. Lincoln TM: Effects of nitroprusside and 8-bromo-cyclic GMP on the contractile activity of the rat aorta. J Pharmacol Exp Ther 224-1: 100–107, 1983
122. Lincoln TM: Cyclic GMP and mechanisms of vasodilation. Pharmacol Ther 41: 479–502, 1989
123. Hardman JG: Cyclic nucleotides and regulation of vascular smooth muscle. J Cardiovasc Pharmacol 6 Suppl 4: S639–S645, 1984
124. Furuya M, Yoshida M, Hayashi Y, Ohnuma N, Minamino N, Kangawa K, Matsuo H: C-type natriuretic peptide is a growth inhibitor of rat vascular smooth muscle cells. Biochem Biophys Res Commun 177: 927–931, 1991
125. Murad F: Cyclic guanosine monophosphate as a mediator of vasodilation. J Clin Invest 78: 1–5, 1986
126. Cornwell TL, Lincoln TM: Regulation of intracellular Ca^{2+} levels in cultured vascular smooth muscle cells. J Biol Chem 264: 1146–1155, 1989
127. Lincoln TM, Cornwell TL: Towards an understanding of the mechanism of action of cyclic AMP and cyclic GMP in smooth muscle relaxation. Blood Vessels 28: 129–137, 1991
128. Dey NB, Boerth NJ, Murphy-Ullrich JE, Chang PL, Prince CW, Lincoln TM: Cyclic GMP-dependent protein kinase inhibits osteopontin and thrombospondin production in rat aortic smooth muscle cells. Circ Res 82: 139–146, 1998
129. Redondo J, Bishop JE, Wilkins MR: Effect of atrial natriuretic peptide and cyclic GMP phosphodiesterase inhibition on collagen synthesis by adult cardiac fibroblasts. Br J Pharmacol 124: 1455–1462, 1998
130. Burnett JC Jr, Kao PC, Hu DC, Heser DW, Heublein D, Granger JP, Opgenorth TJ, Reeder GS: Atrial natriuretic peptide elevation in congestive heart failure in the human. Science 231: 1145–1147, 1986
131. Oliver PM, Fox JE, Kim R, Rockman HA, Kim HS, Reddick RL, Pandey KN, Milgram SL, Smithies O, Maeda N: Hypertension, cardiac hypertrophy, and sudden death in mice lacking natriuretic peptide receptor A. Proc Natl Acad Sci USA 94: 14730–14735, 1997
132. Tamura N, Ogawa Y, Chusho H, Nakamura K, Nakao K, Suda M, Kasahara M, Hashimoto R, Katsuura G, Mukoyama M, Itoh H, Saito Y, Tanaka I, Otani H, Katsuki M: Cardiac fibrosis in mice lacking brain natriuretic peptide. Proc Natl Acad Sci USA 97: 4239–4244, 2000
133. Masciotra S, Picard S, Deschepper CF: Cosegregation analysis in genetic crosses suggests a protective role for atrial natriuretic factor against ventricular hypertrophy. Circ Res 84: 1453–1458, 1999
134. Deschepper CF, Masciotra S, Zahabi A, Boutin-Ganache I, Picard S, Reudelhuber TL: Functional alterations of the Nppa promoter are linked to cardiac ventricular hypertrophy in WKY/WKHA rat crosses. Circ Res 88: 223–228, 2001
135. Hutchinson HG, Trindade PT, Cunanan DB, Wu CF, Pratt RE: Mechanisms of natriuretic-peptide-induced growth inhibition of vascular smooth muscle cells. Cardiovasc Res 35: 158–167, 1997
136. Pedram A, Razandi M, Kehrl J, Levin ER: Natriuretic peptides inhibit G protein activation. Mediation through cross-talk between cyclic GMP-dependent protein kinase and regulators of G protein-signaling proteins. J Biol Chem 275: 7365–7372, 2000
137. Suhasini M, Li H, Lohmann SM, Boss GR, Pilz RB: Cyclic-GMP-dependent protein kinase inhibits the Ras/Mitogen-activated protein kinase pathway. Mol Cell Biol 18: 6983–6994, 1998
138. Erhardt P, Troppmair J, Rapp UR, Cooper GM: Differential regulation of Raf-1 and B-Raf and Ras-dependent activation of mitogen-activated protein kinase by cyclic AMP in PC12 cells. Mol Cell Biol 15: 5524–5530, 1995
139. Lemay J, Hou Y, Tremblay J, Hamet P, deBlois D: Angiotensin 1-converting enzyme activity and vascular sensitivity to angiotensin I in rat carotid artery. Eur J Pharmacol 394: 301–309, 2000
140. Furuya M, Aisaka T, Honbou N, Kawashima K, Ohno T, Tanaka S, Minamino N, Kangawa K, Matsuo H: C-type natriuretic peptide inhibits intimal thickening after vascular injury. Biochem Biophys Res Commun 193: 248–253, 1993

141. Morishita R, Gibbons GH, Pratt RE, Tomita N, Kaneda Y, Ogihara T, Dzau VJ: Autocrine and paracrine effects of atrial natriuretic peptide gene transfer on vascular smooth muscle and endothelial cellular growth. J Clin Invest 94: 824–829, 1994
142. Louzier V, Eddahibi S, Raffestin B, Deprez I, Adam M, Levame M, Eloit M, Adnot S: Adenovirus-mediated atrial natriuretic protein expression in the lung protects rats from hypoxia-induced pulmonary hypertension. Hum Gene Ther 12: 503–513, 2001
143. Hamet P, Pausova Z, Adarichev S, Adaricheva K, Tremblay J: Hypertension: Genes and environment. J Hypertens 16: 397–418, 1998
144. Cusson JR, Tremblay J, Larochelle P, Schiffrin EL, Gutkowska J, Hamet P: Clinical relationships of cyclic GMP. In: F. Murad (ed), Cyclic GMP: Synthesis, Metabolism, and Function. Academic Press, San Diego, 1994, pp 305–319
145. Cusson JR, Hamet P, Gutkowska J, Kuchel O, Genest J, Cantin M, Larochelle P: Effects of atrial natriuretic factor on natriuresis and cGMP in patients with essential hypertension. J Hypertens 5: 435–443, 1987
146. Pang SC, Hoang MC, Tremblay J, Cantin M, Garcia R, Genest J, Hamet P: Effect of natural and synthetic atrial natriuretic factor on arterial blood pressure, natriuresis and cyclic GMP excretion in spontaneously hypertensive rats. Clin Sci 69: 721–726, 1985
147. Cusson JR, Thibault G, Kuchel O, Hamet P, Cantin M, Larochelle P: Cardiovascular, renal and endocrine responses to low doses of atrial natriuretic factor in mild essential hypertension. J Human Hypertens 3: 89–96, 1989
148. Dahl LK, Heine M, Tassinari L: Effects of chronic excess salt ingestion – evidence that genetic factors play an important role in susceptibility to experimental hypertension. J Exp Med 115: 1173–1190, 1962
149. Dahl LK, Knudsen KD, Heine M, Leith GJ: Effects of chronic excess salt ingestion: Modification of experimental hypertension in rat by variations in the diet. Circ Res 22: 11–18, 1968
150. Snajdar RM, Rapp JP: Atrial natriuretic factor in Dahl rats – atrial content and renal and aortic responses. Hypertension 7: 775–782, 1985
151. Hirata Y, Ganguli M, Tobian L, Iwai J: Dahl S rats have increased natriuretic factor in atria but are markedly hyporesponsive to it. J Hypertens 6 (suppl I): I148–I155, 1984
152. Takayanagi R, Imada T, Grammer RT, Misono KS, Naruse M, Inagami T: Atrial natriuretic factor in spontaneously hypertensive rats: Concentration changes with the progression of hypertension and elevated formation of cyclic GMP. J Hypertens 4 (suppl 3): S303–S307, 1986
153. Marsh EA, Seymour AA, Haley AB, Whinnery MA, Napier MA, Nutt RF, Blaine EH: Renal and blood pressure responses to synthetic atrial natriuretic factor in spontaneously hypertensive rats. Hypertension 7: 386–391, 1985
154. Kurihara M, Gutkind JS, Saavedra JM: Alteration of atrial natriuretic peptide binding sites in spontaneously hypertensive rats. Am J Hypertens 1 (suppl): 12S–14S, 1988
155. Saito H, Inui KI, Matsukawa Y, Okano T, Maegawa H, Nakao K, Morii N, Imura H, Makino S, Hori R: Specific binding of atrial natriuretic polypeptide to renal basolateral membranes in spontaneously hypertensive rats (SHR) and stroke-prone SHR. Biochem Biophys Res Commun 137: 1079–1085, 1986
156. Swithers SE, Stewart RE, McCarty R: Binding sites for atrial natriuretic factor (ANF) in kidneys and adrenal glands of spontaneously hypertensive (SHR) rats. Life Sci 40: 1673–1681, 1987
157. Tremblay J, Huot C, Willenbrock RC, Bayard F, Gossard F, Fujio N, Koch C, Kuchel O, Debinski W, Hamet P: Increased cyclic guanosine monophosphate production and overexpression of atrial natriuretic peptide A-receptor mRNA in spontaneously hypertensive rats. J Clin Invest 92: 2499–2508, 1993
158. Deng AY, Dene H, Rapp JP: Congenic strains for the blood pressure quantitative trait locus on rat chromosome 2. Hypertension 30: 199–202, 1997
159. Tremblay J, Dutil J, Hamet P, Deng AY: Dissection of rat chromosome 2 with congenic strains support the ANP-receptor, GC-A as a candidate gene of hypertension in the Dahl rat. J Hypertens 19 (suppl 2): S163, 2001
160. Lopez J, Wong SKF, Kishimoto I, Dubois S, Mach V, Friesen J, Garbers DL, Beuve A: Salt-resistant hypertension in mice lacking the guanylyl cyclase-A receptor for atrial natriuretic peptide. Nature 378: 65–68, 1995
161. Kishimoto I, Dubois SK, Garbers DL: The heart communicates with the kidney exclusively through the guanylyl cyclase-A receptor: Acute handling of sodium and water in response to volume expansion. Proc Natl Acad Sci USA 93: 6215–6219, 1996
162. Brandt RR, Wright RS, Redfield MM, Burnett JC Jr: Atrial natriuretic peptide in heart failure. J Am Coll Cardiol 22: 86A–92A, 1993
163. Pettersson A, Hedner J, Hedner T: Renal interaction between sympathetic activity and ANP in rats with chronic ischaemic heart failure. Acta Physiol Scand 135: 487–492, 1989
164. Abassi Z, Haramati A, Hoffman A, Burnett JC Jr, Winaver J: Effect of converting-enzyme inhibition on renal response to ANF in rats with experimental heart failure. Am J Physiol 259: R84–R89, 1990
165. Abassi ZA, Kelly G, Golomb E, Klein H, Keiser HR: Losartan improves the natriuretic response to ANF in rats with high-output heart failure. J Pharmacol Exp Ther 268: 224–230, 1994
166. Yechieli H, Kahana L, Haramati A, Hoffman A, Winaver J: Regulation of renal glomerular and papillary ANP receptors in rats with experimental heart failure. Am J Physiol 265: F119–F125, 1993
167. Sakamoto A, Ono K, Abe M, Jasmin G, Eki T, Murakami Y, Masaki T, Toyo-Oka T, Hanaoka F: Both hypertrophic and dilated cardiomyopathies are caused by mutation of the same gene, delta-sarcoglycan, in hamster: An animal model of disrupted dystrophin-associated glycoprotein complex. Proc Natl Acad Sci USA 94: 13873–13878, 1997
168. Levin ER, Frank HJL, Chaudhari A, Kirschenbaum MA, Bandt A, Mills S: Decreased atrial natriuretic factor receptors and impaired cGMP generation in glomeruli from the cardiomyopathic hamster. Biochem Biophys Res Commun 159: 807–814, 1989
169. Tsunoda K, Mendelsohn FA, Sexton PM, Chai SY, Hodsman GP, Johnston CI: Decreased atrial natriuretic peptide binding in renal medulla in rats with chronic heart failure. Circ Res 62: 155–161, 1988
170. Cachofeiro V, Schiffrin EL, Cantin M, Garcia R: Glomerular and vascular atrial natriuretic factor receptors in cardiomyopathic hamsters: Correlation with the peptide biological effects. Cardiovasc Res 24: 843–850, 1990
171. Legault L, Cernacek P, Levy M: Attempts to alter the heterogeneous response to ANP in sodium-retaining caval dogs. Can J Physiol Pharmacol 70: 897–904, 1992
172. Bianchi C, Thibault G, Wrobel-Konrad E, de Lean A, Genest J, Cantin M: Atrial natriuretic factor binding sites in experimental congestive heart failure. Am J Physiol 257: F515–F523, 1989
173. Isnard R, Carayon A, Eurin J, Maistre G, Bouanani N, Barthelemy C, Crozatier B, Komajda M, Legrand JC: Glomerular atrial natriuretic factor receptors in experimental congestive heart failure. Am J Physiol 265: H923–H928, 1993
174. Garcia R, Bonhomme MC, Schiffrin EL: Divergent regulation of atrial natriuretic factor receptors in high-output heart failure. Am J Physiol 263: H1790–H1797, 1992
175. Cohen AM, Gelvan A, Winaver J, Kahna L, Floru S, Djaldetti M: Atrial natriuretic peptide in relative polycythemia and polycythemia vera. Haematologica 74: 351–353, 1989

176. Koller KJ, Lowe DG, Bennett GL, Minamino N, Kangawa K, Matsuo H, Goeddel DV: Selective activation of the B natriuretic peptide receptor by C-type natriuretic peptide (CNP). Science 252: 120–123, 1991
177. Canaan-Kuhl S, Jamison RL, Myers BD, Pratt RE: Identification of B-receptor for natriuretic peptide in human kidney. Endocrinology 130: 550–552, 1992
178. Mukaddam-Daher S, Gutkowska J, Tremblay J, Dam TV, Quillen EW Jr: Regulation of renal atrial natriuretic peptide receptors in pregnant sheep. Endocrinology 136: 4565–4571, 1995
179. Mukaddam-Daher S, Jankowski M, Dam TV, Quillen EW Jr, Gutkowska J: Renal atrial natriuretic factor receptors in hamster cardiomyopathy. Kidney Int 48: 1875–1885, 1995
180. Schiffrin EL: Decreased density of binding sites for atrial natriuretic peptide on platelets of patients with severe congestive heart failure. Clin Sci 74: 213–218, 1988
181. Strom TM, Weil J, Braun F, Stangl K, Timnik AM, Heim JM, Gerzer R: Binding sites for atrial natriuretic peptide on platelets in patients with congestive cardiomyopathy. Eur J Clin Invest 18: 524–528, 1988
182. Galla JH, Schneider G, Kotchen TA, Hayslett JP: Renin and aldosterone in the cardiomyopathic hamster in congestive heart failure. Endocrinology 101: 389–395, 1977
183. Lambert C, Larose P, Ong H, Gutkowska J, du Souich P: Influence of heart failure and sodium content in the diet on the natriuretic response to furosemide in hamsters. Res Commun Chem Pathol Pharmacol 50: 181–199, 1985
184. Quirion R, Dalpe M, de Lean A: Characterization, distribution, and plasticity of atrial natriuretic factor binding sites in brain. Can J Physiol Pharmacol 66: 280–287, 1988
185. Kubo SH, Atlas SA, Laragh JH, Cody RJ: Maintenance of forearm vasodilator action of atrial natriuretic factor in congestive heart failure secondary to ischemic or idiopathic dilated cardiomyopathy. Am J Cardiol 69: 1306–1309, 1992
186. Hirooka Y, Takeshita A, Imaizumi T, Suzuki S, Yoshida M, Ando S, Nakamura M: Attenuated forearm vasodilative response to intra-arterial atrial natriuretic peptide in patients with heart failure. Circulation 82: 147–153, 1990
187. Hirsch AT, Dzau VJ, Creager MA: Baroreceptor function in congestive heart failure: Effect on neurohumoral activation and regional vascular resistance. Circulation 75: IV36–IV48, 1987
188. Vagnetti D, Secca T, Roscani C, Santarella B, Di Nardo P: Atrial natriuretic factor stimulation of particulate guanylate cyclase in the alveolar zone of hamster lung: A cytochemical study. Med Sci Res 19: 47–48, 1991
189. Mukaddam-Daher S, Tremblay J, Fujio N, Koch C, Jankowski M, Quillen EW, Gutkowska J: Alteration of lung atrial natriuretic peptide receptors in genetic cardiomyopathy. Am J Physiol 271: L38–L45, 1996
190. Imamura T, Ohnuma N, Iwasa F, Furuya M, Hayashi Y, Inomata N, Ishihara T, Noguchi T: Protective effect of alpha-human atrial natriuretic polypeptide (alpha-hANP) on chemical-induced pulmonary edema. Life Sci 42: 403–414, 1988
191. Lofton CE, Newman WH, Currie MG: Atrial natriuretic peptide regulation of endothelial permeability is mediated by cGMP. Biochem Biophys Res Commun 172: 793–799, 1990
192. Perreault T, Gutkowska J: Role of atrial natriuretic factor in lung physiology and pathology. Am J Respir Crit Care Med 151: 226–242, 1995
193. Brandt RR, Redfield MM, Aarhus LL, Lewicki JA, Burnett JC Jr: Clearance receptor-mediated control of atrial natriuretic factor in experimental congestive heart failure. Am J Physiol 266: R936–R943, 1994
194. Nussenzveig DR, Lewicki JA, Maack T: Cellular mechanisms of the clearance function of type C receptors of atrial natriuretic factor. J Biol Chem 265: 20952–20958, 1990
195. Gutkowska J, Nemer M, Sole MJ, Drouin J, Sirois P: Lung is an important source of atrial natriuretic factor in experimental cardiomyopathy. J Clin Invest 83: 1500–1504, 1989
196. Lin X, Hanze J, Heese F, Sodmann R, Lang RE: Gene expression of natriuretic peptide receptors in myocardial cells. Circ Res 77: 750–758, 1995
197. Tei M, Horie M, Makita T, Suzuki H, Hazama A, Okada Y, Kawai C: Atrial natriuretic peptide reduces the basal level of cytosolic free Ca^{2+} in guinea pig cardiac myocytes. Biochem Biophys Res Commun 167: 413–418, 1990
198. Brown LA, Nunez DJ, Wilkins MR: Differential regulation of natriuretic peptide receptor messenger RNAs during the development of cardiac hypertrophy in the rat. J Clin Invest 92: 2702–2712, 1993
199. Willenbrock RC, Tremblay J, Garcia R, Hamet P: Dissociation of natriuresis and diuresis and heterogeneity of the effector system of atrial natriuretic factor in rats. J Clin Invest 83: 482–489, 1989
200. Cody RJ, Atlas SA, Laragh JH, Kubo SH, Covit AB, Ryman KS, Shaknovich A, Pondolfino K, Clark M, Camargo MJF, Scarborough RM, Lewicki JA: Atrial natriuretic factor in normal subjects and heart failure patients – plasma levels and renal, hormonal, and hemodynamic responses to peptide infusion. J Clin Invest 78: 1362–1374, 1986
201. Davis D, Baily R, Zelis R: Abnormalities in systemic norepinephrine kinetics in human congestive heart failure. Am J Physiol 254: E760–E766, 1988
202. Levine TB, Francis GS, Goldsmith SR, Simon AB, Cohn JN: Activity of the sympathetic nervous system and renin-angiotensin system assessed by plasma hormone levels and their relation to hemodynamic abnormalities in congestive heart failure. Am J Cardiol 49: 1659–1666, 1982
203. Saito Y, Nakao K, Morii N, Sugawara A, Shiono S, Yamada T, Itoh H, Sakamoto M, Kurahashi K, Fujiwara M, Imura H: Bay K 8644, a voltage-sensitive calcium channel agonist, facilitates secretion of atrial natriuretic polypeptide from isolated perfused rat hearts. Biochem Biophys Res Commun 138: 1170–1176, 1986
204. Burnett JC Jr: Vasopeptidase inhibition: A new concept in blood pressure management. J Hypertens Suppl 17: S37–S43, 1999

Molecular and Cellular Biochemistry **230**: 49–60, 2002.

Natriuretic peptide receptor: Structure and signaling

Kunio S. Misono

Department of Molecular Cardiology, Lerner Research Institute, Cleveland Clinic Foundation, Cleveland, OH, USA

Abstract

The ANP receptor is a single-transmembrane sequence receptor coupled to guanylate cyclase (GCase). It belongs to a family of GCase-coupled receptors that share a common overall molecular configuration. Collectively, theses GCase-coupled receptors belong to a larger family of single-transmembrane sequence receptors that include growth hormone and cytokine receptors. The signal transduction mechanism of these receptors has not been thoroughly understood. Receptor dimerization (or oligomerization) has been suggested as the mechanism. However, at least for the ANP receptor, dimerization has been seen to occur in the absence of the ligand, suggesting that an additional, as yet unknown effect of hormone binding is responsible for receptor activation. To understand the signaling mechanism, some of the functions and subsites of the ANP receptor critical for signaling have been identified, including the binding stoichiometry, receptor self-association, the juxtamembrane hinge structure containing a signature motif critical for GCase signaling, ANP-binding site residues, chloride-dependence of ANP binding, disulfide linkages, and glycosylation structures. These structures and the functional sites have been identified in the crystal structure of dimerized recombinant extracellular domain of the ANP receptor. The intracellular domain contains a kinase-homologous domain that regulates the activity of the GCase domain responding to ANP binding and also to binding of the allosteric effector ATP. Moreover, this regulatory role of the kinase-homologous domain is modulated by its own phosphorylated state. Although considerable data have been accumulated, the mechanism of ANP receptor signaling has not been well defined. Further studies are necessary to understand how ANP binds to the receptor, what conformational effect is caused by ANP binding, how this effect is transduced across the cell membrane, and how this transmembrane effect leads to stimulation of the GCase catalytic activity. (Mol Cell Biochem **230**: 49–60, 2002)

Key words: atrial natriuretic peptide, receptor, signal transduction, protein structure, binding, feedback regulation, protein crystallography

Introduction

Atrial natriuretic peptide (ANP) is a hormone secreted by the atrium of the heart that has potent natriuretic [1] and vasorelaxant activities [2, 3]. In the kidney, ANP increases glomerular filtration rate and inhibits Na^+-reabsorption at the collecting duct. ANP also suppresses secretion of aldosterone, vasopressin, and renin. These actions of ANP stimulate natriuresis and diuresis. In the vasculature, ANP relaxes norepinephrine-, angiotensin II-, and K^+-induced vascular smooth muscle contraction, causing immediate hypotensive effect (reviews [4–6]). More recently, it has been found that ANP suppresses vascular smooth muscle cell growth [7, 8], and induces apoptosis in cardiac myocytes [9] and vascular endothelial cells [10]. Transgenic animal studies have shown that mice lacking ANP gene develop salt-sensitive hypertension [11]. The animals lacking the ANP receptor gene develop salt-insensitive essential hypertension accompanied by severe cardiac hypertrophy, fibrosis, and dilatation, that are reminiscent of hypertensive heart disease in humans [12]. Furthermore, in some animals, sudden death ensues with morphologic evidence of congestive heart failure, and administration of ANP to these animals diminishes the incidence of heart failure. Thus, it is evident that ANP plays important regulatory roles in both the circulation and cardiovascular biology, and that aberration in ANP and ANP receptor system may lead to

Address for offprints: K. Misono, Department of Molecular Cardiology, Lerner Research Institute, NB50, Cleveland Clinic Foundation, 9500 Euclid Avenue, Cleveland, OH 44195, USA (E-mail: misonok@ccf.org)

hypertension, congestive heart failure, and other cardiovascular diseases. Analysis of the structure of the ANP receptor and the mechanism of its signal transduction will provide better understanding of the physiological and pathophysiological roles of ANP. It will also promote development of drugs targeted to this receptor system.

Natriuretic peptides

ANP is a 28-residue peptide that contains an intra-chain disulfide bond. The carboxyl-terminal sequence –Asn-Ser-Phe-Arg [2], the disulfide bond [13], and residues in the peptide loop [14], and are critical for the biological activity (a comprehensive review in [14]). Peptides homologous to ANP have been found. Brain natriuretic peptide (BNP, originally isolated from porcine brain) is produced and secreted mainly by the heart atrium and shows activities similar to ANP [15]. C-type natriuretic peptide (CNP) has weak natriuretic and vasorelaxant activities [16]. Physiological functions of CNP in not clearly understood. CNP may be involved in regulation of body fluid balance through the central nervous system [17, 18]. ANP, BNP, and CNP share common structural features such as the disulfide-bond forming a 17-residue peptide ring, and several conserved residues within the ring that apparently are necessary for the activity.

Natriuretic peptide receptors

The biological actions of natriuretic peptides are mediated by the A-type and B-type natriuretic peptide receptor (NPR-A and NPR-B, respectively) that are linked to guanylate cyclase (GCase). NPR-A exhibits ligand selectivity in the order ANP>BNP>>CNP and mediates most of the known actions of the natriuretic peptides. On the other hand, NPR-B shows binding selectivity in the order CNP>>ANP>BNP [17, 19]. NPR-B is thought to mediate central control of body fluid homeostasis [17, 18, 20]. NPR-A and NPR-B receptors share a common overall molecular configuration. The receptors are type-I transmembrane proteins consisting of a single ~ 1,000 residue polypeptide that contains an extracellular ANP-binding domain (ECD), a single transmembrane sequence, and an intracellular domain (ICD) consisting of a kinase-homologous regulatory domain (KHD) and a GCase catalytic domain (GCD) (Fig. 1). Binding of ANP to the ECD stimulates the catalytic activity of the GCD by an as yet unknown mechanism. It is generally accepted that, at the basal state, the KHD interacts with the GCD suppressing its activity and that ANP binding relieves this suppressive effect, thereby elevating the GCase catalytic activity [21]. An allosteric effector ATP binds to the KHD and augments GCase activation by ANP [22, 23]. The precise molecular mechanism mediating this signaling process is not well understood.

The molecular configuration of NPR-A and NPR-B receptors is shared by other membrane-bound receptors coupled to GCase, including sea urchin egg-peptide receptor [24], receptor for heat-stable enterotoxin, guanylin and uroguanylin peptides [25], retinal GCase (ret-GC) [26]; GC-D in olfactory sensory neurons [27]; GC-E and GC-F in the eye [28]; GC-G in lung, intestine, and skeletal muscle [29], and GCY-X1 in sensory neurons of *C. elegans* [30, 31]. The similarity in the molecular configuration suggests a common signal transduction mechanisms.

Collectively, these GCase-couple receptors belong to a larger family of membrane receptors having a single-transmembrane sequence, such as growth hormone and cytokine receptors. The mechanism of signal transduction by these transmembrane receptors has not been thoroughly described. Thus, resolving the structure of the ANP receptor and elucidating its mechanism of action is likely to shed light on the

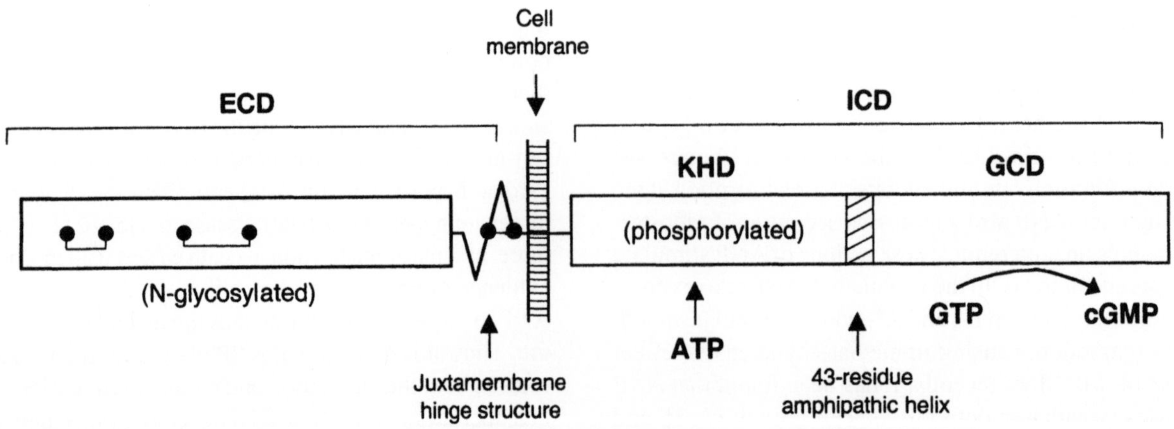

Fig. 1. Diagram showing the overall molecular topology of the ANP receptor. The extracellular ANP binding domain (ECD) is linked through the juxtamembrane hinge structure and single transmembrane sequence to the intracellular domain (ICD) consisting of the kinase-homologous regulatory domain (KHD) and guanylate cyclase catalytic domain (GCD). Disulfide bonds in the ECD are shown with thin lines.

signal transduction mechanisms of not only the GCase-coupled receptor family but also of the larger family of transmembrane receptors in general.

Juxtamembrane hinge structure and GCase signaling motif

During our photoaffinity-labeling studied of the native ANP receptor in adrenal membranes, we observed rapid proteolytic cleavage of the receptor to yield an ANP-binding 65-kDa fragment when the membranes were incubated with an ANP photoligand [32]. In contrast, no cleavage occurred in the absence of the ligand. This observation suggested that ANP binding caused a conformational change at a specific site in the receptor, rendering the site highly susceptible to proteolysis. Concomitant with this receptor cleavage, the GCase activity in the membrane was lost. Based on the size of fragments and the disulfide structure of the ECD [33], it was evident that the cleavage occurred in the peptide loop formed by Cys423-Cys432 disulfide bond near the membrane. Together, these findings suggested that the receptor contains a unique site, termed juxtamembrane hinge structure, near the membrane that undergoes a significant conformational change responding to ANP binding and may mediate transmembrane signal transduction.

The NPR-A sequence near the transmembrane region is well conserved among the species. It contains several closely spaced Pro residues and a pair of Cys residues Cys423 and Cys432 (Fig. 2). These structural features in NPR-A are shared by NPR-B receptors, but not by C-receptors that are not coupled to GCase. Mutation of Pro417 to Ala (P417A) caused uncoupling of the receptor [32]. Namely, the receptor binds ANP but does not stimulate GCase activity. Mutations of other Pro residues had no such effect. A similar uncoupling effect was also observed by F425D mutation. On the other hand, elimination of the disulfide bond by C423S/C432S double-mutation resulted in constitutive receptor activation [32]. These results demonstrate an important contribution of the structure maintained by Pro417, Phe425, and Cys423-Cys432 disulfide bond to the mechanism of receptor signaling. Remarkably, residues Pro417, Phe425, Cys423, and Cys432 are strictly conserved among the classes of GCase-coupled receptors (Fig. 2), being consistent with the notion that the juxtamembrane hinge structure plays a critical role in receptor/GCase signal transduction. We searched protein sequence databases for a similar sequence (e.g. $PX_nCX_nFX_mC$ and PX_nCX_mC) and found no other receptors containing a similar structure in the extracellular juxtamembrane region. Thus, the structure formed by residues Pro417, F425, Cys 423 and Cys432 (PX_nCXFX_nC) appears to represent a signature motif for the GCase-coupled receptor family [32].

Extracellular ANP-binding domain: Expression, purification, and the structure

To facilitate structural and functional studies of the ANP receptor, we produced the ECD of rat NPR-A in a soluble form [34]. The ECD, secreted from transfected COS cells, was purified by a single step of affinity chromatography on ANP-agarose. The purified ECD bound ANP with the affinity (K_d at ~ 2 nM) comparable to that of the native full-length receptor. This material was used to determine the disulfide structure by LC/MS coupled with Edman sequencing [33]. The data showed that the six Cys residues in the ECD form disulfide bonds Cys60-Cys86, Cys164-Cys213, and Cys423-Cys432, forming a 1-2, 3-4, 5-6 disulfide-pairing pattern. No inter-chain disulfide bond was found. The positions of the six Cys residues are strictly conserved in NPR-A from different species. In NPR-B, Cys residues corresponding to 1-2 and 5-6 disulfide-pairs are conserved, suggesting a similar arrangement of disulfide bonds in NPR-B.

The sites of glycosylation and type of bound oligosaccharides were determined as follows: Asn13 (high-mannose), Asn180 (complex; partial occupancy), Asn306 (complex), Asn347

```
                                                 Transmembrane Domain

--SEHKLYWPLGYPPPDVPKCGFDNEDPACNQDHFSTLEVLALVGSLSLISFLIVSFFIYRKMQLEK--   NPR-A
--TGRPIPWVKGVPPLDNPPCAFDMDDPSCDKTPLSTLAIVALGTGITFIMFGVSSFLIFRKLMLEK--   NPR-B
--GTRMHFPRGGSAPGPDPSCWFDPNNI-CGGGLEPGLVFLGFLLVVGMGLAGAFLAHYVRHRLLHM--   retGC
--LGTAVHFPGGSPPAHDASCWFDPNTL-CIRGVQPLGSLLTLTITCVLALVGGFLAYFIRLGLQQL--   GC-D
--GTPVHFPRGAPAPGPDPSCWFDPDVI-CNGGVEPGLVFVGFLLVIVVGLTGAFLAHYLRHRLLHM--   GC-E
--RGTPIHFPGGRPTSADAKCWFAQGKI-CQGGIDPALAMMVCFALLLALLSINGFAYFIRRRINKI--   GC-F
--VLKPVWHNRDEPPLDMPPCGFHGEL--CTNWALYLGASIPTFLIIFGGLIGFFIYRKRAYEAALD--   GC-su
```

Fig. 2. Sequence alignment of the region containing the hinge structure and transmembrane domain. The residues Pro417, Phe425, Cys423, and Cys432, constituting a GCase signaling motif, are shown shaded along with the transmembrane region. The sequences of A- and B-type natriuretic peptide receptor (NPR-A and NPR-B, respectively), retinal GCase (retGC), olfactory receptors (GC-D and GC-E), retinal GCase type-2 (GC-F) and sea urchin speract receptor (GC-su) are shown.

(complex; partial occupancy), and Asn395 (high-mannose and hybrid) [35]. Enzymatic deglycosylation of the ECD by endoglycosidases F and H had no effect on the ANP binding activity, suggesting that the oligosaccharide structures are not involved in hormone binding. On the other hand, it has been reported that glycosylation is essential for expression of ANP binding activity [36–38]. Therefore, it is evident that glycosylation is required for correct folding and/or transport of the receptor polypeptide to the cell surface, as it is the case for most glycoproteins, but is not involved in ANP binding [35].

The HPLC/MS data obtained in determining the disulfide structure and glycosylation sites described above accounted for the entire 435-residue sequence of the ECD and showed no evidence for any other post-translational modifications. The purified ECD was crystallized and the 3-dimensional structure of the *apo*ECD was determined by X-ray crystallography at 2.0 Å resolution [39] (see blow).

ANP binding stoichiometry and receptor dimerization

It has been proposed that GCase-activation may be mediated by ANP-induced receptor dimerization [40, 41] in a manner similar to the signaling mechanisms of growth hormone and cytokine receptors. Dimer and oligomers of ANP receptor were found in cells overexpressing the receptor [42, 43] as well as in detergent-solubilized adrenal membrane preparation [44]. However, in all such studies, the receptor dimers and oligomers were found irrespective of the presence of ANP. It was also found that the purified ECD, in the absence of ANP, undergoes spontaneous dimerization, while ANP binding strongly enhances this dimerization [34]. These findings suggest that receptor dimerization (or oligomerization) alone is not sufficient to cause receptor activation. It is apparent that ANP binding induces, in addition to dimerization, an additional, as yet unidentified effect on the receptor that mediates receptor activation.

de Lean and his associates [44] have reported a receptor-to-ANP binding ratio of 2:1 and have suggested a signaling mechanism similar to that of the growth hormone receptor, where a single hormone molecule binds two receptor molecules, thereby causing receptor dimerization and activation. On the other hand, receptor-to-ANP binding ratio of approximately 1 has been reported previously by studies with purified receptor preparations [45–47]. Our studies with the purified recombinant ECD showed that ECD binds ANP at 1:1 ratio and that this equimolar binding induces formation of 2:2 complexes [34]. The ANP-induced dimerization of the ECD has been further confirmed by SEC-LS and sedimentation equilibrium analysis (unpublished data). The stoichiometry also appears to be consistent with the crystal structure of the *apo*ECD dimer [39], in which two putative ANP binding sites in the dimeric structure are not geometrically overlapping (see blow). We hypothesize that the receptor is in a dynamic equilibrium between the monomer and dimer states and this equilibrium is shifted to dimer by ANP binding (Fig. 3). The linkage between ANP binding and dimerization implies that the ANP binding affinity changes with the association state of the ECD. We are currently investigating this ANP binding and receptor dimerization equilibrium by sedimentation equilibrium analysis, the preliminary data of which suggest strong enhancement (> 30-fold) of ECD dimerization by ANP. How this enhancement of dimerization relates to the signal transduction mechanism remains to be determined.

ANP-binding site

The structure of ANP binding site was probed by the stepwise-affinity-labeling method that gives high crosslinking yields [48]. Affinity labeling was performed using ANP peptide analogs containing an electrophile, iodoacetyl-group (IAc-, ^{14}C-labeled), at position 4 (N_{4a}-[^{14}C]IAc-ANP(4-28)), position 18 (N_{4a}-acetyl-N_{18e}-[^{14}C]IAc-[Lys_{18}]ANP(4-28)), or position 29 (N_{4a}-acetyl-N_{29e}-[^{14}C]IAc-[Lys_{29}]ANP(4-29)). Stepwise affinity labeling followed by HPLC peptide-mapping and sequencing allowed direct determination of labeled receptor residues [49]. The affinity reagents with the IAc-group attached at position 4 or 18 crosslinked to residue Met173 of the receptor, and that with the IAc-group attached

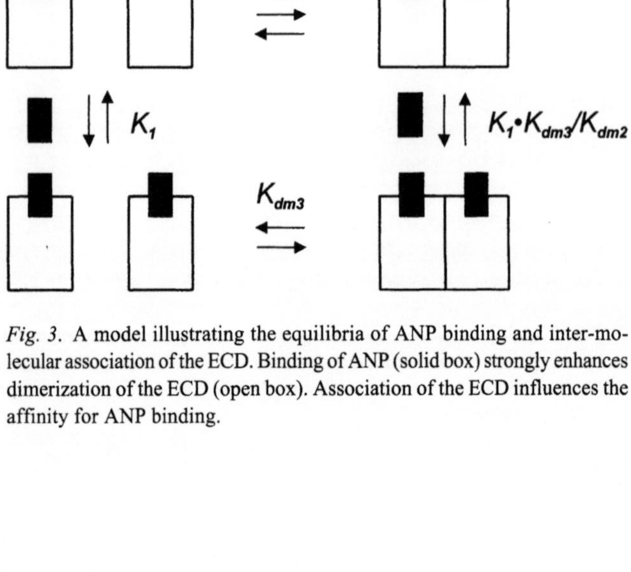

Fig. 3. A model illustrating the equilibria of ANP binding and inter-molecular association of the ECD. Binding of ANP (solid box) strongly enhances dimerization of the ECD (open box). Association of the ECD influences the affinity for ANP binding.

at position 29 crosslinked to Glu194. These results demonstrate localization of Met173 and Glu194 proximal to the ANP binding site. The results also suggest the relative orientation of the bound ANP molecule to the receptor residues. McNicoll *et al.* [50] studied the ANP binding site using ANP analogs containing photoreactive moieties and identified sequences Met173-Phe188 and Asp191-Arg198 that were labeled. The specific sites of crosslinking were not determined. In the crystal structure of the ECD (see below), these sequences extend across the molecule and cover large geometric areas and, hence, were insufficient for mapping the ANP binding site.

The binding site was probed further by mutagenesis experiments directed to the residues in sequence encompassing Met173 and Glu194. Among the mutations produced, E169K and H185D caused marked loss of ANP binding, suggesting involvement of Glu169 and His185 in hormone binding. Mutants F166A, R176L, and N180I showed ANP binding comparable to the wild-type but significantly reduced cGMP stimulation, suggesting that these residues play a role in initiating or transferring the signal of hormone binding to activate GCase catalytic domain. D160A, E190A, and D194A mutations had no effect on ANP binding or cGMP levels, suggesting that these residues are not involved in binding or receptor activation. The results of affinity labeling and mutagenesis studies allowed assignment of the ANP binding site in the *apo*ECD structure (see below).

We also probed the binding site structure by differential chemical modification, in which the ECD was chemically modified by a group-specific reagent in the presence and absence of ANP. By comparative peptide-mapping, residues protected from chemical modification upon ANP binding were identified. Tyr88 and Tyr120 were protected from modification by tetranitromethane by ANP. Similarly, His99 was protected from modification by diethylpyrocarbonate by ANP. Interestingly, in the crystal structure of the ECD, these residues are localized away from the ANP binding site determined by the affinity labeling and site-directed mutagenesis studies described above. The protection of Tyr88, Tyr120 and His99 by ANP may have resulted from ANP-induced conformational change. Alternatively, these residues may be localized at the interface of ECD dimerization or oligomerization induced by ANP binding (manuscript in preparation).

Crystal structure of the *apo*ECD dimer

The crystal structure of the ECD has been solved in collaboration with the Structural Biology Group at the Cleveland Clinic Foundation [39]. The ECD crystalized as a dimer (*apo*ECD dimer), being consistent with its tendency to spontaneously dimerize at high protein concentrations [34]. The ECD monomer contains two subdomains, each comprised of a central β-sheet flanked by α-helices (Fig. 4a). The dimerization interface consists of two pairs of juxtaposed parallel helices, one helix of each pair being presented by each monomer ECD. Dimerization through this interface brings the two protruding C-terminal hinge structures to a close proximity. The juxtamembrane hinge structure, containing a disulfide-linked 8-residue peptide loop, is anchored to the base of the membrane-proximal domain through Pro417 and Phe425 (Fig. 4b). This structure is consistent with the finding that either mutation P417A or F427D, which presumably distorts the anchored hinge structure, causes receptor uncoupling [32]. On the other hand, C423S/C432S double mutation causes constitutive GCase activation [32], indicating that the disulfide bond is necessary to maintain the basal state of the ANP receptor.

The binding site residues Met173 and Glu194 identified by affinity labeling, and Glu169 and His185 identified by mutagenesis map the ANP binding site to the side of the membrane-proximal domain extending closely to the dimer interface (Fig. 4c). Tyr88, Tyr120, and His99 implicated in ANP binding by differential chemical modification are localized near the inter-subdomain hinge region but away from the ANP binding site. It is probable that these residues are protected from modification by ANP as a result of ANP-induced change in conformation or alteration in the associated state rather than a direct steric shielding by bound ANP. The two ANP binding sites in the *apo*ECD dimer are not geometrically overlapping, apparently being consistent with the 2:2 receptor:ANP binding stoichiometry.

The ECD contains structural similarity to the type-I bacterial periplasmic binding fold protein family. This family includes extracellular solute-binding proteins such as the D-ribose binding and the Leu/Ile/Val binding proteins (LIVBP) [51] and bacterial repressors such as the amidase operon repressor AmiC [52]. These proteins are allosterically regulated by small molecules, such as amino acids, sugars and ions, that bind at the subdomain interface. The structural similarity of the ECD to these bacterial proteins suggests possible allosteric regulation of the receptor by an as yet unidentified small molecule(s). In AmiC, binding of acetamide causes a change in the inter-subdomain angle and releases the regulatory protein AmiR [52]. The ANP binding site in the ANP receptor structurally corresponds to the AmiR protein binding site in AmiC. This suggests a possibility that an as yet unidentified small ligand may bind to a site corresponding to the acetamide site in AmiC and regulate ANP binding.

A conserved chloride-binding site is found in the membrane-distal domain (Fig. 4d). The chloride ion is hydrogen bonded to hydroxyl group of Ser53 and backbone N-H moiety of Gly85 and Cys86. The presence of protein-bound chloride ion is consistent with the biochemical data that show chloride-dependence of ANP binding [53]. We have found

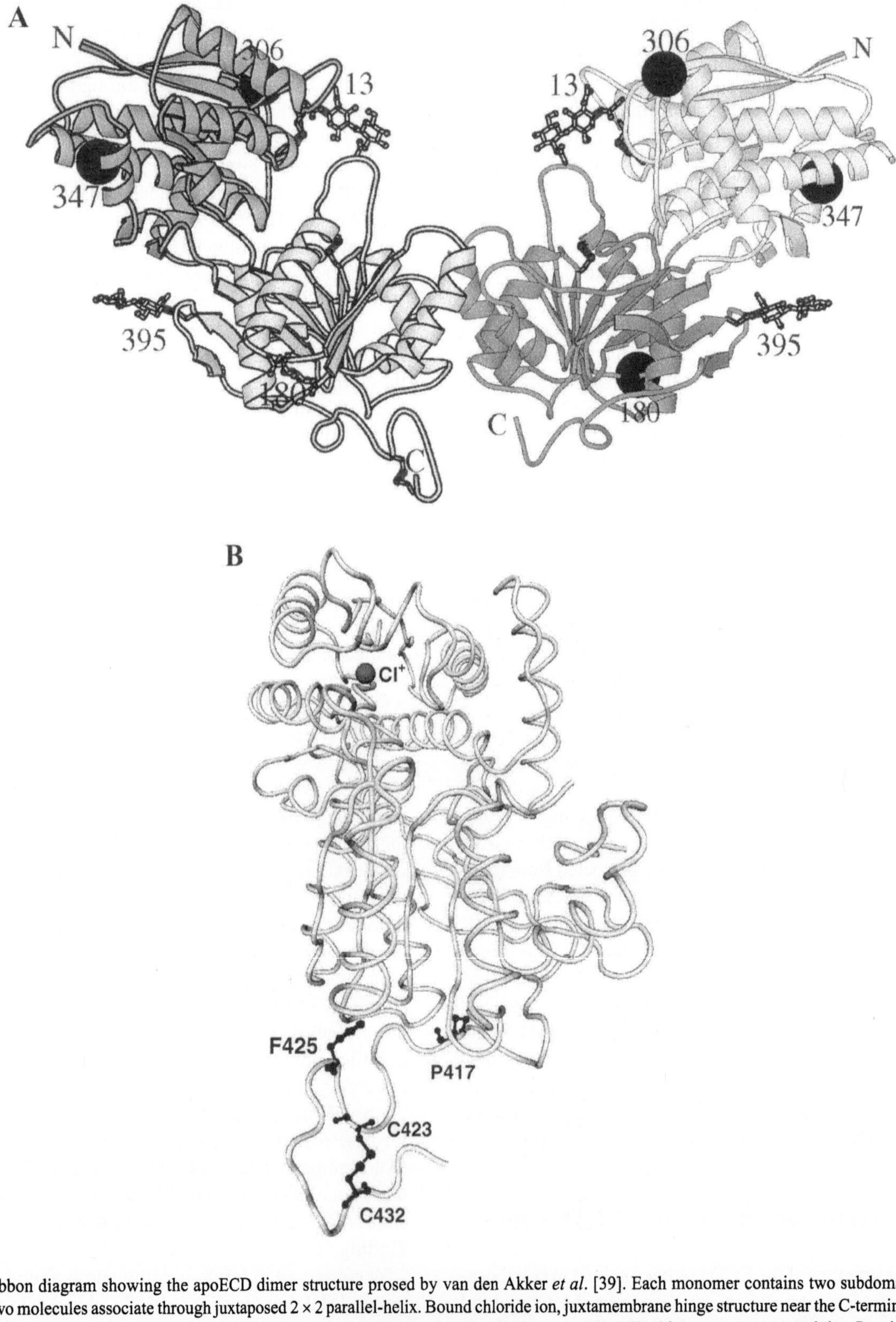

Fig. 4. (A) Ribbon diagram showing the apoECD dimer structure prosed by van den Akker *et al.* [39]. Each monomer contains two subdomains (shaded differently). Two molecules associate through juxtaposed 2 × 2 parallel-helix. Bound chloride ion, juxtamembrane hinge structure near the C-terminus, disulfide bonds, and N-glycosylation sites are shown. (B) Diagram showing the juxtamembrane hinge structure. The hinge structure, containing 7-residues peptide loop formed by Cys423-Cys432 disulfide bond, is anchored to the base of the membrane-proximal subdomain through Pro417 and Phe425. For clarity, only a monomer molecule is shown. (C) Identification of amino acid residues involved in ANP binding. The binding site residues M173 and D194 identified by the affinity labeling are shown in blue. E169 and H185 identified by site-directed mutagenesis are shown in red. Residues Y88, Y120, and H99 identified by differential chemical modification are shown in purple. Residues F166, R176 and N180 whose mutations did no affect ANP binding but abolished cGMP

C

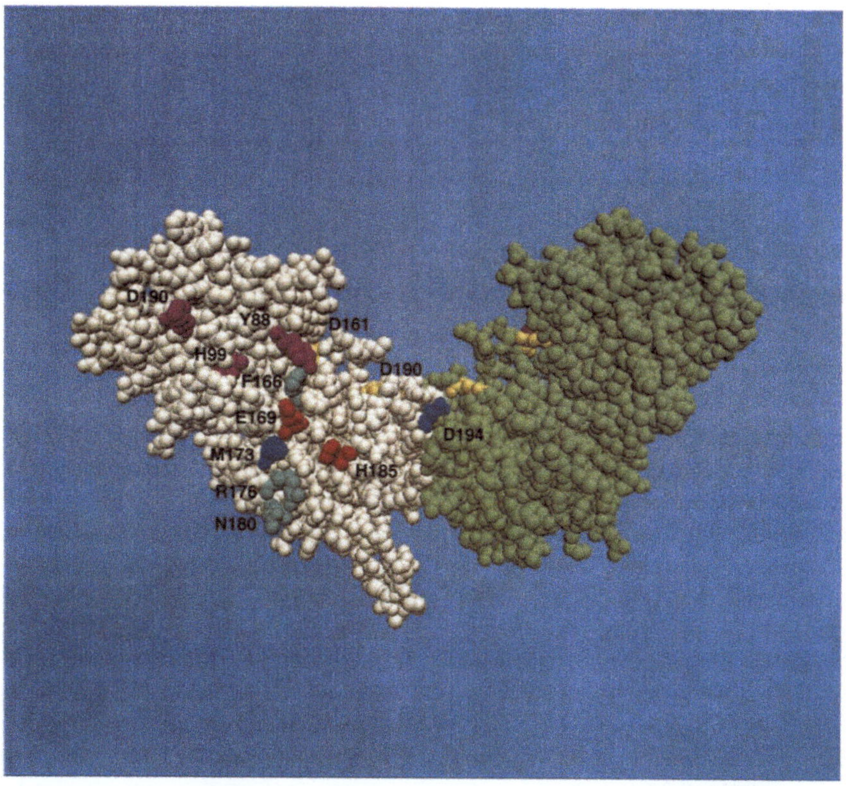

D

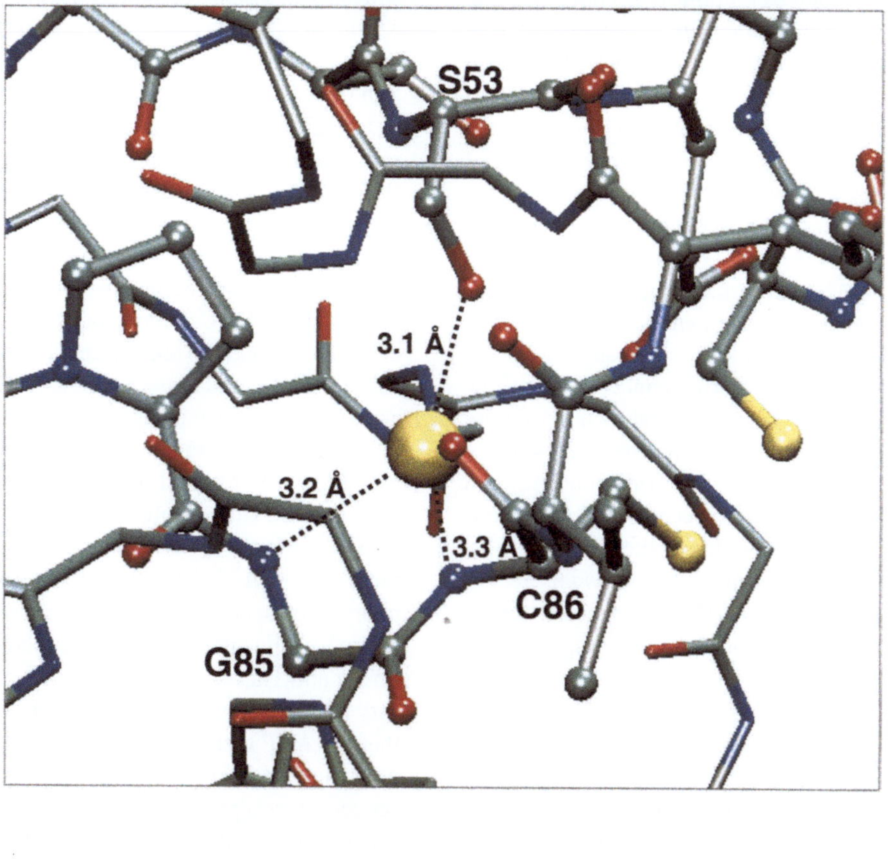

Fig. 4. (Continued)
stimulation are shown in light blue, and residues D161 and E190 whose mutations affected neither ANP binding or cGMP stimulation are shown in yellow. (D) Chloride-binding site. The chloride ion is hydrogen-bonded to hydroxyl group of Ser53 and backbone amide moieties of Gly85 and Cys86.

recently that S53T mutation retains chloride-dependence of ANP binding, while S53A mutation abolishes binding (unpublished data). These findings demonstrate critical contribution of the structure of the chloride binding site to hormone binding.

Chloride-dependent ANP binding and possible chloride-mediated feedback mechanism

In a separate line of studies, it was found that ANP binding to its receptor required the presence of chloride [53]. Binding was chloride concentration-dependent over a range from 0.05–10 mM equivalent chloride concentration with a half-maximum at ~ 0.6 mM. Binding assays at several fixed concentrations of NaCl showed that lowering chloride concentration caused a decrease in maximum binding but did not alter K_d values, suggesting that a loss of chloride blocks ANP binding rather than reducing the affinity for ANF. Excess ANP could not overcome loss of binding caused by the absence of chloride.

The chloride-dependence of ANP binding suggests possible regulation of the receptor by chloride concentration at the target sites. It has been proposed that chloride-mediated ANP receptor regulation may function as a feedback mechanism modulating ANP natriuresis[53]. In the loop of Henle and distal tubule, sodium reabsorption is mediated by Na^+-Cl^- cotransporters and, therefore, the tubular concentration of chloride changes in parallel with that of Na^+. Under certain physiological or pathophysiological conditions, the chloride concentration may decrease to mM ranges in the collecting duct (the main site of ANP action) and suppress ANP binding. For example, aldosterone treatment has been shown to lower tubular chloride concentration to mM ranges [54, 55]. Diuretics also reduce tubular chloride concentration. Low tubular chloride may also occur in congestive heart failure where both renin-angiotensin-aldosterone system and vasopressin secretion are elevated. Low chloride concentration may then inhibit ANP binding to the receptor, allowing salt-retention irrespective of the presence of the hormone. This feedback inhibition mechanism may explain widely recognized yet perplexing phenomena of ANP-resistance in certain disease states such as congestive heart failure [56], nephrotic syndrome [57], and hepatic cirrhosis [58]. In these disease states, salt is retained despite high plasma ANP leading to edematous conditions. ANP resistance observed in ANP-overexpressing transgenic animals [59] and animals on low-salt diet [60] may also be explained by this chloride-mediated inhibition mechanism. Chloride-dependent ANP binding [53] and the physical presence of receptor-bound chloride [39] strongly support the hypothesis that this chloride-mediated control of ANP receptor operates *in vivo*. In light of this possible feedback mechanism in ANP natriuresis, the physiological and pathophysiological roles of ANF in salt and volume homeostasis need to be reevaluated.

Intracellular domain

The intracellular domain of the ANP receptor consists of two interacting subdomains KHD and GCD. Deletion of the KHD causes constitutive activation of the receptor, suggesting repressive action of the KHD on the GCase catalytic activity of the GCD [21]. ATP augments GCase activation by ANP [22]. Later studies with purified recombinant NPR-A have shown that ATP is absolutely required for the responsiveness to ANP [61, 62]. The KHD is thought to be the site for ATP binding. By a series of site-directed and deletion mutagenesis analyses, the sequence region Gly-Arg-Gly-Ser-Asn-Tyr-Gly (residues 503 through 509) has been suggested to form an ATP-binding regulatory site, referred to as the ATP-regulatory module (ARM). ATP binding to this site regulates both signal transduction and receptor down regulation [23, 63, 64]. The potential ATP-binding site has been modeled based on the structures of insulin receptor kinase domain and hematopoietic cell kinase [65]. It has been reported that the ANP receptor does not require any additional proteins for signal transduction [66].

Jewett *et al.* [67] observed transient binding of ANP to the receptor stably expressed on human embryonic kidney cells (HEK cells). This transient binding is thought to have resulted from a low affinity state of the receptor induced by receptor occupancy. Applying a parallelism to the signaling mechanism of G-protein coupled receptors, a mechanism has been proposed for the ANP receptor, in which the following three events are thought to take place: (1) ANP binding to the ECD causes ATP to bind to the KHD, (2) ATP binding to the KHD activates the GCase catalytic activity, and (3) ATP bound to the KHD induces the low-affinity state of the receptor, facilitating release of bound ANP.

Potter and Hunter found that the ANP receptor expressed in HEK cells occurs as a phosphoprotein [68]. Incubation of these cells with ANP caused time-dependent receptor dephosphorylation accompanied with receptor desensitization. Furthermore, phosphorylation of the receptor, occurring mostly on Ser and Thr residues in the KHD, was shown to be necessary for the responsiveness to ANP [69]. It has bee suggested that the ANP receptor in its basal (and responsive) state is phosphorylated and that dephosphorylation occurring subsequent to receptor activation by ANP leads to desensitization (review by Potter and Hunt [70]). The phosphorylation sites in both NPR-A [69] and NPR-B have been mapped [71].

Wilson and Chinkers [72] showed that the ICD expressed in insect cells forms a dimer and exhibits low GCase activity and that deletion of the KHD resulted in increased GCase activity. By generating a series of deletion mutants of the ICD by removing varying length of the sequence from its amino-terminus, a 43-residue sequence between the KHD and GCD was shown to be required for both dimerization and the GCase activity. By molecular modeling, it has been suggested that this 43-residue sequence forms an amphipathic helix structure that mediates ICD dimerization. Direct evidence for the existence of this amphipathic helix is yet to be provided. Nevertheless, it appears that the ICD has an intrinsic capacity to dimerize and that the KHD suppresses GCase activity in the dimerized ICD. It remains to be answered how ANP bound to the ECD removes this suppressive effect of the KHD on the GCD and how the allosteric effector ATP and KHD phosphorylation may modulate this process.

Toward understanding the signal transduction mechanism

While considerable progress has been made, the signal transduction mechanism of the ANP receptor remains unresolved. At the basal state, the ANP receptor appears to be in a dynamic equilibrium between the monomer and dimer, and this equilibrium is strongly shifted to dimer by hormone binding. Yet, spontaneous dimerization of unoccupied receptors implies that dimerization alone is not sufficient for receptor activation. It is apparent that binding of ANP causes, in addition to enhancing dimerization, an as yet undetermined conformational effect that leads to receptor activation (Fig. 5). Such a conformational effect may be clarified in part by solving the crystal structure of the ECD:ANP complex in the future. This ANP-induced conformational effect caused in the ECD, in turn, needs to be transferred across the cell membrane to effect GCase activation. Transmembrane signaling may occur by a change in the distance, angle, and/or rotational relationship between the two transmembrane helices in the receptor dimer by a manner analogous to those proposed for the erythropoietin receptor [73, 74] and aspartate receptor [75]. Further, it also remains unknown as to how this transmembrane conformational effect conveyed from the ECD activates the GCase catalytic activity and how this activation process may be regulated (or switched on and off) by the KHD and its putative allosteric ligand ATP. Additionally, the mechanism by which phosphorylation in KHD regulates this signaling process also remains to be resolved. Clearly, further studies are necessary to answer all these questions. While a number of insightful studies by mutagenesis approach have provided valuable insights and hypotheses, direct structure-function studies at the level of protein by biochemical, biophysical and structural analyses seem to be particularly warranted.

Note added to proof

After this review was completed, the crystal structures of the ECD of natriuretic peptide clearance receptor (NPR-C) with and without bound C-type peptide have been reported (He *et al.* Science 293: 1657–1662, 2001). The monomer struc-

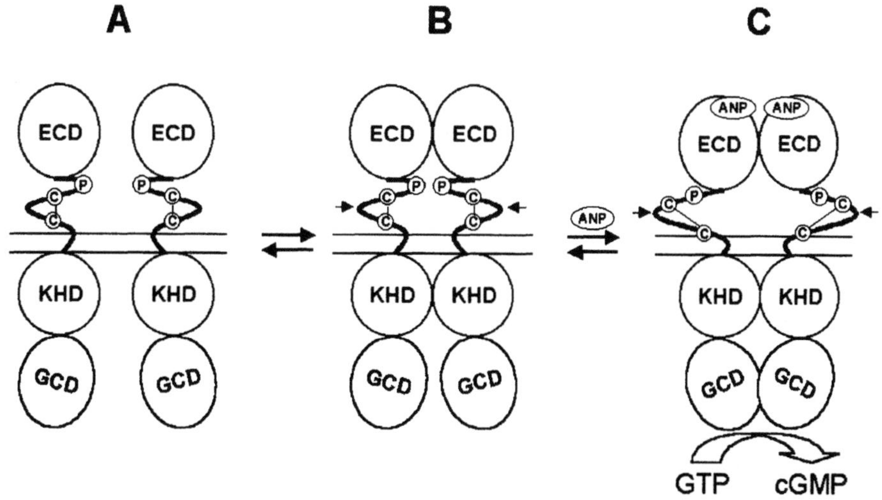

Fig. 5. Hypothetical model of signal transduction by the ANP receptor. In the absence of the hormone, the receptor is in an equilibrium between its monomer (A) and dimer states (B). ANP binding shift equilibrium strongly to the dimer and, at the same time, causes a conformational change that is transferred through the juxtamembrane hinge structure to the ICD (C). This effect then reorients the GCD such that its GCase catalytic activity becomes manifested.

ture of NPR-C ECD shows a close similarity to that of the GCase-coupled A-type receptor (NPR-A) that was reported earlier (van den Akker *et al.* Nature 406: 101–104, 2000) and is discussed here. However, the dimer configuration of the NPR-C differs distinctly from that proposed for NPR-A. He *et al.* have proposed that dimerization of NPR-C occurs through the membrane-distal subdomain (referred to as head-to-head dimerization) and suggested that dimerized structure through the membrane-proximal subdomain (tail-to-tail dimerization) proposed for NPR-A by van den Akker *et al.* may not be physiologically relevant. As discussed in the text, residues Tyr88, Tyr120 and His99 in NPR-A are protected from chemical modification by ANP binding, yet these residues are localized away from the ANP binding site mapped by affinity labeling and site-directed mutagenesis. Protection of these residues by ANP could not be explained by the tail-to-tail dimer structure at the time when we reported the NPR-A structure with our crystallgrapher group. The above protecting effect by ANP is better accounted for by the head-to-head dimer structure (figure below), in which residues Tyr88, Tyr120 and His99 in NPR-A indeed occur at the dimer interface.

Acknowledgements

I thank members of my lab and collaborators for their contributions to the work discussed in this review.

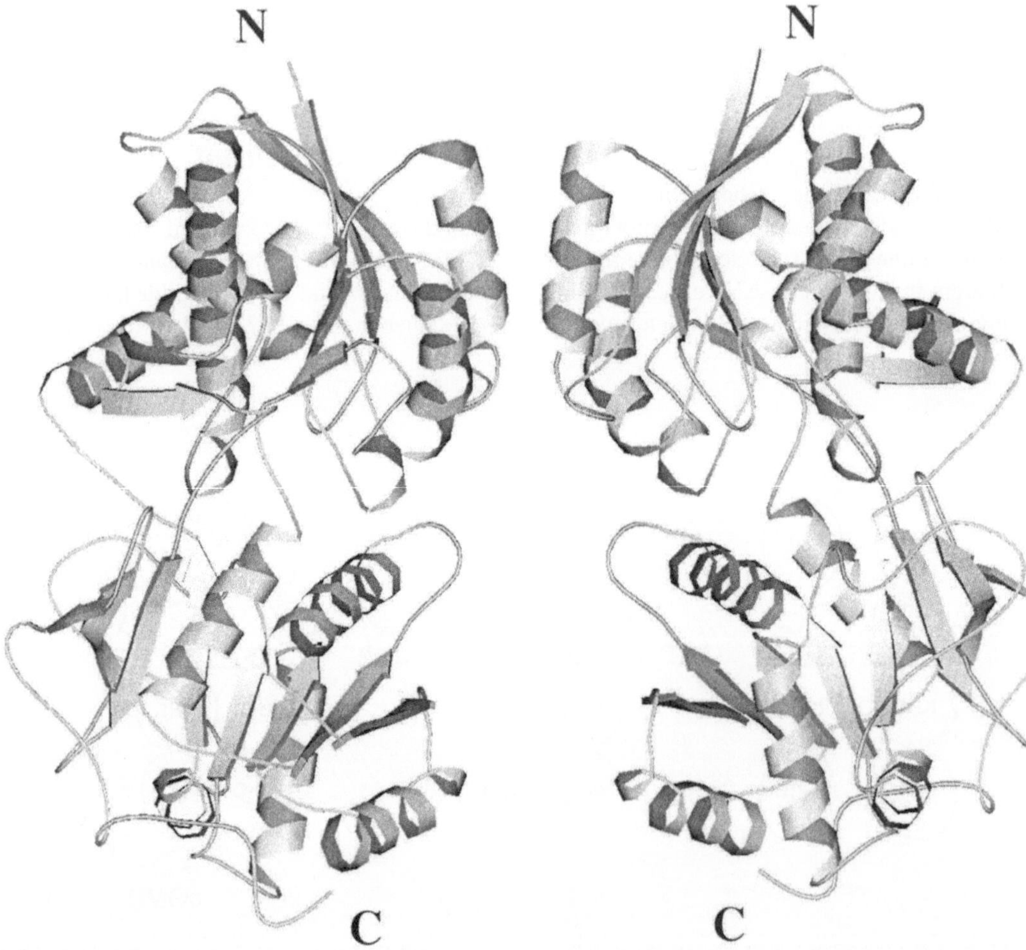

A ribbon diagram showing the structure of dimerized NPR-A receptor ECD modeled based on the C-receptor ECD dimer structure of He *et al.* (Science 293: 1657–1662, 2001).

References

1. de Bold J, Borenstein HB, Veress AT, Sonnenberg H: A rapid and potent natriuretic response to intravenous injection of atrial myocardial extract in rats. Life Science 28: 89–94, 1981
2. Currie MG, Geller DM, Cole BR, Boylan JG, YuSheng W, Holmberg SW, Needleman P: Bioactive cardiac substances: potent vasorelaxant activity in mammalian atria. Science 221: 71–73, 1983
3. Grammer RT, Fukumi H, Inagami T, Misono KS: Rat atrial natriuretic factor. Purification and vasorelaxant activity. Biochem Biophys Res Commun 116: 696–703, 1983
4. Cantin M., Gutkowska J, Thibault G, Garcia R, Anand-Srivastava M, Hamet P, Schiffrin E, Genest J: The heart as an endocrine gland. J Hypertens(suppl 2): S329–S331, 1984
5. Sonnenberg H: The physiology of atrial natriuretic factor. Can J Physiol Pharmacol 65: 2021–2023, 1987
6. Gerzer R, Heim JM, Schutte B, Weil J: Cellular mechanisms of action of atrial natriuretic factor. Klin Wochenschr 65: 109–114, 1987
7. Itoh H, Pratt RE, Dzau VJ: Atrial natriuretic polypeptide inhibits hypertrophy of vascular smooth muscle cells. J Clin Invest 86: 1690–1697, 1990
8. Furuya M, Aisaka K, Miyazaki T, Honbou N, Kawashima K, Ohno T, Tanaka S, Minamino N, Kangawa K, Matsuo H: C-type natriuretic peptide inhibits intimal thickening after vascular injury. Biochem Biophys Res Commun 193: 248–253, 1993
9. Wu CF, Bishopric NH, Pratt RE: Atrial natriuretic peptide induces apoptosis in neonatal rat cardiac myocytes. J Biol Chem 272: 14860–14866, 1997
10. Suenobu N, Shichiri M, Iwashina M, Marumo F, Hirata Y: Natriuretic peptides and nitric oxide induce endothelial apoptosis via a cGMP-dependent mechanism Arterioscler Thromb Vasc Biol 19: 140–146, 1999
11. John SW, Krege JH, Oliver PM, Hagaman JR, Hodgin JB, Pang SC, Flynn TG, Smithies O: Genetic decreases in atrial natriuretic peptide and salt-sensitive hypertension. Science 267: 679–678, 1995
12. Oliver PM, Fox JE, Kim R, Rockman HA, Kim HS, Reddick RL, Pandey KN, Milgram SL, Smithies O, Maeda N: Hypertension, cardiac hypertrophy, and sudden death in mice lacking natriuretic peptide receptor A. Proc Natl Acad Sci USA 94: 14730–14735, 1997
13. Misono KS, Fukumi H, Grammer RT, Inagami T: Rat atrial natriuretic factor: complete amino acid sequence and disulfide linkage essential for biological activity. Biochem Biophys Res Commun 119: 524–529, 1984
14. Bovy PR: Structure activity in the atrial natriuretic peptide (ANP) family. Med Res Rev 10: 115–142, 1990
15. Nakao K, Ogawa Y, Suga S, Imura H: Molecular biology and biochemistry of the natriuretic peptide system. I: Natriuretic peptides. J Hypertens 10: 907–912, 1992
16. Sudoh T, Minamino N, Kangawa K, Matsuo H: C-type natriuretic peptide (CNP): A new member of natriuretic peptide family identified in porcine brain. Biochem Biophys Res Commun 168: 863–870, 1990
17. Koller KJ, Lowe DG, Bennett GL, Minamino N, Kangawa K, Matsuo H, Goeddel DV: Selective activation of the B natriuretic peptide receptor by C-type natriuretic peptide (CNP). Science 252: 120–123, 1991
18. Imura H, Nakao K, Itoh H: The natriuretic peptide system in the brain: Implications in the central control of cardiovascular and neuroendocrine functions. Front Neuroendocrinol 13: 217–249, 1992
19. Schulz S, Singh S, Bellet RA, Singh G, Tubb DJ, Chin H, Garbers DL: The primary structure of a plasma membrane guanylate cyclase demonstrates diversity within this new receptor family. Cell 58: 1155–1162, 1989
20. Nakao K, Ogawa Y, Suga S, Imura H: Molecular biology and biochemistry of the natriuretic peptide system. II: Natriuretic peptide receptors. J Hypertens 10: 1111–1114, 1992
21. Chinkers M, Garbers DL: The protein kinase domain of the ANP receptor is required for signaling. Science 245: 1392–1394, 1989
22. Kurose H, Inagami T, Ui M: Participation of adenosine 5′-triphosphate in the activation of membrane-bound guanylate cyclase by the atrial natriuretic factor. FEBS Lett 219: 375–379, 1987
23. Goraczniak RM, Duda T, Sharma RK: A structural motif that defines the ATP-regulatory module of guanylate cyclase in atrial natriuretic factor signalling. Biochem J 282: 533–537, 1992
24. Thorpe DS, Garbers DL: The membrane form of guanylate cyclase. Homology with a subunit of the cytoplasmic form of the enzyme. J Biol Chem 264: 6545–6549, 1989
25. Schulz S, Green CK, Yuen PS, Garbers DL: Guanylyl cyclase is a heat-stable enterotoxin receptor. Cell 63: 941–948, 1990
26. Lowe DG, Dizhoor AM, Liu K, Gu Q, Spencer M, Laura R, Lu L, Hurley JB: Cloning and expression of a second photoreceptor-specific membrane retina guanylyl cyclase (RetGC), RetGC-2. Proc Natl Acad Sci USA 92: 5535–5539, 1995
27. Fulle HJ, Vassar R, Foster DC, Yang RB, Axel R, Garbers DL: A receptor guanylyl cyclase expressed specifically in olfactory sensory neurons. Proc Natl Acad Sci USA 92: 3571–3575, 1995
28. Yang RB, Foster DC, Garbers DL, Fulle HJ: Two membrane forms of guanylyl cyclase found in the eye. Proc Natl Acad Sci USA 92: 602–606, 1995
29. Schulz S, Wedel BJ, Matthews A, Garbers DL: The cloning and expression of a new guanylyl cyclase orphan receptor. J Biol Chem 273: 1032–1037, 1998
30. Baude EJ, Arora VK, Yu S, Garbers DL, Wedel BJ: The cloning of a Caenorhabditis elegans guanylyl cyclase and the construction of a ligand-sensitive mammalian/nematode chimeric receptor. J Biol Chem 272: 16035–16039, 1997
31. Yu S, Avery L, Baude E, Garbers DL: Guanylyl cyclase expression in specific sensory neurons: a new family of chemosensory receptors. Proc Natl Acad Sci USA 94: 3384–3387, 1997
32. Huo X, Abe T, Misono KS: Ligand binding-dependent limited proteolysis of the atrial natriuretic peptide receptor: Juxtamembrane hinge structure essential for transmembrane signal transduction. Biochemistry 38: 16941–16951, 1999
33. Miyagi M, Misono KS: Disulfide bond structure of the atrial natriuretic peptide receptor extracellular domain: Conserved disulfide bonds among guanylate cyclase-coupled receptors. Biochim Biophys Acta 1478: 30–38, 2000
34. Misono KS, Sivasubramanian N, Berkner K, Zhang X: Expression and purification of the extracellular ligand-binding domain of the atrial natriuretic peptide (ANP) receptor: monovalent binding with ANP induces 2:2 complexes. Biochemistry 38: 516–523, 1999
35. Miyagi M, Zhang X, Misono KS: Glycosylation sites in the atrial natriuretic peptide receptor oligosaccharide structures are not required for hormone binding. Eur J Biochem 267: 5758–5768, 2000
36. Lowe DG, Fendly BM: Human natriuretic peptide receptor-A guanylyl cyclase. Hormone cross- linking and antibody reactivity distinguish receptor glycoforms. J Biol Chem 267: 21691–21697, 1992
37. Fenrick R, McNicoll N, De Lean A: Glycosylation is critical for natriuretic peptide receptor-B function. Mol Cell Biochem 165: 103–109, 1996
38. Fenrick R, Bouchard N, McNicoll N, De Lean A: Glycosylation of asparagine 24 of the natriuretic peptide receptor-B is crucial for the formation of a competent ligand binding domain. Mol Cell Biochem 173: 25–32, 1997
39. van den Akker F, Zhang X, Miyagi M, Huo X, Misono KS, Yee VC: Structure of the dimerized hormone-binding domain of a guanylyl-cyclase- coupled receptor. Nature 406: 101–104, 2000

40. Koesling D, Bohme E, Schultz G: Guanylyl cyclases, a growing family of signal-transducing enzymes. FASEB J 5: 2785–2791, 1991
41. Schulz S, Yuen PS, Garbers DL: The expanding family of guanylyl cyclases. Trends Pharmacol Sci 12: 116–120, 1991
42. Chinkers M, Wilson EM: Ligand-independent oligomerization of natriuretic peptide receptors. Identification of heteromeric receptors and a dominant negative mutant. J Biol Chem 267: 18589–18597, 1992
43. Lowe DG: Human natriuretic peptide receptor-A guanylyl cyclase is self- associated prior to hormone binding. Biochemistry 31: 10421–10425, 1992
44. Rondeau JJ, McNicoll N, Gagnon J, Bouchard N, Ong H, De Lean A: Stoichiometry of the atrial natriuretic factor-R1 receptor complex in the bovine zona glomerulosa. Biochemistry 34: 2130–2136, 1995
45. Kuno T, Andresen JW, Kamisaki Y, Waldman SA, Chang LY, Saheki S, Leitman DC, Nakane M, Murad F: Co-purification of an atrial natriuretic factor receptor and particulate guanylate cyclase from rat lung. J Biol Chem 261: 5817–5823, 1986
46. Takayanagi R, Inagami T, Snajdar RM, Imada T, Tamura M, Misono KS: Two distinct forms of receptors for atrial natriuretic factor in bovine adrenocortical cells. Purification, ligand binding, and peptide mapping. J Biol Chem 262: 12104–12113, 1987
47. Meloche S, McNicoll N, Liu B, Ong H, De Lean A: Atrial natriuretic factor R1 receptor from bovine adrenal zona glomerulosa: Purification, characterization, and modulation by amiloride. Biochemistry 27: 8151–8158, 1988
48. He X, Nishio K, Misono KS: High-yield affinity alkylation of the atrial natriuretic factor receptor binding site. Bioconjug Chem 6: 541–548, 1995
49. Misono KS: Determination of binding-site sequences of the atrial natriuretic peptide receptor by stepwise-affinity alkylation. In: Y. Shimonishi (ed). Peptide Science – Present and Future. Kluwer Academic Publishers, Dordrecht, Netherlands, 1999, pp 163–166.
50. McNicoll N, Gagnon J, Rondeau JJ, Ong H, De Lean A: Localization by photoaffinity labeling of natriuretic peptide receptor-A binding domain. Biochemistry 35: 12950–12956, 1996
51. Saper MA, Quiocho FA: Leucine, isoleucine, valine-binding protein from *Escherichia coli*. Structure at 3.0-A resolution and location of the binding site. J Biol Chem 258: 11057–11062, 1983
52. O'Hara BP, Norman RA, Wan PT, Roe SM, Barrett TE, Drew RE, Pearl LH: Crystal structure and induction mechanism of AmiC-AmiR: A ligand- regulated transcription antitermination complex. Embo J 18: 5175–5186, 1999
53. Misono KS: Atrial natriuretic factor binding to its receptor is dependent on chloride concentration: A possible feedback-control mechanism in renal salt regulation. Circ Res 86: 1135–1139, 2000
54. Hanley MJ, Kokko JP: Study of chloride transport across the rabbit cortical collecting tubule. J Clin Invest 62: 39–44, 1978
55. Moe OW, Berry CA, Rector FC: Renal transport of glucose, amino acid, sodium, chloride, and water. In: B.M. Brenner (ed). The Kidney, 6th edn. 2000, pp 375–415
56. Burnett JC Jr, Kao PC, Hu DC, Heser DW, Heublein D, Granger JP, Opgenorth TJ, Reeder GS: Atrial natriuretic peptide elevation in congestive heart failure in the human. Science 231: 1145–1147, 1986
57. Koepke JP, DiBona GF: Blunted natriuresis to atrial natriuretic peptide in chronic sodium- retaining disorders. Am J Physiol 252: F865–F871, 1987
58. Warner L, Skorecki K, Blendis LM, Epstein M: Atrial natriuretic factor and liver disease. Hepatology 17: 500–513, 1993
59. Field LJ, Veress AT, Steinhelper ME, Cochrane K, Sonnenberg H: Kidney function in ANF-transgenic mice: Effect of blood volume expansion. Am J Physiol 260: R1–R5, 1991
60. Honrath U, Chong CK, Wilson DR, Sonnenberg H: Dietary salt extremes and renal function in rats: Effect of atrial natriuretic factor. Clin Sci (Colch) 87: 525–531, 1994
61. Marala RB, Sitaramayya A, Sharma RK: Dual regulation of atrial natriuretic factor-dependent guanylate cyclase activity by ATP. FEBS Lett 281: 73–76, 1991
62. Chinkers M, Singh S, Garbers DL: Adenine nucleotides are required for activation of rat atrial natriuretic peptide receptor/guanylyl cyclase expressed in a baculovirus system. J Biol Chem 266: 4088–4093, 1991
63. Duda T, Goraczniak RM, Sharma RK: Core sequence of ATP regulatory module in receptor guanylate cyclases. FEBS Lett 315: 143–148, 1993
64. Duda T, Sharma RK: ATP modulation of the ligand binding and signal transduction activities of the type C natriuretic peptide receptor guanylate cyclase. Mol Cell Biochem 152: 179–183, 1995
65. Duda T, Yadav P, Jankowska A, Venkataraman V, Sharma RK: Three dimensional atomic model and experimental validation for the ATP-Regulated Module (ARM) of the atrial natriuretic factor receptor guanylate cyclase. Mol Cell Biochem 214: 7–14, 2000
66. Wong SK, Ma CP, Foster DC, Chen AY, Garbers DL. The guanylyl cyclase-A receptor transduces an atrial natriuretic peptide/ATP activation signal in the absence of other proteins. J Biol Chem 270: 30818–30822, 1995
67. Jewett JR, Koller KJ, Goeddel DV, Lowe DG: Hormonal induction of low affinity receptor guanylyl cyclase. Embo J 12: 769–777, 1993
68. Potter LR, Garbers DL: Dephosphorylation of the guanylyl cyclase-A receptor causes desensitization. J Biol Chem 267: 14531–14534, 1992
69. Potter LR, Hunter T: Phosphorylation of the kinase homology domain is essential for activation of the A-type natriuretic peptide receptor. Mol Cell Biol 18: 2164–2172, 1998
70. Potter LR, Hunter T: Guanylyl cyclase-linked natriuretic peptide receptors: structure and regulation. J Biol Chem 276: 6057–6060, 2001
71. Potter LR, Hunter T: Identification and characterization of the major phosphorylation sites of the B-type natriuretic peptide receptor. J Biol Chem 273: 15533–15539, 1998
72. Wilson EM, Chinkers M: Identification of sequences mediating guanylyl cyclase dimerization. Biochemistry 34: 4696–4701, 1995
73. Livnah O, Stura EA, Middleton SA, Johnson DL, Jolliffe LK, Wilson IA: Crystallographic evidence for preformed dimers of erythropoietin receptor before ligand activation Science 283: 987–990, 1999
74. Remy I, Wilson IA, Michnick SW: Erythropoietin receptor activation by a ligand-induced conformation change. Science 283: 990–993, 1999
75. Ottemann KM, Xiao W, Shin YK, Koshland DE Jr: A piston model for transmembrane signaling of the aspartate receptor. Science 285: 1751–1754, 1999

Molecular and Cellular Biochemistry **230**: 61–72, 2002.

Intracellular trafficking and metabolic turnover of ligand-bound guanylyl cyclase/atrial natriuretic peptide receptor-A into subcellular compartments

Kailash N. Pandey

Department of Physiology, Tulane University School of Medicine and Health Sciences Center, New Orleans, LA, USA

Abstract

Atrial natriuretic peptide (ANP) is the first described member of the natriuretic peptide hormone family. ANP elicits natriuretic, diuretic, vasorelaxant and antiproliferative effects, important factors in the control of blood pressure homeostasis. One of the principal loci involved in the regulatory action of ANP is the guanylyl cyclase-linked ANP-receptor which has been designated as NPRA, also referred to as GC-A, whose ANP-binding efficiency and guanylyl cyclase activity vary remarkably in different target tissues. However, the cellular and molecular basis of these activities and the functional expression and regulation of NPRA are not well understood. The mature form of receptor resides in the plasma membrane and consists of an extracellular ligand-binding domain, a single transmembrane-spanning region, and intracellular protein kinase-like homology and guanylyl cyclase catalytic domains. In this review, emphasis has been placed on the interaction of ANP with NPRA, the ligand-mediated endocytosis, trafficking, and subcellular distribution of ligand-receptor complexes from cell surface to the intracellular compartments. Furthermore, it is implicated that after internalization, the ANP/NPRA complexes dissociate into the subcellular compartments and a population of receptor recycles back to the plasma membrane. This is an interesting area of research in the natriuretic peptide receptor field because there is currently debate over whether ANP/NPRA complexes internalize at all or whether cell utilizes some other mechanisms to release ANP from the bound receptor molecules. Indeed, controversy exist since it has been previously reported by default that among the three natriuretic peptide receptors only NPRC internalizes with bound ligand. Hence, from a thematic standpoint it is clearly evident that there is a current need to review this subject and provide a consensus forum that establishes the cellular trafficking, sequestration and processing of ANP/NPRA complexes in intact cells. Towards this aim the cellular life-cycle of NPRA will be described in the context of ANP-binding, internalization, metabolic processing, and/or inactivation, down-regulation, and degradation of ligand-receptor complexes in model cell systems. (Mol Cell Biochem **230**: 61–72, 2002)

Key words: atrial natriuretic peptide (ANP), guanylyl cyclase/natriuretic peptide receptor-A (NPRA), ligand binding, receptor internalization and trafficking

Introduction

Atrial natriuretic peptide (ANP) is the first described member in the natriuretic peptide hormone family [1]. Two other natriuretic peptide hormones closely related to ANP have been identified and isolated with molecular and biochemical characteristics similar to ANP but each derived from separate genes [2]. The nomenclatures of these natriuretic peptides are referred to as: A-type natriuretic peptide (ANF or ANP), primarily synthesized in the granules of heart atrium; B-type natriuretic peptide (BNP), initially isolated from the brain, but is predominantly present in heart ventricle; which displays the most variability in primary structure; and C-type natriuretic peptide (CNP), isolated from brain but largely synthesized in the vascular endothelial cells, and highly conserved among the species. All three types of natriuretic peptides contain a highly conserved 17-member disulfide-bonded ring structure but deviate from each other in the

Address for offprints: K.N. Pandey, Department of Physiology SL39, Tulane University School of Medicine and Health Sciences Center, 1430 Tulane Avenue, New Orleans, LA 70112, USA (E-mail: kpandey@tulane.edu)

flanking sequences. Membrane-bound forms of guanylyl cyclase (GC) represent biologically active natriuretic peptide receptors (NPRs), which have been cloned and sequenced from rat brain [3, 4], human placenta [5, 6] and mouse testis [7]. The molecular cloning and expression of cDNAs have identified three different forms of NPRs. Two of these receptors contain intrinsic GC activity that have been designated as NPRA and NPRB, also referred to as GC-A and GC-B, respectively [8–11]. A third receptor lacks the GC catalytic domain and has been termed as a natriuretic peptide clearance receptor (NPRC) which is considered to clear natriuretic peptides from the circulation [12]. Both ANP and BNP selectively stimulate NPRA, whereas CNP activates primarily NPRB and all three natriuretic peptides indiscriminately bind to NPRC [13, 14]. The NPRA is a 135-kDa transmembrane protein which generates second messenger cGMP that mediates the functional effects of ANP [15–19]. The general topological structure of NPRA is consistent with GC receptor family with at least four distinct regions comprising; ligand-binding, transmembrane, protein kinase-like homology (KHD) and GC catalytic domains. The NPRB has the overall domain structure similar to that of NPRA with the binding selectivity to CNP. NPRB is localized mainly in the brain and is thought to mediate the actions of CNP in the central nervous systems and also in vascular systems. Comparison of the amino acid sequence indicates a 62% identity among NPRA and NPRB and the intracellular regions appear to be more highly conserved than the extracellular domains of these two receptors (78 vs. 43%). The extracellular domain of NPRA is homologous to the NPRC, which contains a short (35-residues) cytoplasmic tail [20], apparently not coupled to the guanylyl cyclase activation. NPRA is the dominant form of the natriuretic peptide receptors found in peripheral organs and mediates most of the known actions of ANP. This review concern mainly the cellular itinerary of NPRA.

The binding of ANP to NPRA occurs at the plasma membrane, the receptor is synthesized on the polyribosomes of the endoplasmic reticulum, and is presumably degraded within the lysosomes. Therefore, it is apparent that NPRA is a dynamic cellular macromolecule that traverses through different compartments of the cell through its lifetime. This review describes the experiments which address the interaction of ANP with NPRA, the receptor-mediated internalization, stoichiometric distribution of ANP/NPRA complexes from cell surface to cell interior, and the release of degraded products into culture media. Furthermore, it is implicated that after internalization, the ligand-receptor complexes dissociate inside the subcellular compartments and a population of NPRA recycles back to plasma membrane. Subsequently, some of the dissociated ligand molecules escape the lysosomal degradative pathway and released intact into culture media, which reenter the cell by retroendocytotic mechanisms. By utilizing the pharmacologic and physiologic perturbants, the emphasis has been placed on the cellular regulation and processing of ligand-receptor complexes in intact cells. The discussion is concluded by examining the data available on the utilization of deletion mutations of NPRA cDNA, which has afforded experimental insights into the mechanisms, the cell utilizes in modulating the expression and functioning of NPRA in target cells.

Structural components and domains of NPRA

The primary structure of the guanylyl cyclase receptor family is consistent with at least four distinct regions which include ligand-binding, transmembrane, protein kinase-like homology, and GC catalytic domains. The GC catalytic region of NPRA has been assigned to an approximately 250 amino acid residues that presumably constitute the catalytic active site of the receptor [21–23]. Although the transmembrane GC receptors contain a single cyclase catalytic active site per polypeptide molecule, however, based on the structure modeling data [24] two polypeptide chains seem to be required to activate NPRA, and receptor functions as a homodimer [25–27]. The protein kinase-like homology domain (KHD) is a region of approximately 280 amino acids that immediately follows the transmembrane spanning domain of the receptor. The dimerization region of the receptor has been suggested to be located between the KHD and GC domains that has been predicted to form an amphipathic alpha helix structure [28]. The integrity of these regions of NPRA are conserved among human, mouse and rat [3, 5, 7]. The KHD of NPRA is more closely related to protein tyrosine kinases than protein serine/threonine kinases. In fact, it is largely similar to the protein kinase domain of the platelet-derived growth factor receptor with approximately 31% amino acid sequence identity between the comparable regions of the kinase domains [3, 5, 7]. It has been demonstrated that the KHD of NPRA serves as an important mediatory role in transducing the ligand-induced signals to activate the GC catalytic domain of the receptor [8, 29, 30]. It has been suggested that an intervening step involving the KHD is necessary to the cyclase catalytic activation process [9, 31, 32]. Several studies have suggested that ATP serves as an intracellular allosteric regulator of KHD for the activation of NPRA [30, 33–36].

Initially, ATP was considered to function by interacting with KHD because this region contains a glycine-rich nucleotide-binding motif and was postulated to provide the ATP-regulatory module for ANP signaling [30]. Indeed, previous studies as well as the recent data have indicated that KHD seems to be important for ANP-dependent activation of NPRA [37, 38]. It has been also suggested that the binding of ANP to NPRA activates the ATP binding to the KHD in the intracellular cytoplasmic space which, in turn, activates GC catalytic domain of the receptor [39]. However, the exact mechanisms of activation and the relay of signals from KHD to GC catalytic active site of the receptor remains to be established. It has been suggested that transmembrane GC receptors seem to function as homodimers coupled with intracellular domains of the two receptor molecules [25, 26, 40]. Previous studies have proposed that NPRA exists as a dimer and that one molecule of ANP binds to a receptor dimer suggesting a receptor-to-ANP binding stoichiometry of 2:1 [40]. In contrast recent observations have indicated that an equimolar binding stoichiometry between the extracellular domain of NPRA and ANP as evidenced for ligand-induced dimerization was 1:1 [24, 41]. Nevertheless, the validity of these schemes remains to be firmly established and the interactive role of receptor dimers with bound ligands in the internalization process have yet to be determined.

Ligand binding and internalization of ANP/NPRA complexes

ANP binds to NPRA in a specific and saturable manner with a kD value of $1 \times 10^{-10} - 1 \times 10^{-9}$ M [18]. ANP binding to intact cells containing endogenous NPRA or to purified receptor preparations evaluated by binding analysis, have revealed a single class of binding sites with intrinsic guanylyl cyclase activity [42–49]. The kinetics of internalization and metabolic processing of ANP through NPRC which does not contain GC catalytic activity has been extensively studied by us as well as by other investigators [11, 47, 50–57]. Early studies of the post-binding events of NPRC were greatly facilitated due to its predominant presence in vascular smooth muscle cells which are one of the best known target cells for ANP. In contrast, the studies on the post-binding events of NPRA were hampered due to the lack of suitable target cells containing predominantly this receptor protein. Nevertheless, our initial studies in Leydig tumor (MA-10) cell line [11] as well as report by others in PC-12 cells [58] indicated that ANP/NPRA complexes were internalized and sequestered inside the cells. On the other hand, studies by Maack *et al.*, however, suggested that in renomedullary interstitial as well as in mesangial cells, ANP/NPRA complexes were not processed intracellularly [59]. These authors suggested that a rapid dissociation of ligand-receptor complexes seems to take place upon ANP binding to NPRA at 37°C and intact ligand is released into culture medium. However, it is difficult to contemplate the findings of those previous studies since the dissociation of ligand was carried out in a medium containing very high concentrations of unlabeled ANP (1 μM) to preclude the rebinding of dissociated ligand to receptors which might have produced artifactual results. The cell lines utilized in those previous studies contained more than one ANP receptor subtypes; including both NPRA and NPRC.

Our studies utilizing MA-10 cell line containing predominantly high density of endogenous NPRA, as well as COS-7 cells overexpressing recombinant NPRA have firmly established that ANP/NPRA complexes are rapidly internalized, processed intracellularly, and degraded products released into culture medium [60, 61]. We performed the kinetics of ligand-binding, internalization, and intracellular sequestration of NPRA to distinguish between cell-surface associated, internalized and degraded ligand-receptor complexes [18, 60, 61]. As shown in Fig. 1, the results of those previous studies indicated that acid treatments at 4°C, removed a greater than

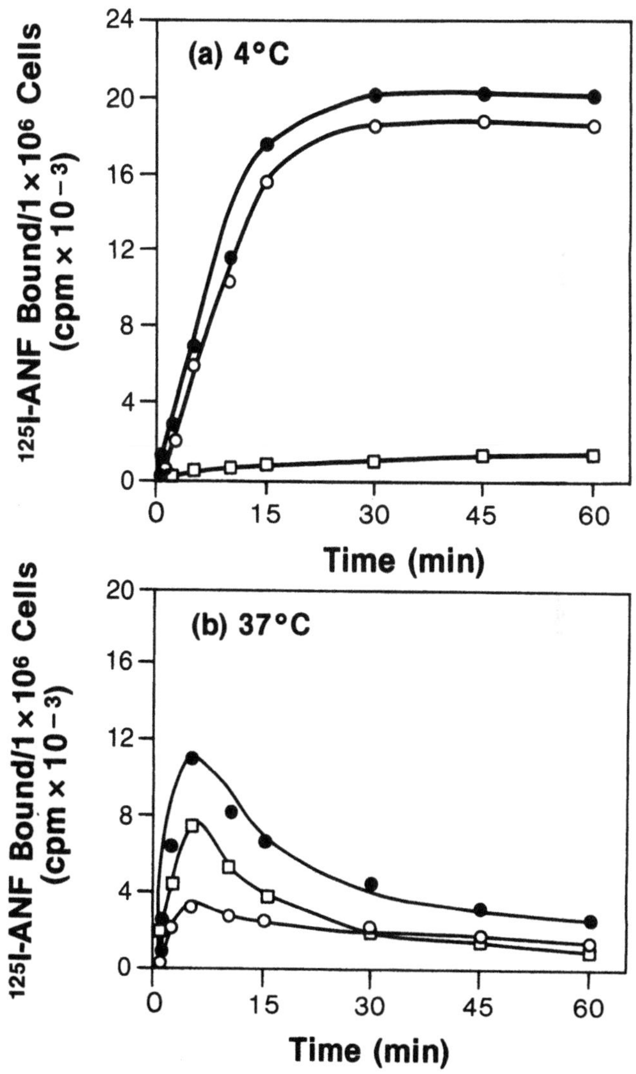

Fig. 1. Specific binding of ^{125}I-ANP to MA-10 cells at 4 and 37°C as a function of time. Confluent cells in 6-cm² culture dishes were washed with assay medium (culture medium containing 0.1% bovine serum albumin; BSA) and incubated with ^{125}I-ANP at 4°C (a) or 37°C (b) in 2 ml assay medium. At the indicated times, cells were washed with assay medium and the acid-sensitive (○) and acid resistant (□) ^{125}I-ANP radioactivity was determined. The total cell-associated radioactivity (●) is indicated as the sum of the acid-sensitive and acid-resistant radioactivities. Adapted from [60] with permission.

95% cell-surface-associated ^{125}I-ANP radioactivity whereas at 37°C, a less than 50% radioactivity was removed [60]. These observations suggested that after binding of ^{125}I-ANP to NPRA, the ligand-receptor complexes were internalized at physiological temperatures and both the degraded and intact ligands were released into culture media. Chloroquine profoundly inhibited the intracellular degradation of ^{125}I-ANP in MA-10 cells. The acid wash procedure allowed the quantitative determination of cell surface-associated, internalized and released ^{125}I-ANP radioactivity in chloroquine-treated and control cells (Figs 2a and 2b). After 15 min incubation at 37°C, approximately 40% of the total ^{125}I-ANP radioactivity was cell surface-associated in controls whereas only 5–10% ^{125}I-ANP radioactivity remained at cell surface in chloroquine-treated cells. The acid-resistant radioactivity decreased to a level of 40% and remained constant for at least 2 h. The cells which did not receive chloroquine exhibited only 10–20% acid-resistant ^{125}I-ANP radioactivity.

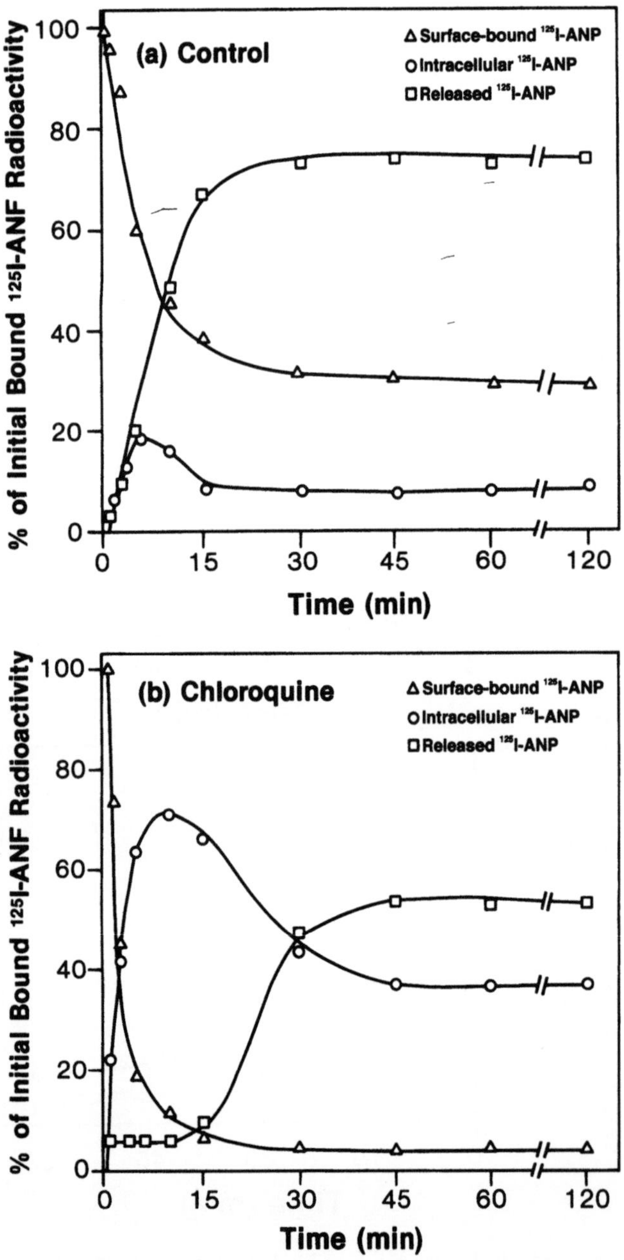

Fig. 2. Quantitative analysis of ^{125}I-ANP radioactivity in the absence or presence of chloroquine treatment. The control (a) and chloroquine pretreated (b) cells were allowed to bind ^{125}I-ANP at 4°C for 30 min, after which cells were washed 4 times with cold assay medium to remove the unbound ligand. Cells were reincubated at 37°C, and at indicated intervals, the cell surface-associated (Δ), internalized (O), and released (□) ^{125}I-ANP radioactivities were determined in acid eluates, cell extracts, and culture medium, respectively. Adapted from [60] with permission.

The distribution of ^{125}I-ANP radioactivity on the cell-surface, in the intracellular compartments and into culture media of MA-10 cells provided a dynamic equilibrium between the rates of ^{125}I-ANP uptake, its degradation, and extrusion [60]. The quantitative analysis of intact and degraded ligand was determined by measuring the solubility of ^{125}I-ANP products in 10% trichloroacetic acid (TCA). The precipitates (intact ligand) and supernatants (degraded ligand) were separated by centrifugation. It was observed that a major portion of the internalized ^{125}I-ANP was released into culture media which consisted approximately 70–75% degraded products and about 25–30% intact ligand. During the initial incubation period, the release of both the degraded and intact ligand was blocked to a greater extent by the lysosomotropic agent chloroquine, however, after longer incubation period, the effect of chloroquine was only partially effective in blocking the release of both the degraded and intact ligands [60, 61]. Overall, the available evidence support the notion that a majority of internalized ^{125}I-ANP is processed through the degradative compartments. Nevertheless, an alternate mechanism also seems to exist for the release of intact ligand. Dual pathways for the intracellular processing of ligand-receptor complexes of insulin have been previously suggested [62, 63]. Our previous studies have also provided a definitive evidence for the recycling of internalized NPRA from cell interior to the plasma membrane [60, 61]. Those previous studies have also shown that majority of the internalized ligand (>70%) is degraded in the lysosomes and released into culture medium. However, 25–30% ligand-receptor complexes escape lysosomal pathway and extruded intact into cell exterior. It is expected that these intact ligand can rebind the recycled receptors on the cell surface and reenter the cell via repeated retroendocytotic mechanisms. We predict that the homeostatic regulation of NPRA and its cellular sensitivity to ANP would be very much dependent on dynamic equilibrium of endocytotic mechanisms and intracellular processing of ANP/NPRA complexes in a ligand-regulated manner. The rates of both the internalization and breakdown of ^{125}I-ANP in MA-10 cells containing endogenous receptors as well as in COS-7 cells expressing recombinant receptors, were markedly inhibited in the presence of metabolic inhibitors such as chloroquine and dinitrophenol [60, 61].

Metabolic processing and degradation of internalized ligand receptor complexes of NPRA

Stoichiometric analyses of the metabolic processing of ANP/NPRA complexes in MA-10 cell line containing endogenous receptors and COS-7 cells expressing recombinant receptors, provided the evidence that a large population of the bound ligand-receptor complexes entered into lysosomes and released into culture media [60, 61]. Lysosomotropic agent chloroquine inhibited the degradation of ANP, suggesting that ANP was metabolized in lysosomes. ANP-specific endopeptidase is known to degrade ANP, however, its location in the endosomal compartment has not been established. The endosomal system requires ATP for the maintenance of pH gradient across the endosomal membrane. Since acidification is believed to augment the release of ligand from the receptors within the endosomes, the question can be raised as if the ANP-endopeptidase is active on ANP, bound to the receptor or ANP acting as a substrate for endopeptidase, has to be free in the lumen of the endosome. An earlier study has indicated that in neuroblastoma cells the bound-ligand to NPRA was degraded and released by a neutral metalloendopeptidase on plasma membrane [64], however, further studies have not been carried out to confirm these results. The localization of ANP-endopeptidase-like activity into cytosol, plasma membrane, and endosomes would be of a great significance. If the physiologically relevant location of the endopeptidase is endosomal compartment and if the enzymes in all three compartment are identical, the question can be raised as by what mechanism the cytosolic enzyme is inserted into the endosome. Furthermore, if this insertion process is locus for cellular regulation of the degradation of ANP, remains to be seen.

Our previous studies indicated that at 37°C, the bound ANP/NPRA complexes are rapidly internalized and processed intracellularly in target cells [60, 61]. The kinetic analysis of internalization, subcellular distribution of ANP/NPRA complexes were determined in intact cells. Both ^{125}I-ANP binding assays and photoaffinity labeling procedures were utilized to label the cell surface NPRA and to determine the post-binding kinetics and cellular fate of the labeled ligand-receptor complexes. At 37°C, almost 90% of the acid-resistant intracellular radioactivity consisted of intact ^{125}I-ANP molecules during the initial incubation period (15 min); however, it decreased to approximately 75% level after longer incubation period (30–45 min). A rapid release of ^{125}I-ANP radioactivity was detected in control culture medium, almost immediately which accounted for 75–80% of total radioactivity. The ^{125}I-ANP radioactivity released into the culture medium of control cells consisted 75–85% degraded products and 25–30% intact molecules. However, in chloroquine-treated cells, the released ^{125}I-ANP radioactivity was minimal during the initial incubation period but after 15 min the radioactivity began to appear in culture medium and steadily increased to a level of 50% ($t_{1/2}$ = 6 min). In chloroquine-treated cells, the released ^{125}I-ANP accounted for approximately 60% of total radioactivity in culture medium which consisted of 55–60% degraded products and 35–40% intact molecules.

Recycling of internalized NPRA from cell interior to plasma membrane

The tryptic proteolysis of photoaffinity labeled receptors was utilized as a means to distinguish between NPRA that remained on the cell surface and those that were internalized. The MA-10 cells which contain predominantly NPRA, were incubated with photoaffinity ligand azidobenzoyl ^{125}I-ANP (AZB-^{125}I-ANP) at 37°C followed by tryptic proteolysis to digest the photoaffinity-labeled intact 135-kDa receptor on the cell surface. Cells were incubated with AZB ^{125}I-ANP at 4°C for 10 min and photolyzed. The labeled 135-kDa NPRA was visualized by sodium dodecyl sulfate-polyacrylamide gel electrophoresis (SDS-PAGE) and autoradiography (Fig. 3A, lane a). If cells were photoaffinity-labeled at 4°C and then treated with trypsin at 16°C for 15 min, all most all the 135-kDa labeled receptors became sensitive to trypsin which generated a proteolytic fragment corresponding to the 68-kDa receptor band (Fig. 3A, lane b). The internalization of photoaffinity-labeled NPRA in MA-10 cells was examined as follows: After photoaffinity labeling, culture dishes were preincubated at 37°C for 10 min to allow the majority of ligand-and-receptor complexes to enter the intracellular compartments. The cells were then further incubated for 10 min at 37°C to assess the trypsin-sensitive receptor molecules. Cells from one set of dishes were solubilized to measure the intact 135-kDa receptors (Fig. 3B, lane a) and cells from the second set were treated with trypsin at 16°C for 5 min. During the 20 min exposure at 37°C, approximately 50–60% of labeled receptors became sensitive to trypsin and produced the 68-kDa protelytic fragment (Fig. 3B, lane b). Thus trypsin sensitivity of the photoaffinity-labeled NPRA allowed cells surface receptors to be distinguished from those internalized. It was assumed that trypsin cannot enter the cell and it will have an access only to those receptor molecules which will be present on the cell surface. Ligand binding and photoaffinity labeling studies in the intact cells demonstrated that internalized NPRA recycles back to the plasma membrane [18, 60]. The recycling of endocytosed NPRA was analyzed by incubating MA-10 cells with photoaffinity ligand AZB-^{125}I-ANP at 37°C followed by tryptic proteolysis to digest the photoaffinity-labeled intact 135-kDa receptor on the cell surface. The internalized NPRA became trypsin-resistant and

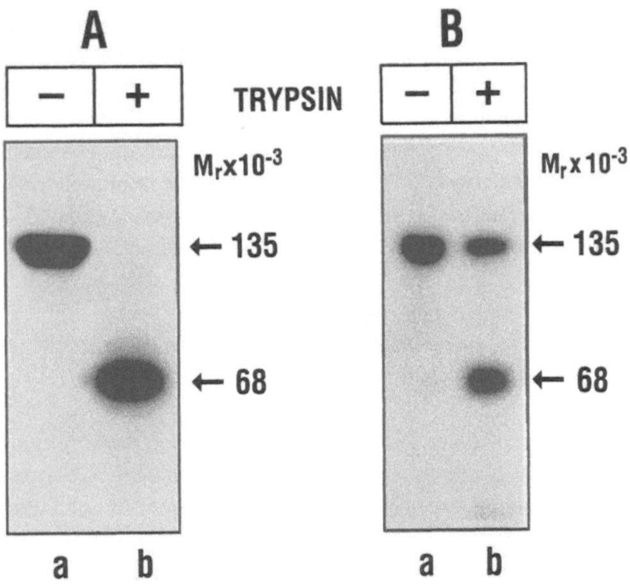

Fig. 3. Limited tryptic digestion of photoaffinity-labeled MA-10 cells to assess the internalization of NPRA. Confluent MA-10 cells were labeled with AZB ^{125}I-ANP at 4°C. After photolysis, in A, cells were treated in the absence (lane a) or presence (lane b) of trypsin (100 μg/ml) at 16°C for 5 min and then solubilized to evaluate trypsin-resistant and trypsin-sensitive cell surface receptors. In B, cells were preincubated at 37°C for 10 min to permit the internalization of ligand-receptor complexes inside the cell. Following this internalization preincubation period, cells were further incubated for 10 min at 37°C, after which cells from one set of dishes were solubilized to measure the intact 135-kDa receptors (lane a) and cells from the second set of dishes were treated with trypsin (100 μg/ml) at 16°C for 5 min (lane b). Both the 135-kDa intact receptor and the 68-kDa proteolytic fragment were separated by the SDS-PAGE and autoradiography. Arrows indicate the Mr and position of radiolabeled receptor bands. Adapted from [60] with permission.

those which returned to the cell surface became susceptible to trypsin and produced the 68-kDa tryptic fragments. The control cells which were not exposed to trypsin showed exclusively the 135-kDa intact receptor band [60]. After trypsin treatment of the photoaffinity-labelled cells, the amount of intracellular 135-kDa tryptic fragment increased in a time-dependent manner. Densitometric scanning of the radioactive-labeled bands indicated that approximately 60% of the intact receptor pool became sensitive to trypsin and produced 68-kDa proteolytic fragments after 10 min of internalization-incubation process. The subsequent decrease in the 68-kDa tryptic fragments of the receptor suggested that the internalized NPRA recycles from intracellular compartments to the plasma membrane [60]. The results of those previous studies indicated that after initiation of the internalization process with increasing time periods, the internalized receptors recycled from the intracellular compartments to the plasma membrane. The quantity of the intact 135-kDa receptor decreased in the cell interior and the amount of the 68-kDa tryptic fragment of NPRA increased. These observations provided the strong evidence that internalized receptor molecules (those once were trypsin-resistant) became trypsin-sensitive after returning to the cell surface [18, 60].

It should be noted that using trypsin protocols a decrease in the receptor population from the intracellular pool was impaired by known inhibitors of recycling such as chloroquine and dinitrophenol which further substantiated the notion that a decrease in the intracellular pool of NPRA occurs via the receptors returning to cell surface and not by trypsin entering inside the cell. After 20 min incubation at 37°C, the density of intracellular receptor decreased, indicating that approximately 10–20% of the total number of photoaffinity-labeled receptors seem to be degraded. This reflected only a small fraction of receptor loss through the degradation process that must have occurred over a period of time through internalization and recycling pathways. The reinsertion of photoaffinity-labeled receptors in the plasma membrane was very rapid at 37°C and reached to completion in 10 min ($t_{1/2}$ = 5 min). After initiation of the internalization of ligand-receptor complexes, approximately 50–60% labeled receptors recycled from the cell interior to the plasma membrane in 10 min [60]. The recycling of NPRA was slower at low temperature (22°C) as compared to 37°C and essentially negligible at 10°C. In this context it is important to note that the incubation of adrenal membranes at pH 3.5–5.6, resulted in the degradation of the 135-kDa intact NPRA and the generation of the 68–70-kDa receptor fragment [65]. Those previous findings led to the proposal that the 70-kDa receptor fragment might be generated through an endogenous proteolytic degradation of the 135-kDa receptor protein in adrenal cortex and possibly in other target cells.

Since recycling process of NPRA was completely inhibited by energy depleter dinitrophenol and lysosomotropic agent chloroquine, suggesting that the recycling of NPRA is energy-dependent process and probably requires alkalization [42, 60]. Both chloroquine and dinitrophenol are known to impair the recycling of various ligand-receptor complexes [66–68]. The results of our previous studies as well as recent experimental data are consistent with the notion that if recycling is inhibited, the internalization of the receptor continues suggesting that the receptor will not be reinserted into the plasma membrane and a rapid loss in the cell surface receptor will occur [60, 61]. The ANP binding studies clearly demonstrated that in the presence of chloroquine the amount of bound ^{125}I-ANP to cell surface receptors was significantly decreased. Similarly in the chloroquine- and dinitrophenol-treated cells, the return of internalized photoaffinity labeled ligand-receptor complexes to the plasma membranes was severely impaired, and the production of 68-kDa tryptic receptor fragment was drastically inhibited. It has been suggested that the internalized hormone-receptor complexes of the low density lipoprotein (LDL) enter the acidic vesicular compartments where the ligand dissociates from the receptor. The dissociated LDL receptors recycle back to the plasma membrane and the ligand is degraded by lysosomal pathways [69]. The results of our studies in MA-10 cells suggested that the dissociation of bound ANP from the receptor is not a prerequisite for the recycling of NPRA [60]. It is indicative that the recycling of NPRA probably also occurs constitutively in addition to the ligand-dependent regulatory mechanisms.

Alternatively to examine the recycling of NPRA, MA-10 cells were treated with trypsin at 4°C for 10 min which abolished the cell-surface receptors. However, after washing the cells free of trypsin and incubating in fresh medium at 37°C, a return in ^{125}I-ANP was observed [60]. In parallel, one group of trypsin-treated cells were also exposed to cycloheximide. Interestingly, in cycloheximide-treated cells, the binding of ^{125}I-ANP was 25–30% lower as compared with control cells without cycloheximide treatment. These observations provided further evidence that a return in ^{125}I-ANP binding was due to recycling of NPRA. Although, a complete return in ANP-binding did not occur, suggesting that a new protein synthesis may also be required. Those previous studies suggested that both lysosomotropic agent chloroquine and metabolic inhibitor dinitrophenol which depletes cellular ATP, disrupted internalization and recycling process of NPRA [60, 61]. However, it has been reported that ATP is not required for internalization of insulin receptor [70, 71], nevertheless, it seems to be essential for internalization of epidermal growth factor (EGF) receptor [72].

Signals for internalization and metabolic degradation of NPRA

Using deletion mutagenesis, we determined the specific regions in the intracellular domain of NPRA relevant to the receptor function namely; internalization, sequestration, and recycling in transfected COS-7 cells expressing NPRA. The transfection studies have relied on the loss of function of deletion mutations to identify the regions within the KHD and GC catalytic domain of NPRA [61]. The findings of those studies have suggested that the truncation of NPRA at the carboxyl-terminal end significantly reduced the hydrolysis of ligand-receptor complexes compared with wild-type receptor. For example, the complete deletion of both KHD and GC catalytic domains abolished the internalization of NPRA. The deletion of 170 amino acids at the carboxyl-terminal end of NPRA (Δ937-NPRA, Δ916-NPRA, Δ887-NPRA) reduced the internalization of ligand-receptor complexes by 60%. Cells expressing full length (wild-type) NPRA released 40–45% ^{125}I-ANP radioactivity into culture medium, but only 10–15% ^{125}I-radioactivity was released from the cells that expressed mutant NPRA with a deletion of 422 aa (Δ635-NPRA). Furthermore, 35–40% ^{125}I-ANP radioactivity was detected into the intracellular compartments of cells that expressed the wild-type NPRA and only 5–10% ^{125}I-ANP radioactivity was observed in cells expressing the Δ635-NPRA. These results suggested that specific regions within the intracellular domains of NPRA determine the extent of ligand binding efficiency, endocytosis, and intracellular sequestration of ligand-receptor complexes in transfected COS-7 cells expressing NPRA [61]. Furthermore, those previous studies indicated that with increasing deletions of amino acid residues at the carboxyl-terminal end of receptor, a large proportion of ligand-receptor complexes did not internalize and remained on the cell surface (Figs 4a and 4b). The quantitative assessment of the intact and degraded ligand was determined by measuring the solubility of ^{125}I-ANP products in 10 % TCA. As described above, the TCA precipitates (intact ligand) and supernatants (degraded ligand) were separated by centrifugation. The release of ^{125}I-ANP consisted of higher amounts of degraded products and a lesser amount of intact ligand in culture media of the cells expressing either wild-type or carboxyl-terminal truncated mutant receptors (Figs 5a and 5b).

Interestingly, most of the internalization signals have been reported to reside in the cytoplasmic domains of endocytosed receptors [73–76]. Evidence suggest that a majority of the receptors that undergo endocytosis, contain internalization signals such as NPXY sequence motif in the cytoplasmic portion, near the transmembrane domain of the receptor. However, NPRA including other members of GC-A receptor family do not seem to contain an intracellular NPXY sequence motif near the transmembrane domain. Nevertheless, examples exist where this motif is not essential for endocytosis and processing of ligand-receptor complexes such as in case of insulin-receptor complexes in CHO cells [77]. Therefore, it is indicative that the internalization of ligand-receptor complexes may not be solely dependent on NPXY motif located near the transmembrane domain of the receptor proteins. Although KHD of NPRA has been suggested to play a critical role in the functional ability of the receptor, however, the exact mechanisms by which it controls GC catalytic activity and other function of NPRA are not well understood. The search for specific sequences within the cytoplasmic as well as in the extracellular regions of receptors required for endocytosis and sequestration of NPRA are currently in progress. After internalization and sequestration from the plasma membrane, NPRA with bound ligand enters the milieu of cell interior. A variety of structures supposed to be involved in the uptake of extracellular substances, is collectively termed as the endosomal apparatus. This endosomal apparatus is non-lysosomal in nature and is characterized by subcellular fractionation, morphological markers, and enzyme activities. It has been suggested that endosomal apparatus may contain three separate compartments; (i) early stage can be found close to plasma membrane, (ii) the larger intracellular structures contain endocytosed ligand and/or receptor which is

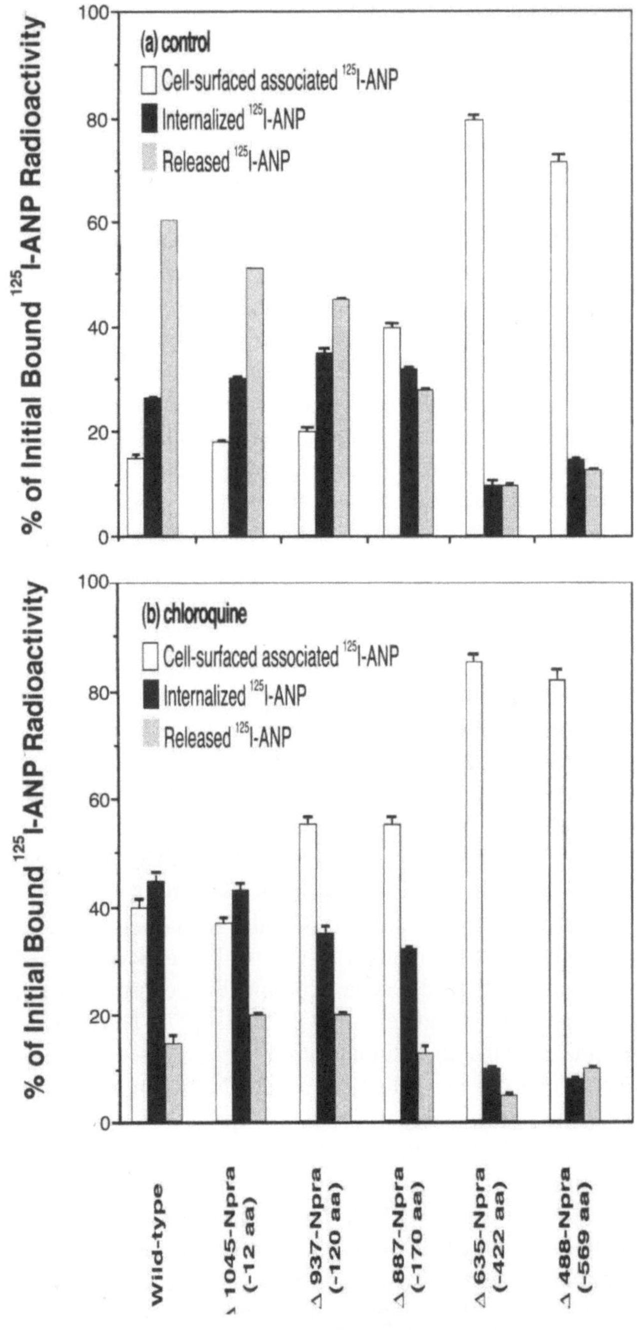

Fig. 4. Quantitative assessment of cell-surface associated, internalized and released ^{125}I-ANP radioactivity in control and chloroquine-treated COS-7 cells expressing the wild-type and carboxyl-terminal deleted mutant NPRA: Forty-eight hours after transfection, cells were washed and pretreated in the presence or absence of 200 μM chloroquine at 37°C for 30 min. ^{125}I-ANP binding was performed at 4°C for 1 h, after which cells were washed with ice-cold assay medium to remove the unbound ligand and then reincubated at 37°C for 15 min in the absence (a) or presence (b) of chloroquine. The cell surface associated, internalized and released ^{125}I-ANP was determined. Adapted from [61] with permission.

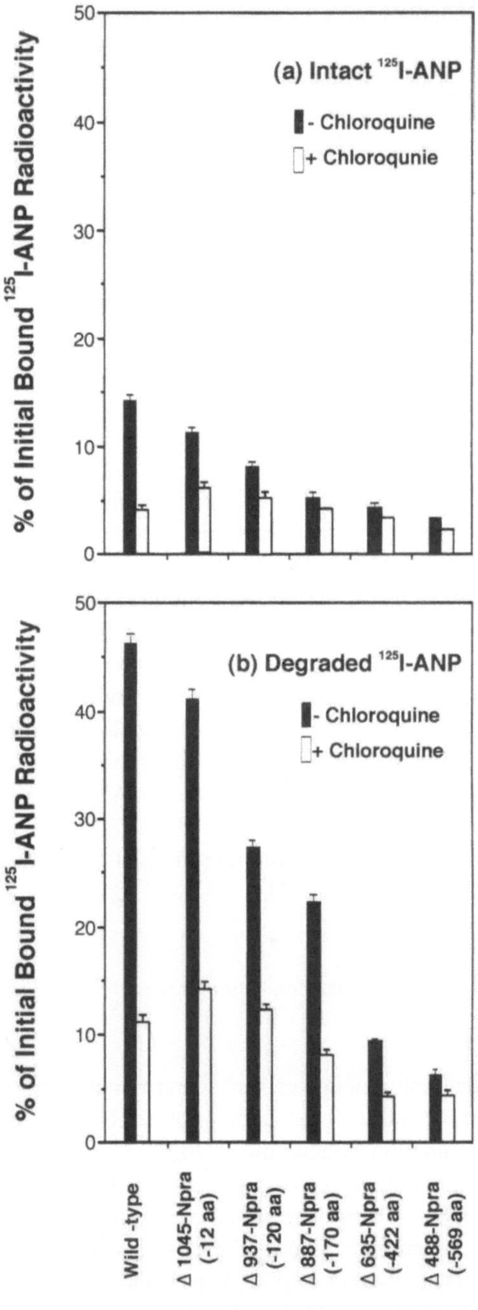

Fig. 5. Quantitative analysis of degraded and intact ^{125}I-ANP released into culture medium of COS-7 cells expressing the wild-type and carboxyl terminal deleted mutant NPRA. Forty-eight hours after transfection, COS-7 cells, were washed and preincubated in the absence (a) or presence (b) of the lysosomotropic agent chloroquine (200 μM) at 37°C for 30 min. At the end of preincubation period, cells in culture dishes were allowed to bind ^{125}I-ANP at 4°C for 1 h, after which cells were washed with ice-cold assay medium at 37°C for 15 min. At the end of incubation periods, the culture media were quantified by established methods. Adapted from [61] with permission.

vesicular and tubular in nature, and (iii) a structural compartment that contains acid phosphatase activity, similar to lysosomes [78]. Most likely, it is this third compartment which accumulates the weak base chloroquine, maintains low pH, and contains an ATP-dependent pump. Presumably, the discharge of ANP most likely may occur in this compartment. At present, no linker or adaptor proteins are known to be identified which could provide associations of NPRA with coated pit proteins essential for receptor endocytosis.

Internalization, down-regulation, and desensitization of NPRA

In large part, the mechanisms which lead to homologous down-regulation of receptors involve the complete removal of ligand-bound receptors from cell surface by an endocytic process. There could be several variations in this theme and that can result in receptor internalization, ligand degradation and then recycling of the receptor to the surface or alternatively both the receptor and the ligand can be degraded. If receptor recycles to the surface, to some extent, the internalization can be compensated by the reappearance of the receptors in the plasma membrane and thus down-regulation can be delayed, until receptors begin to be degraded. Receptors that are degraded following internalization can have important physiological and pathophysiological implications. In situation where circulating hormone levels are maintained above normal, the cells of peripheral tissues are exposed to unusually high levels of the hormone and, thus the proportion of cell surface receptors which contain bound ligand will be increased under resting conditions. Consequently, this would have an effect on promoting receptor-ligand internalization that would lead to the degradation of both species. In the course of time the increased rate of degradation can exceed the rate at which the receptors are replaced by *de novo* synthesis and the total receptor population is correspondingly reduced. This in turn, renders the cell refractory to the hormone and results in a state of hormone resistance whereby the prevailing concentrations is not sufficient to effectively cause the biological responsiveness. After binding of ANP to NPRA, ligand-receptor complexes are internalized and sequestered into the intracellular compartments and degraded products are released into culture medium. The phenomenon of internalization of ANP/NPRA complexes seem to play important role in receptor down-regulation [79]. This type of down-regulation by receptor degradation is usually homologous in nature since it only involves the receptor of one particular ligand and even though other receptors may be present on the cell surface, the rate of internalization of these is unaffected. This mechanism differs from that of sequestration since it involves the total loss of cellular receptors rather than their redistribution into a compartment which may be inaccessible to extracellular ligands.

On the other hand, heterologous desensitization is a slowly developing response which is only a manifest after continuous exposure of a cell to an agent for several hours. An example of this mechanism can be provided by the demonstration that agents which artificially elevate cell cAMP levels can provoke the heterologous desensitization followed by down-regulation of several hormone receptors linked to an effector system for example adenylyl cyclase. A change in the receptor phosphorylation state has been implicated in the process of heterologus desensitization. Interestingly, the heterologous desensitization does not require the large scale removal of receptors from the cell surface and is probably achieved by a combination of both receptor phosphorylation and degradation. Desensitization of receptor can be defined as a process by which the function of receptor is lost, whether it be ANP-binding affinity, autoregulatory efficacy of KHD, and/or GC catalytic activity. In our metabolic processing studies of NPRA, ANP-binding has been utilized as the index of NPRA activity. Degradation on the other hand, is considered as the actual loss of the receptor protein from the cell by proteolysis into amino acids. Temporally, inactivation may precede degradation or both events may occur simultaneously. An invaluable experimental tool in the elucidation of NPRA inactivation would be to use ANP-induced receptor down-regulation. Essentially down-regulation may result in a loss of cellular NPRA by means of an accelerated rate constant for receptor internalization and presumably inactivation. Desensitization and/or inactivation of NPRA has been suggested by ANP-dependent dephosphorylation of receptor protein [80, 81]. However, the exact mechanisms of the dephosphorylation-dependent inactivation of NPRA is not well understood. Previous studies as well as recent data have indicated that the NPRA phosphorylation seems to occur in a number of cell types [30, 37, 38, 82–84]. More recently, it has been suggested that the phosphorylation of kinase homology domain of NPRA is essential for its activation process [38]. We anticipate that the inactivation of receptor occurs intracellularly and inhibition of internalization should prevent this process. However, the biochemical nature of the inactivation process and subcellular location of the event have yet to be determined. These questions await further studies on the metabolic processing and turn-over of NPRA.

Our previous studies have demonstrated that after endocytosis of NPRA, some of the internalized pool of receptors could be diverted to degradation while the remainder could be subject to regulated recycling [60, 61]. Clearly, further studies would be required to clarify the mechanisms that control the endocytosis and recycling of NPRA. The consideration should be given to the fact that internalized NPRA may provide signals that contribute to the regulation of receptor gene expression. The ligand-evoked internalization and partial degradation of GABA receptors are known to both enhance as well as repress the receptor gene-expression [85–88]. Intriguing was the finding that agonist-dependent endocytosis of β2-adrenergic receptors is a necessary step in the activation of MAP kinases and mitogenic signals [89]. Those previous findings support the view that certain class of receptors may regulate their own biosynthesis through putative intracellular signals by ligand-dependent endocytosis. Indeed, at present this is a speculative notion, however, it may provide new direction for future investigations on the trafficking and functional regulation of NPRA as well as GC family of receptors.

Summary

In conclusion, evidence to date suggest that NPRA is rapidly internalized with bound ligand, processed and redistributed into subcellular compartments. The measurements of the stoichiometric distribution of ^{125}I-ANP bound to NPRA from cell surface to the cell interior and into culture media have provided the definitive means to directly examine the dynamics of ANP-mediated translocation and redistribution of biologically active receptor in intact cells containing either endogenous or recombinant NPRA. Thus it has become apparent that the cellular regulation of NPRA expression and its activity involves the movement and routing of the active receptor protein through the multiple subcellular compartments of the cell. The evidence to date further indicate that at least two major components seem to influence the cellular trafficking and the activity of NPRA. The first component is the NPRA itself and which must be properly modulated for directed routing and trafficking of the receptor protein. The second component is the interaction of the NPRA with ANP that modifies not only the rates at which the NPRA moves through the intracellular compartments and that may also influence the course of routing and trafficking pathways. The cellular trafficking and the aspects of the path followed by the receptor may be cell type-specific. The routing pathway followed by the NPRA as well as the rate with which the receptor traverses into the subcellular compartments that influences the sensitivity of the cell to ANP remains to be determined. The molecular basis of this process, as well as the clarification of the role of NPRA phosphorylation and its internalization, requires additional investigations. The possible role of NPRA internalization seems to mediate the accumulation of ANP in its target tissues, but its possible role in sustained signal generation and delivery of ANP to the subcellular compartments remains to addressed. Indeed, future studies of the biosynthetic assemblies of NPRA will be important for our understanding of the mechanisms responsible for its internalization, signaling, thereof, desensitization, down-regulation, and degradation processes of this receptor protein.

Acknowledgements

This work was supported by the grants from the National Institutes of Health (HL57531 and HL62147).

References

1. de Bold AJ: Atrial natriuretic factor: A hormone produced by the heart. Science 230: 767–770, 1985
2. Rosenzweig A, Seidman CE: Atrial natriuretic factor and related peptide hormones. Annu Rev Biochem 60: 229–255,1991
3. Chinkers M, Garbers DL, Chang MS, Lowe DG, Chin H, Goeddel DV, Schulz S: A membrane form of guanylate cyclase is an atrial natriuretic peptide receptor. Nature 338: 78–83, 1989
4. Schulz S, Singh S, Bellet RA, Singh G, Tubb DJ, Chin, H, Garbers DL: The primary structure of a plasma membrane guanylate cyclase demonstrates diversity within this new receptor family. Cell 58: 1155–1162, 1989
5. Lowe DG, Chang M-S, Hellmis R, Chen E, Singh S, Garbers D, Goddel DV: Human atrial natriuretic peptide receptor defines a new paradigm for second messenger signal transduction. EMBO J 8: 1377–1384, 1989
6. Chang MS, Lowe DG, Lewis M, Hellmiss R, Chen E, Goeddel DV: Differential activation by atrial and brain natriuretic peptides of two different receptor guanylate cyclases. Nature 341: 68–72, 1989
7. Pandey KN, Singh S: Molecular cloning and expression of murine guanylate cyclase/atrial natriuretic factor receptor cDNA. J Biol Chem 265: 12342–12348, 1990
8. Garbers DL: Guanylyl cyclase receptors and their endocrine, paracrine and autocrine ligands. Cell 71: 1–4, 1992
9. Koller KJ, de Sauvage FJ, Lowe DG, Goeddel DV: Conservation of the kinase-like regulatory domain is essential for activation of the natriuretic peptide receptor guanylyl cyclase. Mol Cell Biol 12: 2581–2590, 1992
10. Drewett JG, Garbers DL: The family of guanylyl cyclase receptors and their ligands. Endocr Rev 15: 135–162, 1994
11. Pandey KN: Vascular action: Natriuretic peptide receptor. In: J.R. Sowers (ed), Contemporary Endocrinology: Endocrinology of the Vasculature. Humana Press, Totawa, NJ, 1996, pp 255–267
12. Maack T: The role of atrial natriuretic factor in volume control. Kidney Int 49: 1732–1737, 1996
13. Koller KJ, Lowe DG, Bennett GL, Minamino N, Kangawa K, Matsuo H, Goeddel DV: Selective activation of the B natriuretic peptide receptor by C-type natriuretic peptide (CNP). Science 252: 120–123, 1991
14. Suga S-I, Nakao K, Hosda K, Mukoyama M, Ogawa Y, Shirakami G, Arai H, Saito Y, Kambayashi Y, Inoue K, Imura H: Receptor selectivity of natriuretic peptide family, atrial natriuretic peptide, brain natriuretic paeptide and C-type natriuretic peptide. Endocrinology 130: 229–239, 1992
15. Brenner BM, Ballerman BJ, Gunning ME, Zeidel ML: Diverse biological action of atrial natriuretic peptide. Physiol Rev 70: 665–699, 1990
16. Anand-Srivastava MB, Trachte GJ: Atrial natriuretic factor receptors and signal transduction mechanism. Pharmacol. Rev. 45: 455–497, 1993
17. Espiner EA: Physiology of natriuretic peptides: Cardiovascular function. In: W.K. Samson, E.R. Levin (eds), Contemporary Endocrinology:

Natriuretic Peptides in Health and Disease. Humana Press, Totowa, NJ, 1997, 123–146
18. Pandey KN: Physiology of the natriuretic peptides: Gonadal Function. In: W.K. Samson, E.R. Levin (eds), Contemporary Endocrinology: Natriuretic Peptides in Health and Disease. Humana Press, Totowa, NJ, 1997, 171–191
19. Levin ER, Gardner DG, Samson WK: Natriuretic peptides. N Engl J Med 339: 321–328, 1998
20. Fuller F, Porter JG, Arfsten AE, Miller J, Schilling JW, Scarborough RM, Levicki JA, Schenk DB: Atrial natriuretic peptide clearance receptor. Complete sequence and functional expression of cDNA clones. J Biol Chem 263: 9395–9401, 1988
21. Liu Y, Ruoho ER, Rao VD, Hurley, JH: Catalytic mechanism of the adenylyl cyclase: Modeling and mutational analysis. Proc Natl Acad Sci USA 94: 13414–13419, 1997
22. Sunahara RK, Beuve A, Tesmer JJG, Sprang SR, Garbers DL, Gilman AG: Exchange of substrate and inhibitor specificities between adenylyl and guanylyl cyclase. J Biol Chem 273: 16332–16338, 1998
23. Tucker CL, Hurley JH, Miller TR, Hurly JB: Two amino acid substitutions convert a guanylyl cyclase, Ret GC-1 into an adenylyl cyclase. Proc Natl Acad Sci USA 95: 5993–5997, 1998
24. van den Akker F, Zang X, Miyagi , Huo X, Misono KS, Yee VC: Structure of the dimerized hormone-binding domain of a guanylyl cyclase-coupled receptor. Nature 406: 101–104, 2000
25. Wilson EM, Chinkers M: Identification of sequences mediating guanylyl cyclase dimerization. Biochemistry 34: 4696–4701, 1995
26. Yang RB, Garbers DL: Two eye guanylyl cyclases are expressed in the same photoreceptor cells and form homomers in preference to heteromers. J Biol Chem 272: 13738–13742, 1997
27. Labrecque J, McNicoll N, Marquis M, De Lean A: A disulfide-bridged mutant of natriuretic peptide receptor-A displays constitutive activity. Role of receptor dimerization in signal transduction. J Biol Chem 274: 9752–9759, 1999
28. Garbers DL, Lowe DG: Guanylyl cyclase receptor. J Biol Chem 269: 30741–30744, 1994
29. Chinkers M, Garbers DL: The protein kinase of the ANP receptor is required for signaling. Science 245: 1392–1394, 1989
30. Duda T, Goraczniak RM, Sharma RK: Core sequence of ATP regulatory module in receptor guanylate cyclases. FEBS Lett 315: 143–148, 1993
31. Goraczniak RM, Duda T, Sharma RK: A structural motif that defines the ATP-regulatory module of guanylate cyclase in atrial natriuretic factor signaling. Biochem J 282: 533–537, 1992
32. Koller KJ, Lipari MT, Goeddel DV: Proper glycosylation and phosphorylation of the type A natriuretic peptide receptor are required for hormone-stimulated guanylyl cyclase activity. J Biol Chem 268: 5997–6003, 1993
33. Kurose H, Inagami T, Ui M: Participation of adenosine 5′-triphosphate in the activation of membrane-bound guanylate cyclase by the atrial natriuretic factor. FEBS Lett 219: 375–379, 1987
34. Chinkers M, Singh S, Garbers DL: Adenine nucleotides are required for activation of rat atrial natriuretic peptide receptor/guanylyl cyclase expressed in a baculovirus system. J Biol Chem 266: 4088–4093, 1991
35. Larose L, McNicoll N, Ong H, De Lean A: Allosteric modulation by ATP of the bovine adrenal natriuretic factor R1 receptor functions. Biochemistry 30: 8990–8995, 1991
36. Wong SK, Ma CP, Foster DC, Chen AY, Garbers DL: The guanylyl cyclase-A receptor transduces an atrial natriuretic peptide/ATP activation signal in the absence of other proteins. J Biol Chem 270: 30818–30822, 1995
37. Larose L, Rondeau J-J, Ong H, DeLean A: Phosphorylation of atrial natriuretic factor R1 receptor by serine/threonine protein kinase: Evidence for receptor regulation. Mol Cell Biochem 115: 203–211, 1992
38. Potter LR, Hunter T: Phosphorylation of the kinase homology domain is essential for activation of the A-type natriuretic peptide receptor. Mol Cell Biol 18: 2164–2172, 1998
39. Jewett JRS, Koller KJ, Goeddel DV, Lowe DG: Hormonal induction of low affinity receptor guanylyl cyclase. EMBO J 12: 769–777, 1993
40. Rondeau JJ, McNicoll N, Gagnn J, Bouchard N, Ong H, De Lean A: Stoichiometry of the atrial natriuretic factor-R1 receptor complex in the bovine zona glomerulosa. Biochemistry 34: 2130–2136, 1995
41. Misono KS, Sivasubramanian N, Berkner K, Zhang X: Expression and purification of the extracellular ligand-binding domain of the atrial natriuretic peptide (ANP) receptor: Monovalent binding with ANP induces 2:2 complexes. Biochemistry 38: 516–523, 1999
42. Pandey KN, Inagami T, Misono KS: Atrial natriuretic factor receptor on cultured Leydig tumor cells: Ligand binding and photoaffinity labeling. Biochemistry 25: 8467–8472, 1986
43. Kuno T, Andresen JW, Kamisaki Y, Waldman SA, Chang LY, Saheki S, Leitman DC, Nakane M, Murad F: Co-purification of an atrial natriuretic factor receptor and particulate guanylate cyclase from rat lung. J Biol Chem 261: 5817–5823, 1986
44. Takayanagi R, Snajdar RM, Imada T, Tamura, M, Pandey KN, Misono KS, Inagami T: Purification and characterization of two types of atrial natriuretic factor receptors from bovine adrenal cortex: Guanylate cyclase-linked and cyclase-free receptors. Biochem Biophys Res Commun 144: 244–250, 1987
45. Paul AK, Marala RB, Jaiswal RK, Sharam RK: Coexistence of guanylate cyclase and atrial natriuretic factor recetpor in a 180-kD protein. Science 235: 1224–1226, 1987
46. Meloche S, McNicoll N, Liu B, Ong H, DeLean A: Atrial natriuretic factor R1 receptor from bovine adrenal glomerulosa: Purification, characterization and modulation by amiloride. Biochemistry 27: 8151–8158, 1987
47. Pandey KN, Pavlou SN, Inagami T: Identification and characterization of three distinct atrial natriuretic factor receptors: Evidence for tissue-specific heterogeneity of receptor subtypes in vascular smooth muscle, kidney tubular efithelium and Leydig tumor cells by ligand binding, photoaffinity labeling and tryptic proteolysis. J Biol Chem 263: 13406–13413, 1988
48. Khurana ML, Pandey KN: Receptor-mediated stimulatory effect of atrial natriuretic factor, brain natriuretic peptide and C-type natriuretic peptide on stimulation of testosterone production in purified mouse Leydig cells: Activation of cholesterol side-chain cleavage enzyme. Endocrinology 133: 2141–2149, 1993
49. Kanungo J, Pandey KN: Kinasing PCR products for efficient blunt-end cloning and linker addition. BioTechniques 14: 912–913, 1993
50. Hirata Y, Takata S, Tomita M, Takaichi S: Binding, internalization, and degradation of atrial natriuretic peptide in cultured vascular smooth muscle cells of rat. Biochem Biophys Res Commun 132: 976–984, 1985
51. Napier M, Arcuri K, Vandlen R: Binding and internalization of atrial natriuretic factor by high-affinity receptors in A10 smooth muscle cells Arch Biochem Biophys 248: 516–522, 1986
52. Murthy KK, Thibault G, Cantin M: Binding and intracellular degradation of atrial natriuretic factor by cultured vascular smooth muscle cells. Mol Cell Endocrinol 67: 195–206, 1989
53. Nussenzveig DR, Lewicki JA, Maack T: Cellular mechanisms of the clearance function of type-C receptors of atrial natriuretic factor. J Biol Chem 265: 20952–20958, 1990
54. Cahill PA, Redmond EM, Keenan AK: Vascular atrial natriuretic factor receptor subtypes are not independently regulated by atrial peptides. J Biol Chem 265: 21896–21906, 1990
55. Cohen D, Koh GY, Nikonova LN, Porter, JG, Maack T: Molecular determinants of the clearance function of type-C receptor of natriuretic peptides. J Biol Chem 271: 9863–9869, 1996

56. Pandey KN: Kinetic analysis of internalization, recycling and redistribution of atrial natriuretic factor-receptor complex in cultured vascular smooth-muscle cells. Biochem J 288: 55–61, 1992
57. Anand-Srivastava MB: Down-regulation of atrial natriuretic peptide ANP-C receptor is associated with alterations in G-protein expression in A10 smooth muscle cells. Biochemistry 39: 6503–6513, 2000
58. Rathinavelu A, Isom GE: Differential internalization and processing of atrial natriuretic factor B and C receptors in PC 12 cells. Biochem J 276: 493–497, 1991
59. Koh GY, Nussenzweig DR, Okolicany J, Price DA, Maack T: Dynamics of atrial natriuretic factor-guanylate cyclase receptors and receptor-ligand complexes in cultured glomerular mesangial and renomedullary interstitial cells. J Biol Chem 267: 11987–11994, 1992
60. Pandey KN: Stoichiometric analysis of internalization, recycling and redistribution of photoaffinity labeled guanylyl cyclase/atrial natriuretic factor receptors in cultured murine Leydig tumor cells. J Biol Chem 268: 4382–4390, 1993
61. Pandey KN, Kumar R, Li M, Nguyen H: Functional domains and expression of truncated atrial natriuretic peptide receptor-A: The carboxyl-terminal regions direct the receptor internalization and sequestration in COS-7 cells. Mol Pharm 57: 259–267, 2000
62. Tietze C, Schlesinger P, Stahl P: Mannose-specific endocytosis receptor of alveolar macrophages: Demonstration of two functionally distinct intracellular pools of receptor and their roles in receptor recycling. J Cell Biol 92: 417–424,1982
63. Marshall S: Dual pathways for the intracellular processing of insulin: Relationship between retroendocytosis of intact hormone and the recycling of insulin receptors. J Biol Chem 260: 13524–13531, 1985
64. Delporte C, Poloczek P, Tastenoy M, Winard J, Christopher J: Atrial natriuretic peptide binds to ANP-R 1 receptors in neuroblastoma cells or is degraded extracellularly at the Ser-Phe bond. Eur J Pharmacol 227: 247–256, 1992
65. Misono KS: Acidic pH- and metal ion (Zn^{++} or Mn^{++})-dependent proteolysis of 140 kDa atrial natriuretic factor receptor in bovine adrenal cortex plasma membranes: Evidence for membrane-bound acidic metallo-endopeptidase. Biochem Biophys Res Commun 152: 658–667, 1988
66. Ganzalez-Noriega A, Grubb A, Talkad JT, Sly WS: Chloroquine inhibits lysosomal enzyme pinocytosis and enhances lysosomal enzyme secretion by impairing receptor recycling. J Cell Biol 85: 839–852, 1980
67. Tietze C, Schlesinger P, Stahl P: Mannose-specific endocytosis receptor of alveolar macrophages: Demonstration of two functionally distinct intracellular pools of receptor and their roles in receptor recycling. J Cell Biol 92: 417–424, 1982
68. Van Leuvan F, Cassiman JJ, Van-Den Berghe H: Primary amines inhibit recycling of $\alpha 2$ M receptors in fibroblasts. Cell 20: 37–43, 1980
69. Brown MS, Anderson RGW, Goldstein JL: Recycling receptors: The round-trip itinerary of migrant membrane proteins. Cell 32: 663–667, 1983
70. Backer JM, Kahn CR, White MF: Tyrosine phosphorylation of the insulin receptor is not required for receptor internalization: Studies in 2,4-dinitrophenol-treated cells. Proc Natl Acad Sci USA 86: 3201–3213, 1989
71. Smith RM, Jarett L: Differences in adenosine triphosphate dependency of receptor-mediated endocytosis of α-2 macroglobulin and insulin correlate with separate routes of ligand-receptor complex internalization. Endocrinology 126: 1551–1560, 1990
72. Carpenter G, Cohen S: ^{125}I-labeled human epidermal growth factor: Binding, internalization, and degradation in human fibroblasts. J Cell Biol 71: 159–171, 1976
73. Parez HD, Holmes R, Bilander LR, Adams RR, Manzan W, Jolley D, Andrews WH: Formyl peptide receptor chimeras define domains involved in ligand binding. J Biol Chem 268: 2292–2295, 1993
74. Sorokin A, Mohammadi M, Huang J, Slessinger J: Internalization of fibroblast growth factor receptor is inhibited by a point mutation at tyrosine 766. J Biol Chem 269: 17056–17061,1994
75. Haft SK, Quinn AM, Taylor SI: Involvement of dileucine motif in the internalization and degradation of the insulin receptor. J Biol Chem 269: 26286–26294, 1994
76. Huang Z, Chen Y, Nissenson RA: The cytoplasmic tail of the G-protein-coupled receptor for parathyroid hormone and parathyroid hormone-related protein contains positive and negative signals for endocytosis. J Biol Chem 270: 151–156, 1995
77. Berhanu P, Ibrahim-Schneck RH, Anderson C, Wood WM: The NPEY sequence is not necessary for endocytosis and processing of insulin receptor complexes. Mol Endocrinol 5: 1827–1835, 1991
78. Bergeron JJM, Cruz J, Khan MN, Posher BI: Uptake of insulin and other ligands in receptor rich endocytic components of target cells: The endosomal apparatus. Ann Rev Physiol 47: 383–403, 1985
79. Pandey KN, Nguyen H: Ligand-dependent down-regulation of atrial natriuretic peptide receptor-A in HEK-293 cells. FASEB J 14: A784, 2000
80. Potter LR, Garbers DL: Dephosphorylation of the guanylyl cyclase-A receptor causes desensitization. J Biol Chem 267: 14531–14534, 1992
81. Potter LR, Garbers DL: Protein kinase C-dependent desensitization of the atrial natriuretic peptide receptor is mediated by dephosphorylation. J Biol Chem 269: 14636–14642, 1994
82. Pandey KN: Stimulation of protein phosphorylation by atrial natriuretic factor in plasma membranes of bovine adrenal cortical cells. Biochem Biophys Res Commun 163: 988–994, 1989
83. Sharma RK, Duda T: Plasma membrane guanylate cyclase: A multimodule transduction system. Adv Exp Med Biol 407: 271–279, 1997
84. Duda T, Yadav P, Jankowska A, Venkataraman V, Sharma RK: Three dimensional atomic model and experimental validation for the ATP-regulated module (ARM) of the atrial natriuretic factor receptor guanylate cyclase. Mol Cell Biochem 217: 165–172, 2001
85. Kim HY, Sapp DE, Olsen RW, Tobin AJ: GABA alters GABAA receptor mRNAs and increases ligand binding. J Neurochem 61: 2334–2337, 1993
86. Elster L, Hansen GH, Belhage BO, Fritschy JM, Mohler H, Schousboe A: Differential distribution of GABAA receptor subunits in soma and processes of cerebellular granule cells: Effects of maturation and a GABA agonist. J Dev Neurosci 13: 417–428, 1995
87. Barnes EM: Intracellular trafficking of GABAA receptors. Int Rev Neurobiol 39: 53–76, 1996
88. Miranda JD, Barnes EM: Repression of γ-aminobutyric acid type A receptor a 1 polypeptide biosynthesis requires chronic agonist exposure. J Biol Chem 272: 16288–16294, 1997
89. Lefkowitz RJ: G protein-coupled receptors. J Biol Chem 273: 18677–18680, 1998

Molecular and Cellular Biochemistry **230**: 73–83, 2002.

Structure and function of the heat-stable enterotoxin receptor/guanylyl cyclase C

Arie B. Vaandrager

Department of Biochemistry and Cell Biology, Faculty of Veterinary Medicine, Utrecht University, Utrecht, The Netherlands

Abstract

Guanylyl cyclase C (GC-C) was found to function as the principal receptor for heat-stable enterotoxins (STa), major causative factors in *E. coli*-induced secretory diarrhea. GC-C is enriched in intestinal epithelium, but was also detected in other epithelial tissues. The enzyme belongs to the family of receptor guanylyl cyclases, and consists of an extracellular receptor domain, a single transmembrane domain, a kinase homology domain, and a catalytic domain. GC-C is modified by N-linked glycosylation and, at least in the small intestine, by proteolysis, resulting in a STa receptor that is coupled non-covalently to the intracellular domain. So far two endogenous ligands of mammalian GC-C have been identified i.e. the small cysteine-rich peptides guanylin and uroguanylin. The guanylins are released in an auto- or paracrine fashion into the intestinal lumen but may also function as endocrine hormones in gut-kidney communication and as regulators of ion transport in extra-intestinal epithelia. They are thought to activate GC-C by inducing a conformational change in the extracellular portion of the homotrimeric GC-C complex, which allows two of the three intracellular catalytic domains to dimerize and form two active catalytic clefts. In the intestine, activation of GC-C results in a dual action: stimulation of Cl and HCO_3 secretion, through the opening of apical CFTR Cl channels; and inhibition of Na absorption, through blockade of an apical Na/H exchanger. The principal effector of the GC-C effect on ion transport is cGMP dependent protein kinase type II, which together with GC-C and the ion transporters, may form a supramolecular complex at the apical border of epithelial cells. (Mol Cell Biochem **230**: 73–83, 2002)

Key words: heat-stable enterotoxin, secretory diarrhea, guanylate cyclase, cGMP, intestine, guanylin

Introduction

Heat-stable enterotoxins (STa), small peptides secreted by some pathogenic strains of *Escherichia coli*, were the first natural compounds found to activate a particulate guanylyl cyclase [1, 2]. The target of STa was located primarily on the apical or brush border membrane of intestinal epithelial cells, already known at that time as a rich source of particulate guanylyl cyclase [3]. The discovery that STa actually served as an extracellular activator of this enzyme, in combination with its known causative role in severe secretory diarrhea, comparable to that caused by activation of the adenylyl cyclase by cholera toxin, strongly suggested a major function of the guanylyl cyclase as a regulator of intestinal salt and water transport [1, 2, 4]. Subsequently, it was shown that the intestinal form of guanylyl cyclase was different from the particulate isoforms present in most other tissues, which could be stimulated by natriuretic peptides and were shown to be covalently coupled to a receptor domain [5, 6]. In contrast, initially studies suggested the existence of a variety of different receptors for STa in the intestine, as well as a non-covalent linkage between the receptor and the intestinal guanylyl cyclase [5–8]. The cloning of an intestinal guanylyl cyclase comprising an STa receptor within the same poly-peptide, which showed a considerable homology to the natriuretic peptide receptors/guanylyl cyclases A and B, came therefore as a surprise [9]. The cloned intestinal STa receptor, designated guanylyl cyclase C, was subsequently shown by biochemical characterization and targeted deletion to be responsible for most, if not all, of the particulate guanylyl cyclase activity and enterotoxin binding sites in intestinal brush border membranes and for the physiological effects of

Address for offprints: A.B. Vaandrager, Department of Biochemistry and Cell Biology, Faculty of Veterinary Medicine, Utrecht University, P.O. Box 80.176, 3508 TD Utrecht, The Netherlands (E-mail: A.B.Vaandrager@vet.uu.nl)

STa on intestinal water secretion and absorption in whole animals [10–15]. Shortly after the cloning of GC-C, an endogenous activator peptide named guanylin, was isolated from intestinal mucosa [16]. Guanylin together with the related peptide uroguanylin are now believed to function as the physiological ligands for GC-C [17, 18].

Structure of the heat-stable enterotoxin receptor/guanylyl cyclase C

Gene structure

In humans, the gene for GC-C (GUCY2C) is located on chromosome 12 and is much larger than the gene encoding the related natriuretic peptide receptor GC-A (85 kb vs. 17 kb; it also contains more exons (27 vs. 22) (human genome project; NCBI; [19]). A splice variant, resulting in a shorter non-functional transcript, was detected in some colon and colorectal cancer cells [20]. The 5′ upstream promoter region of the GC-C gene was found to contain several potential binding elements for transcription factors, which were conserved between human and mouse. Both hepatocyte nuclear factor-4 (HNF-4) and the homeodomain transcription factor family member Cdx2 were implicated in basal and/or intestine specific expression of GC-C [21, 22].

Primary structure

The primary structure of the enterotoxin receptor/guanylyl cyclase C has now been deduced from its molecular cloning from various mammalian species, including humans [23, 24], rat [9], pig [25], guinea pig (M. Kruhoeffer *et al.*, EMBL/Genbank/DDDJ accession number P70106], cow (R.M. Goraczniak *et al.*, EMBL/Genbank/DDDJ accession number AF081464) and opossum, a marsupial [26]. Fish and amphibians are also shown to express homologs of GC-C ([27]; R.T. MacFarland, EMBL/Genbank/DDDJ accession number BAA08638) which probably interact with STa/guanylin-like peptides detected recently in lower vertebrates [28, 29].

Its primary sequence clearly places GC-C among the family of receptor guanylyl cyclases, together with the natriuretic peptide receptors GC-A and B, the retinal and olfactory cyclases GC-D, GC-E and GC-F, the orphan receptor GC-G and a number of cyclases from non-vertebrates, including the receptors for egg peptides in sea urchins and olfactory receptors in *C. elegans* [19, 30].

All members of the family of receptor guanylyl cyclases are membrane-bound and are composed of an extracellular N-terminal receptor domain, a single transmembrane domain, and a cytosolic region comprising a kinase homology domain, which is linked by an alpha-helical putative dimerization domain to the C-terminal catalytic domain (see Fig. 1). Whereas the intracellular portion of GC-C is fairly homologous to the other receptor cyclases, GC-C contains an unique extracellular receptor domain, and an unique C-terminal extension to the catalytic domain [9, 19, 23–27, 30].

Enterotoxin receptor/extracellular domain

The extracellular domain of GC-C comprises approximately 40% of the total protein and functions as the binding site for (uro)guanylin and STa [9, 19]. It can be expressed independently from the C-terminal part of GC-C as a soluble protein with similar binding properties as the holo-enzyme [31, 32]. It was predicted to be composed of a more rigid N-terminal α/β rich region, stabilized by multiple disulfide bridges, and a more flexible β-rich region [33]. This latter region, which is close to the transmembrane domain, is thought to form the binding site for the ligands, possibly in combination with a more proximal highly conserved region around Arg136 [33, 34] (see Fig. 1). Furthermore, the extracellular domain contains 8–10 N-linked glycosylation sites, depending on the species, which are probably not absolutely required for binding of ligand [35] (see also 'Post translational modifications'). The receptor domain presumably also contains one or more protease sensitive sites [11, 12, 36], which might result in the formation of the various STa receptors with molecular masses ranging from 50–80 kDa detected in intestinal preparations by chemical cross-linking studies [5–8] (see also 'Post translational modifications').

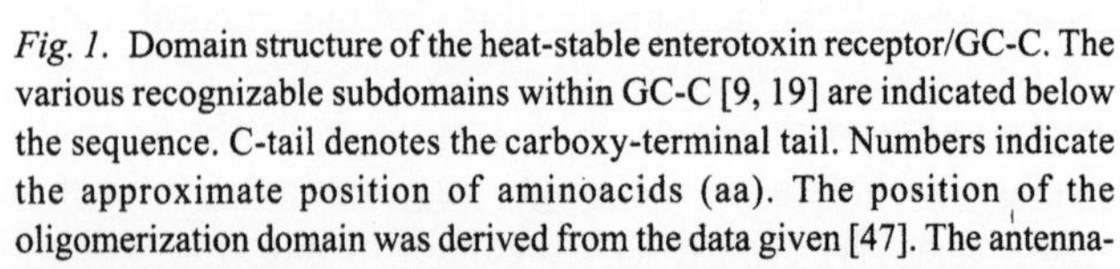

Fig. 1. Domain structure of the heat-stable enterotoxin receptor/GC-C. The various recognizable subdomains within GC-C [9, 19] are indicated below the sequence. C-tail denotes the carboxy-terminal tail. Numbers indicate the approximate position of aminoacids (aa). The position of the oligomerization domain was derived from the data given [47]. The antenna-like symbols indicate 7 of the most conserved glycosylation sites within the GC-C orthologs [35]. The secondary structure of the extracellular/receptor domain is indicated as predicted [33]. The positions of the ligand (STa) binding sites and the protease sensitive sites were deduced from refs [33, 34, 12, 36, 55], respectively and are discussed in the text.

Kinase homology domain

The kinase homology domain (KHD) is situated between the receptor and the catalytic domain and has a clear homology to the catalytic site of protein kinases, in particular to that of the receptor tyrosine kinases like the PDGF receptor [9, 19]. However, in GC-C this domain is thought to be devoid of a protein kinase activity as it lacks a conserved glycine-rich region and an Asp residue required for positioning of the gamma-phosphate of ATP and the transfer of the phosphate group to the substrate, respectively [9, 19]. In line with its position within the protein structure, the KHD is thought to transduce the signal from the ligand-occupied receptor domain to the catalytic domain, analogous to the role of G-proteins in the signal cascade of the heptahelical receptors to the adenylyl cyclase. As predicted by this model, deletion of the KHD resulted in a cyclase, which could not longer be activated by STa [37, 38]. This GC-C KHD-deletion mutant was found to have basal cyclase activity [38; Vaandrager, unpublished results], but was also reported to represent a constitutionally fully active form, similar to GC-A constructs lacking the KHD [37]. In GC-A and GC-B the KHD is suggested to account for the strict adenine nucleotide requirement of the activation of these guanylyl cyclases by natriuretic peptides [19, 39]. Although the activation of GC-C by STa does not depend on the presence of ATP [11, 40–43], the KHD is thought to mediate the observed potentiation of GC-C by adenine nucleotides in *in vitro* assays [11, 41–43]. Under these conditions occupation of the KHD by ATP might stabilize activated GC-C and protect it against rapid desensitization [42].

The catalytic domain

The catalytic domain of GC-C is highly homologous to the catalytic domain of the other receptor GC's. The structure of the catalytic domain of the retinal GC has been modeled to the resolved structure of adenylyl cyclases [44]. It was postulated to exists of a homo-dimer of two catalytic domains, containing two catalytic sites in the cleft between the two domains. So each catalytic site was composed of two complementary half sites residing on the different monomers [19, 44, 45]. As judged by the sequence similarity between the catalytic domains of GC-C and retinal GC, and by their kinetic similarities, as is apparent from the stimulation by Mn and the positive cooperation between the GTP binding sites [46, 47], a very similar model will probably apply to GC-C.

Carboxy terminal tail

Unlike the natriuretic receptors, but similar to the membrane-bound GCs from sensory organs, GC-C contains an approximately 60 amino acid long extension distal to the catalytic domain with unknown function. Deletion of this domain rendered GC-C unresponsive to STa [48]. Residue Ser 1029 in this domain was suggested as the phosphoacceptor for the PK-C-mediated stimulation of GC-C activity [48, 49]. From the correlation between the possession of a carboxy terminal tail and the cytoskeletal localization of GC-C and GC-E, a role of this extension in interaction with the cytoskeleton was proposed [9, 19]. The C-terminal domain was also suggested to play a role in the regulated endocytosis of GC-C in a colonic cell line [50].

Oligomeric state

From a number of studies it is now clear that GC-C is a homo-multimer, both in the absence and presence of an activating ligand. In that respect it is similar to the other members of the receptor guanylyl cyclases studied so far [19, 30, 39]. However, the precise number of monomers forming one GC-C complex under native conditions is still not firmly established. From the point of view of the mechanism of cGMP catalysis, which require two catalytic subunits [19, 45] (also discussed above in 'The catalytic domain'), a dimeric, or tetrameric structure composed of two loosely associated dimers, seems most plausible. A dimeric or possibly tetrameric structure has also been proposed for the natriuretic receptor GC's [39]. However, from actual experimental determination of the molecular mass of solubilized recombinant and native rat GC-C by gel filtration and sucrose density centrifugation, a trimeric structure was deduced [12]. Recently, cross-linking studies with recombinant fragments of GC-C purified from insect cells, suggested that both the extracellular domains and the intracellular domains exists as trimers, although ligand was required to generate or stabilize the trimeric extracellular domain [32, 47]. These data together with the observation that truncated receptor fragments of recombinant GC-C, most probably representing incomplete translation products [51], also form trimers [12], suggest that the putative trimerization of GC-C is mediated by sequences in both the extracellular and the intracellular domain. A region located in the linker between the kinase homology and the catalytic domain, predicted to form a coiled-coil structure, was implied as the intracellular multimerization sequence ([47]; cf. Fig. 1).

Post translational modifications: Glycosylation, phosphorylation and proteolysis

Depending on the species, GC-C contains 8–10 N-linked glycosylation sites, all located in the extra cellular domain [35], which lead to an extensive glycosylation of the receptor [11, 52, 53]. Glycosylation of the receptor is probably not absolutely required for binding of its ligand, as a GC-C receptor domain expressed in *E. coli* showed normal binding characteristics [31], and treatment of an STa receptor fragment purified from pig intestine with endoglycosidase H did not lead to a loss of STa binding capacity [52]. However, other studies observed a reduction in STa binding after removal of the sugar groups by N-glycosidase F [35]. From mutational studies a role for the most distal conserved glycosylation site, which is very close to or part of the binding site for STa, was postulated in both the binding of STa and in the stability of GC-C itself [35].

In contrast to findings with the natriuretic peptide receptors [19, 39], little addition of phosphate was detected in GC-C in the basal state or after addition of ligand [42, 49]. However phorbol esters were found to stimulate the phosphorylation of GC-C in intact cells [42, 49]. This phosphorylation is thought to activate the cyclase, especially in synergy with STa [49, 54]. Residue Ser 1029 was suggested as the phosphoacceptor for the PK-C-mediated phosphorylation [48, 49].

Endogenous GC-C is probably proteolytically modified on the surface of small intestinal cells *in vivo* [11, 12, 36, 55], whereas GC-C expressed in colonic cell lines like CaCo-2 or T84 is relatively intact [55, 56]. The truncated receptor fragments generated by the intestinal proteases apparently gave rise to the suggestion that the STa receptor was not linked to the intestinal guanylyl cyclase moiety [5, 6]. However, interpretation of results from native intestinal preparations are complicated by the existence of multiple potential proteolytic cleavage sites [36] and by the occurrence of differently glycosylated forms of GC-C i.e. mature fully glycosylated GC-C and immature high mannose type GC-C [12, 52, 53]. Most likely, mature GC-C expressed on the brush border membrane of intestinal villus cells is cleaved at a single position in the extra cellular domain resulting in a 50–60 kDa N-terminal and an 85 kDa N-terminal fragment [12, 36]. Although extrapolations from the molecular mass of the fragments to their primary structure are hampered by the extensive glycosylation of the receptor, the cleavage site could be somewhere between the N-terminal α/β rich region and the more flexible β-rich region predicted by Hasagawa *et al.* [33] (see Fig. 1). Under physiologic conditions this cleavage is probably catalyzed by a specific intestinal or pancreatic protease but does not impair the function of GC-C, as the fragments are still associated to each other by non-covalent forces [12, 36]. On the other hand an intracellular cleavage site resulting in an N-terminal 70–75 kDa and a 65 kDa C-terminal GC-C fragment is probably brought about by pancreatic proteases during membrane preparation/solubilization, and may result in STa receptors which are no longer coupled to cyclase domains [36]. It is even conceivable that some of the STa binding proteins with masses between 70–80 kDa belong to a fragment containing the juxtamembrane STa binding site [33] coupled to a large portion of the intracellular domain. This fragment would be formed by the proteolytic removal of both a 50–60 kDa N-terminal domain and a 5–10 kDa C-terminal tail. This latter cleavage would explain the observation that most of the GC-C present in upper villus cells and in brush border caps escape detection by a C-terminal antibody [36; Vaandrager, unpublished results].

The low molecular mass STa receptors observed in intestinal preparations which are generated by proteolysis and are partly located on the luminal side of the brush border membrane, are different from the small STa receptors observed by heterologous expression of GC-C in HEK 293 cells [10–12]. These latter forms were shown to result from incomplete translation and are retained within the cell [51].

Regulation of GC-C activity

Activators and inhibitors of GC-C

The most important activators of GC-C under physiological conditions are thought to be guanylin and uroguanylin, small cysteine-rich peptides derived from larger precursor proteins secreted by various epithelial cells, especially those lining the gastro-intestinal tract [17, 18]. In opossum a third member of the guanylin family, lymphoguanylin, was found to be expressed in lymphoid tissues but also in kidney and heart [57]. Interestingly the genes for guanylin and uroguanylin are arranged in tandem in a tail-to-tail orientation on human chromosome 1, close to the genes for the natriuretic peptides ANP and BNP ([58–60], human genome project, NCBI). Despite their similarity, guanylin and uroguanylin have distinct, although overlapping expression patterns and different potencies to activate GC-C, depending on the pH [17, 18, 58, 61–63]. Heat-stable enterotoxins (STa's) secreted by some pathogenic strains of bacteria, which structurally resemble (uro)guanylin (see Fig. 2), were the first agonists described to activate the intestinal GC. STa's are somewhat more potent than the endogenous (uro)guanylin peptides in activating GC-C [16, 61, 62], and stimulate the intestinal guanylyl cyclase under pathologic conditions, e.g. travelers diarrhea [4]. STa can activate GC-C in the submicromolar range, although it can already bind to the receptor at nanomolar concentrations [9, 64, 65]. This apparent discrepancy was explained by the existence of multiple affinity states of GC-C [64, 65]. The GC-C in the lowest affinity state is apparently associated with ligand-induced cyclase activation [64]. Upon prolonged exposure to STa, GC-C expressed in colonic cell lines, was shown to be desensitized [66, 67]. This homologous desensitization was distinct from the inactivation observed *in vitro*, which could be prevented by ATP, and was not observed in heterologously expressed GC-C in 293 cells [66].

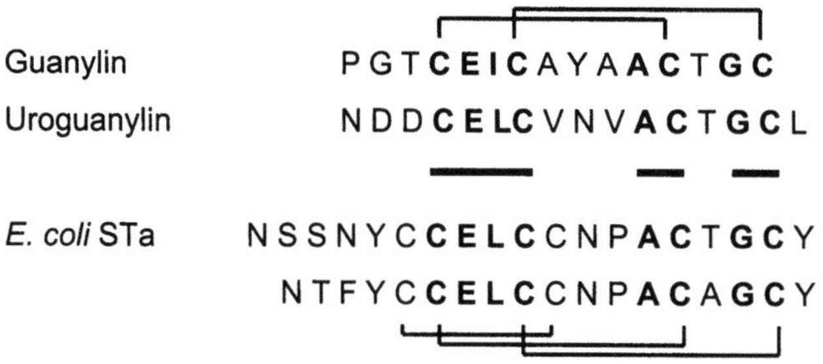

Fig. 2. Primary structure of GC-C ligands. The primary structure of human uroguanylin and guanylin and of two heat-stable enterotoxins (STa) secreted by *E. coli* are given. Conserved aminoacids are printed in bold and indicated by bars between the sequences. The disulfide bridges within the peptides are indicated by accolades.

A number of compounds were identified that could stimulate GC-C in membrane preparations *in vitro*. These compounds were used mainly to characterize the intestinal GC and its molecular mechanism. Especially Mn at milimolar concentrations, replacing the physiological metal Mg, alone or in combination with a detergent, leads to a supra maximal activation of GC-C, which can not be enhanced further by STa [9, 40, 56]. The ligand independent activation by Mn is typical for most GC's and AC's, illustrating the similarity in mechanism of action of the various cyclases [19, 68]. Detergents like Triton and Lubrol alone can also activate GC-C [40, 42, 56]. Also some bivalent metal ions like Cu, Cd and Zn were shown to activate GC-C at micromolar concentrations in the presence of MgGTP [42, 69]. However, when added at higher concentration Zn was shown to inhibit cyclase activity of an intracellular domain of GC-C [47]. A similar biphasic activating-inhibitory behavior was described for hemin [42, 69, 70]. Likewise, adenine nucleotides at submillimolar concentrations can potentiate the activation of GC-C by STa and other compounds, presumably by stabilizing the enzyme and preventing desensitization [11, 41–43, 66], but were shown to inhibit activated GC-C at higher concentrations [40, 42, 43, 47, 71]. Especially 2-substituted adenine nucleotides were found to be potent allosteric inhibitors of GC-C [71, 72]. A membrane-permeable 2-chloroadenosine analog, which could be converted to a (tri)nucleotide in the cell, was shown to inhibit the effect of STa on electrolyte transport in a colonic cell line [72, 73]. Another class of compounds that inhibits GC-C activation are non-specific thiol reactive reagents [6, 69, 74]. Some of these thiol agents can inactivate STa by reducing its internal cysteine bridges [75]. However, despite their obvious clinical benefit in the treatment of diarrheal diseases, so far no specific and potent inhibitors of the STa-GC-C interaction that are effective in intact intestinal epithelium have been described.

Molecular mechanism of GC-C regulation

Evidence from a number of studies, strongly suggests that both adenylyl and guanylyl cyclases have a similar catalytic mechanism for synthesizing cyclic nucleotides [19, 44, 45]. Therefore, GC-C may catalyze the attack of the 3′ hydroxyl of ribose on the α-phosphate of the nucleosidetriphosphate by a two metal-ion mechanism similar to that proposed for adenylyl cyclases and DNA polymerases [68]. According to the model for adenylyl cyclases, the two active sites of GC-C are composed of two half-sites, each formed by structures on both sides of a cleft formed at the interface of two catalytic domains [45, 68]. In order to form cGMP, the two half sites must collapse around the substrate involving a conversion from an open configuration to a closed one [68]. In the basal/inactive state the two half sites should thus be prevented to form a competent active site by either preventing the two half sites to associate or to prevent the conversion of an open site into the closed configuration. Since the intracellular domain of GC-C expressed in insect cells was found to be constitutionally active [47], it seems plausible that in the absence of a ligand, the extracellular receptor domain, possibly in combination with the kinase homology domain, stabilizes the basal/inactive state. Upon binding of (uro)guanylin or STa, a conformational change is induced in the receptor, which releases the inhibition and allows the catalytic sites to become catalytically competent. Radiation inactivation experiments confirmed that GC-C is repressed in the basal state by a relatively large structure [12]. STa seemed to promote clustering of the extracellular domains as assessed by cross-linking [32]. STa was also suggested to stabilize the binding between two GC-C monomers within the GC-C trimer and to change their conformation as judged from their different mobility in SDS-PAGE [11, 12, 42]. It was also reported that binding of STa induced a lower affinity state of the receptor, similar to ANP action on GC-A [19, 38, 64]. This also suggests that STa induces a shift in the conformation of GC-C. Both the activation of the cyclase by STa and the induction of the affinity shift depended on the presence of the intracellular domain including the KHD [37, 38], stressing the importance of the KHD in transducing the conformation-shift of the receptor to the catalytic domains. The model of activation of GC-C depicted above may also explain the activation and inhibition by some of the compounds known to modulate the activity of GC-C. Detergents may partly overcome the repression in the basal state and enable the formation of the active configuration of the catalytic domains by allowing a greater freedom of orientation of the GC-monomers within the GC-C multimer. The bivalent metal ions like Mn, Cd, and Zn might stimulate or inhibit GC-C by acting on one or two of the proposed metal ion binding sites in each catalytic site as suggested for the adenylyl cyclases

[47, 68]. In analogy to the P-site inhibitors in ACs, adenosine triphosphate or analogs thereof, may inhibit the catalytic domain of GC-C in the closed configuration allosterically, since these inhibitors are (i) non-competitive with regard to the substrate GTP, (ii) more effective when substrate is present i.e. when the catalytic site is supposedly in the 'closed' configuration, and (iii) were shown to act directly on the cytoplasmic domain of GC-C [47, 68, 71].

However, knowledge concerning the precise interactions of STa with the receptor domain and the interactions between the extracellular domain and the KHD and catalytic domains in the cytoplasm awaits the determination of the complete 3D-structure of GC-C. Especially the mechanism of the putative STa-induced dimerization of catalytic domains within a trimeric GC-C complex, is intriguing and is likely to be unique to the enterotoxin receptor. A hypothetical model of such a mechanism is depicted in Fig. 3.

Function of the heat-stable enterotoxin receptor/guanylyl cyclase C

The observation that GC-C is specifically activated by heat-stable enterotoxins that provoke a severe secretory diarrhea, immediately implied a function of this cyclase in the regulation of salt and water transport in the intestine. The relative abundance of GC-C and its ligands, guanylin and uroguanylin along the gastrointestinal tract in both crypt, villus and surface cells [76–83] also point to an important role of GC-C in the intestine. GC-C was even proposed as a selective marker for metastatic colorectal tumors [84]. However, (uro)guanylin and/or GC-C have also been found in extra-intestinal tissues, including kidney [85, 86], lung [85, 87], pancreas [88, 89], pituitary [90], adrenal [85, 91] and developing liver [92, 93], suggesting a more general role for GC-C in various epithelial tissues. Especially, in the opossum the (uro)guanylin/GC-C system was found in many epithelia aside the intestine [94]. In the opossum but also in mammals, STa and uroguanylin were shown to stimulate diuresis, natriuresis and kaliuresis by the kidney [95].

From localization studies and from studies directed to the effects of the GC-C ligands *in vivo* and *in vitro*, the following functions for the guanylin-peptides/GC-C system are now emerging;

- regulation of the fluidity of the intestinal contents by stimulation of the secretion of Cl and inhibition of absorption of NaCl, thus preventing dehydration and consequently obstruction of the intestine on one hand and the loss of water (diarrhea) on the other [4, 79].

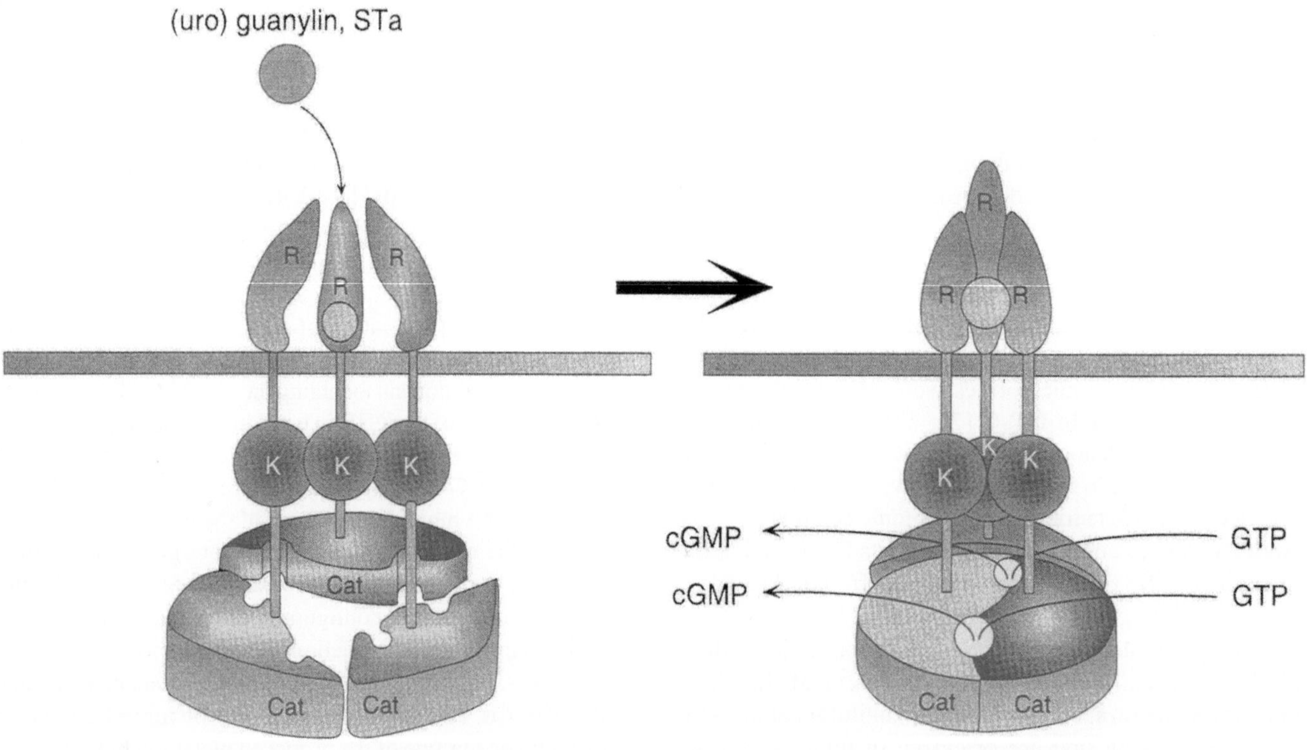

Fig. 3. Mechanism of activation of GC-C. Shown is a hypothetical scheme in which homo-trimeric GC-C is activated by an internal dimerization following ligand binding. R – receptor domain; K – kinase homology domain; Cat – catalytic domain, each containing two catalytic half sites.

- regulation of pH of the intestinal contents by stimulating HCO_3 secretion and inhibiting H extrusion by a Na/H exchanger, and thus stimulating the digestion and absorption of food components [96–98].
- regulation of the Na homeostasis, especially in case of high salt intake, by limiting intestinal Na uptake and by stimulating Na excretion by the kidney. Uroguanylin secreted by intestinal cells into the blood stream was suggested to act as an 'intestinal natriuretic hormone' along a postulated endocrine intestine-kidney axis [95].

Furthermore, the guanylin/GC-C system might contribute to the defense against pathogens by keeping epithelial layers of the intestine, lung or other tissues clean by secretion of mucus and fluid. A disturbance in epithelial Cl secretion is thought to be the main cause of the recurrent lung infections observed in patients with the genetic disease cystic fibrosis [79]. Also, a possible protection by uroguanylin/GC-C against carcinogenesis was proposed from studies with mice, which were sensitive to colon cancer [99].

Despite the forementioned potential physiological roles of GC-C, mice in which the gene for GC-C was deleted appeared generally healthy, although they were clearly insensitive to STa [13–15]. This suggests that many of the functions of GC-C are either of minor importance for a normal physiology, or that they are redundant and can be compensated by other systems. Regarding the last possibility it should be noted that ion transport in both the intestine and the kidney are regulated by many hormones using cAMP and Ca as second messengers in addition to cGMP. In any case, the guanylin/GC-C system might prove to be essential under certain dietary or environmental conditions, yet to be determined. A better knowledge concerning regulation of the expression and secretion of guanylin and uroguanylin would be helpful in that respect. A few studies indicate that the levels of guanylin peptides are indeed under environmental (salt intake) [100], neuronal [101, 102], and circadian [103] control.

Molecular mechanism of GC-C mediated regulation of ion transport

So far the best studied function of GC-C is its regulation of salt and water transport in the intestine. Net intestinal fluid transport results primarily from an electroneutral absorption of NaCl, counteracted by an electrogenic Cl secretion [4, 79]. Activation of GC-C generates cGMP, which leads to a stimulation of electrogenic Cl secretion concomitant with an inhibition of electroneutral Na absorption. The molecular target for the former was shown to be the CFTR chloride channel, which is non-functional in patients with the genetic disease cystic fibrosis [79, 104–107]. The CFTR-Cl channel was also shown to be required for HCO_3 secretion in the intestine [97, 108]. Interestingly, HCO_3 secretion is, as noted above, thought to be another major function of the guanylin/cGMP pathway in the proximal part of the small intestine [97, 98]. The GC-C/cGMP-regulated transporter involved in the observed inhibition of electroneutral Na absorption is less well defined. The Na/H exchanger NHE-3 is a likely candidate, as it is present in the apical membrane of enterocyte [109] and was implicated in intestinal water and salt absorption by gene-knockout [110]. The Na/H exchanger might absorb NaCl in combination with the chloride/anion exchanger 'downregulated in adenoma' (DRA) protein, recently found mutated in congenital chloride diarrhea [111]. However, the regulation of NHE-3 by cGMP might prove to be complex, as the mechanism of the extensively studied cAMP/cAK inhibition of NHE-3 is still unclear and probably involves PDZ domain containing co-factors like NHERF or E3KARP [112, 113]. A specific isoform of cGMP-dependent protein kinase (cGK) designated type II was implicated as the principal mediator of the GC-C/cGMP-provoked Cl secretion and inhibition of Na uptake by pharmacological and gene-knockout methodology [107, 114–116]. Interestingly, cGK II, like GC-C, CFTR and NHE-3, is a membrane and cytoskeletal-bound protein, and could not be functionally replaced by a soluble cGK mutant [117, 118]. This suggests that the proper regulation of the ion transporters by (uro)-guanylin might require a large signal transduction complex (see Fig. 4), as described for the cGMP signaling in the retina [119].

As shown in Fig. 4, a cGK II independent mechanism may also contribute to the effects of GC-C/cGMP in cells expressing a cGMP-inhibitable phosphodiesterase (PDE) III [116]. cGMP may act by inhibiting this phosphodiesterase resulting in a reduced breakdown of cAMP, local cAMP accumulation, and activation of the cAMP/cAK pathway which was shown capable of modulating the same ion transporters, CFTR and NHE-3, as targeted cGK II [79, 114, 116, 118]. An alternative pathway, which may also contribute to a cGK II-independent effect of STa on Cl secretion, was demonstrated in the colon carcinoma cell line T84 [104, 120]. In these cells STa induces very high levels of cGMP, which can directly cross-activate cAK. The cGK II-independent STa pathway is probably more prominent in the large intestine, where the expression of cGK II is relatively low compared with the small intestine [79, 116].

The mechanism by which GC-C exerts its effects in the kidney and other extra-intestinal tissues is largely unknown but some of the key components in the intestine (e.g. cGK II, CFTR, NHE-3) may play a similar role in other epithelia as well. To elucidate these mechanisms and to evaluate their importance for the (patho)physiology of an organism will be important topics for future research.

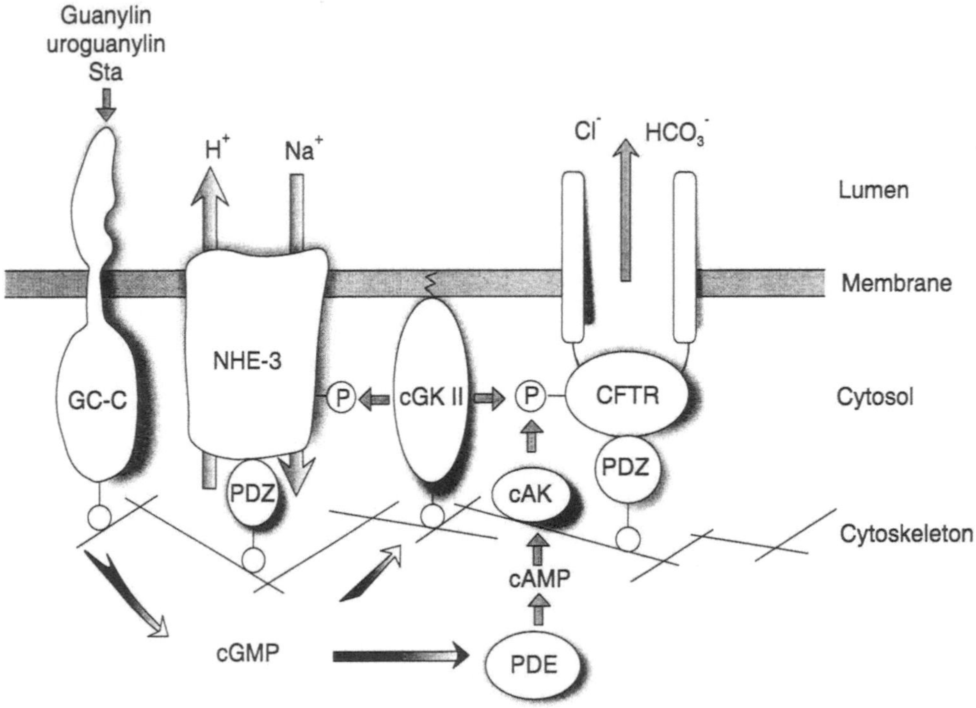

Fig. 4. Regulation of ion transport by GC-C in the apical membrane of intestinal epithelial cells. Activation of GC-C by STa or guanylin-peptides can stimulate both HCO_3 and Cl secretion by opening of the CFTR Cl channel, and reduce Na absorption presumably by inhibiting the Na/H exchanger NHE-3. cGK II – cGMP-dependent protein kinase, which is bound to the membrane by myristoylation and to the cytoskeleton by an unknown mechanism, possibly involving a G-kinase anchoring protein (GKAP); cAK – cAMP-dependent protein kinase, which is thought to co-localize with CFTR and NHE-3 by binding to ezrin; PDZ – PDZ domain containing, ezrin/radixin/moesin (ERM)-binding regulatory proteins like NHERF or E3KARP, which couple CFTR and/or NHE-3 to the cytoskeleton [118]; PDE – cGMP-inhibitable phosphodiesterase, which can convert the cGMP signal into a cAMP signal.

References

1. Field M, Graf LH Jr, Laird WJ, Smith PL: Heat-stable enterotoxin of *Escherichia coli*: *In vitro* effects on guanylate cyclase activity, cyclic GMP concentration and ion transport. Proc Natl Acad Sci USA 75: 2800–2804, 1978
2. Hughes JM, Murad F, Chang B, Guerrant RL: Role of cyclic GMP in the action of heat-stable enterotoxin of *Escherichia coli*. Nature 271: 755–756, 1978
3. De Jonge HR: The localization of guanylate cyclase in rat small intestinal epithelium. FEBS Lett 53: 237–242, 1975
4. Field M, Rao MC, Chang EB: Intestinal electrolyte transport and diarrheal disease. N Engl J Med 321: 800–806, 879–883, 1989
5. Kuno T, Kamisaki Y, Waldman SA, Gariepy J, Schoolnik G, Murad F: Characterization of the receptor for heat-stable enterotoxin from *Escherichia coli* in rat intestine. J Biol Chem 261: 1470–1476, 1986
6. Waldman SA, Kuno T, Kamisaki Y, Chang LY, Gariepy J, O'Hanley P, Schoolnik G, Murad F: Intestinal receptor for heat-stable enterotoxin of *Escherichia coli* is tightly coupled to a novel form of particulate guanylate cyclase. Infect Immun 51: 320–326, 1986
7. Ivens K, Gazzano H, O'Hanley P, Waldman SA: Heterogeneity of intestinal receptors for *Escherichia coli* heat-stable enterotoxin. Infect Immun 58: 1817–1820, 1990
8. Thompson MR, Giannella RA: Different crosslinking agents identify distinctly different putative *Escherichia coli* heat-stable enterotoxin rat intestinal cell receptor proteins. J Recept Res 10: 97–117, 1990
9. Schulz S, Green CK, Yuen PST, Garbers DL. Guanylyl cyclase is a heat-stable enterotoxin receptor. Cell 63: 941–948, 1990
10. de Sauvage FJ, Horuk R, Bennett G, Quan C, Burnier JP, Goeddel DV: Characterization of the recombinant human receptor for *Escherichia coli* heat-stable enterotoxin. J Biol Chem 267: 6479–6482, 1992
11. Vaandrager AB, Schulz S, De Jonge HR, Garbers DL: Guanylyl cyclase-C is an N-linked glycoprotein receptor that accounts for multiple heat-stable enterotoxin binding proteins in the intestine. J Biol Chem 268: 2174–2179, 1993
12. Vaandrager AB, van der Wiel E, Hom ML, Luthjens LH, De Jonge HR: Heat-stable enterotoxin receptor/guanylyl cyclase C is an oligomer consisting of functionally distinct subunits, which are non-covalently linked in the intestine. J Biol Chem 269: 16409–16415, 1994
13. Schulz S, Lopez MJ, Kuhn M, Garbers DL: Disruption of the guanylyl cyclase-C gene leads to a paradoxical phenotype of viable but heat-stable enterotoxin-resistant mice. J Clin Invest 100: 1590–1595, 1997
14. Mann EA, Jump ML, Wu J, Yee E, Giannella RA: Mice lacking the guanylyl cyclase C receptor are resistant to STa-induced intestinal secretion. Biochem Biophys Res Commun 239: 463–466, 1997
15. Charney AN, Egnor RW, Alexander-Chacko JT, Zaharia V, Mann EA, Giannella RA: Effect of *E. coli* heat-stable enterotoxin on colonic transport in guanylyl cyclase C receptor-deficient mice. Am J Physiol 280: G216–G221, 2001
16. Currie MG, Fok KF, Kato J, Moore RJ, Hamra FK, Duffin KL, Smith CE: Guanylin: An endogenous activator of intestinal guanylate cyclase. Proc Natl Acad Sci USA 89: 947–951, 1992

17. Forte LR, Currie MG: Guanylin: A peptide regulator of epithelial transport. FASEB J 9: 643–650, 1995
18. Forte LR: Guanylin regulatory peptides: Structures, biological activities mediated by cyclic GMP and pathobiology. Reg Pep 81: 25–39, 1999
19. Lucas KA, Pitari GM, Kazerounian S, Ruiz-Stewart I, Park J, Schulz S, Chepenik KP, Waldman SA: Guanylyl cyclases and signaling by cyclic GMP. Pharmacol Rev 52: 375–414, 2000
20. Pearlman JM, Prawer SP, Barber MT, Parkinson SJ, Schulz S, Park J, Zook M, Waldman SA: A splice variant of the transcript for guanylyl cyclase C is expressed in human colon and colorectal cancer cells. Dig Dis Sci 45: 298–305, 2000
21. Swenson ES, Mann EA, Jump ML, Giannella RA: Hepatocyte nuclear factor-4 regulates intestinal expression of the guanylin/heat-stable toxin receptor. Am J Physiol 276: G728–G736, 1999
22. Park J, Schulz S, Waldman SA: Intestine-specific activity of the human guanylyl cyclase C promoter is regulated by Cdx2. Gastroenterology 119: 89–96, 2000
23. De Sauvage FJ, Camerato TR, Goeddel DV: Primary structure and functional expression of the human receptor for *Escherichia coli* heat-stable enterotoxin. J Biol Chem 266: 17912–17918, 1991
24. Singh S, Singh G, Heim JM, Gerzer R: Isolation and expression of a guanylate cyclase-coupled heat stable enterotoxin receptor cDNA from a human colonic cell line. Biochem Biophys Res Commun 179: 1455–1463, 1991
25. Wada A, Hirayama T, Kitao S, Fujisawa J, Hidaka Y, Shimonishi Y: Pig intestinal membrane-bound receptor (guanylyl cyclase) for heat-stable enterotoxin: cDNA Cloning, functional expression, and characterization. Microbiol Immunol 38: 535–541, 1994
26. London RM, Eber SL, Visweswariah SS, Krause WJ, Forte LR: Structure and activity of OK-GC: A kidney receptor guanylate cyclase activated by guanylin peptides. Am J Physiol 276: F882–F889, 1999
27. Mantoku T, Muramatsu R, Nakauchi M, Yamagami S, Kusakabe T, Suzuki N: Sequence analysis of cDNA and genomic DNA, and mRNA expression of the medaka fish homolog of mammalian guanylyl cyclase C. J Biochem (Tokyo) 125: 476–486, 1999
28. Comrie MM, Cutler CP, Cramb G: Cloning and expression of guanylin from the European eel (*Anguilla anguilla*). Biochem Biophys Res Commun 281: 1078–1085, 2001
29. Krause WJ, Freeman RH, Eber SL, Hamra FK, Currie MG, Forte LR: Guanylyl cyclase receptors and guanylin-like peptides in reptilian intestine. Gen Comp Endocrinol 107: 229–239, 1997
30. Garbers DL: The guanylyl cyclase receptors. Methods 19: 477–484, 1999
31. Nandi A, Mathew R, Visweswariah SS: Expression of the extracellular domain of the human heat-stable enterotoxin receptor in *Escherichia coli* and generation of neutralizing antibodies. Prot Exp Pur 8: 151–159, 1996
32. Hasegawa M, Kawano Y, Matsumoto Y, Hidaka Y, Fujii J, Taniguchi N, Wada A, Hirayama T, Shimonishi Y: Expression and characterization of the extracellular domain of guanylyl cyclase C from a baculovirus and Sf21 insect cells. Prot Exp Pur 15: 271–281, 1999
33. Hasegawa M, Hidaka Y, Matsumoto Y, Sanni T, Shimonishi Y: Determination of the binding site on the extracellular domain of guanylyl cyclase C to heat-stable enterotoxin. J Biol Chem 274: 31713–31718, 1999
34. Wada A, Hirayama T, Kitaura H, Fujisawa J, Hasegawa M, Hidaka Y, Shimonishi Y: Identification of ligand recognition sites in heat-stable enterotoxin receptor, membrane-associated guanylyl cyclase C by site-directed mutational analysis. Infect Immun 64: 5144–5150, 1996
35. Hasegawa M, Hidaka Y, Wada A, Hirayama T, Shimonishi Y: The relevance of N-linked glycosylation to the binding of a ligand to guanylate cyclase C. Eur J Biochem 263: 338–346, 1999
36. Scheving LA, Chong KM: Differential processing of guanylyl cyclase C along villus-crypt axis of rat small intestine. Am J Physiol 272: C1995–C2004, 1997
37. Rudner XL, Mandal KK, de Sauvage FJ, Kindman LA, Almenoff JS: Regulation of cell signaling by the cytoplasmic domains of the heat-stable enterotoxin receptor: Identification of autoinhibitory and activating motifs. Proc Natl Acad Sci USA 92: 5169–5173, 1995
38. Deshmane SP, Parkinson SJ, Crupper SS, Robertson DC, Schulz S, Waldman SA: Cytoplasmic domains mediate the ligand-induced affinity shift of guanylyl cyclase C. Biochemistry 36: 12921–12929, 1997
39. Potter LR, Hunter T: Guanylyl cyclase-linked natriuretic peptide receptors: Structure and regulation. J Biol Chem 276: 6057–6060, 2001
40. Katwa LC, Parker CD, Dybing JK, White AA: Nucleotide regulation of heat-stable enterotoxin receptor binding and of guanylate cyclase activation. Biochem J 283: 727–735, 1992
41. Gazzano H, Wu HI, Waldman SA: Activation of particulate guanylate cyclase by *Escherichia coli* heat-stable enterotoxin is regulated by adenine nucleotides. Infect Immun 59: 1552–1557, 1991
42. Vaandrager AB, van der Wiel E, de Jonge HR: Heat-stable enterotoxin activation of immunopurified guanylyl cyclase C. Modulation by adenine nucleotides. J Biol Chem 268: 19598–19603, 1993
43. Parkinson SJ, Carrithers SL, Waldman SA: Opposing adenine nucleotide-dependent pathways regulate guanylyl cyclase C in rat intestine. J Biol Chem 269: 22683–22690, 1994
44. Tucker CL, Hurley JH, Miller TR, Hurley JB: Two amino acid substitutions convert a guanylyl cyclase, RetGC-1, into an adenylyl cyclase. Proc Natl Acad Sci USA 95: 5993–5997, 1998
45. Hurley JH: The adenylyl and guanylyl cyclase superfamily. Curr Opin Struct Biol 8: 770–777, 1998
46. De Jonge HR: Properties of guanylate cyclase and levels of cyclic GMP in rat small intestinal villous and crypt cells. FEBS Lett 55: 143–152, 1975
47. Vijayachandra K, Guruprasad M, Bhandari R, Manjunath UH, Somesh BP, Srinivasan N, Suguna K, Visweswariah SS: Biochemical characterization of the intracellular domain of the human guanylyl cyclase C receptor provides evidence for a catalytically active homotrimer. Biochemistry 39: 16075–16083, 2000
48. Wada A, Hasegawa M, Matsumoto K, Niidome T, Kawano Y, Hidaka Y, Padilla PI, Kurazono H, Shimonishi Y, Hirayama T: The significance of Ser1029 of the heat-stable enterotoxin receptor (STaR): Relation of STa-mediated guanylyl cyclase activation and signaling by phorbol myristate acetate. FEBS Lett 384: 75–77, 1996
49. Crane JK, Shanks KL: Phosphorylation and activation of the intestinal guanylyl cyclase receptor for *Escherichia coli* heat-stable toxin by protein kinase C. Mol Cell Biochem 165: 111–120, 1996
50. Urbanski R, Carrithers SL, Waldman SA: Internalization of *E. coli* ST mediated by guanylyl cyclase C in T84 human colon carcinoma cells. Biochim Biophys Acta 1245: 29–36, 1995
51. Rudner XL, Nicchitta C, Almenoff JS: Biogenesis, cellular localization, and functional activation of the heat-stable enterotoxin receptor (guanylyl cyclase C). Biochemistry 35: 10680–10686, 1996
52. Hirayama T, Wada A, Iwata N, Takasaki S, Shimonishi Y, Takeda Y: Glycoprotein receptors for a heat-stable enterotoxin (STh) produced by enterotoxigenic *Escherichia coli*. Infect Immun 60: 4213–4220, 1992
53. Scheving LA, Russell WE, Chong KM: Structure, glycosylation, and localization of rat intestinal guanylyl cyclase C: Modulation by fasting. Am J Physiol 271: G959–G968, 1996
54. Weikel CS, Spann CL, Chambers CP, Crane JK, Linden J, Hewlett EL: Phorbol esters enhance the cyclic GMP response of T84 cells to the heat-stable enterotoxin of *Escherichia coli* (STa). Infect Immun 58: 1402–1407, 1990

55. Cohen MB, Jensen NJ, Hawkins JA, Mann EA, Thompson MR, Lentze MJ, Giannella RA: Receptors for *Escherichia coli* heat stable enterotoxin in human intestine and in a human intestinal cell line (Caco-2). J Cell Physiol 156: 138–144, 1993
56. Visweswariah SS, Ramachandran V, Ramamohan S, Das G, Ramachandran J: Characterization and partial purification of the human receptor for the heat-stable enterotoxin. Eur J Biochem 219: 727–736, 1994
57. Forte LR, Eber SL, Fan X, London RM, Wang Y, Rowland LM, Chin DT, Freeman RH, Krause WJ: Lymphoguanylin: Cloning and characterization of a unique member of the guanylin peptide family. Endocrinology 140: 1800–1806, 1999
58. Magert HJ, Reinecke M, David I, Raab HR, Adermann K, Zucht HD, Hill O, Hess R, Forssmann WG: Uroguanylin: Gene structure, expression, processing as a peptide hormone, and co-storage with somatostatin in gastrointestinal D-cells. Reg Pep 73: 165–176, 1998
59. Whitaker TL, Steinbrecher KA, Copeland NG, Gilbert DJ, Jenkins NA, Cohen MB: The uroguanylin gene (Guca1b) is linked to guanylin (Guca2) on mouse chromosome 4. Genomics 45: 348–354, 1997
60. Miyazato M, Nakazato M, Matsukura S, Kangawa K, Matsuo H: Genomic structure and chromosomal localization of human uroguanylin. Genomics 43: 359–365, 1997
61. Greenberg RN, Hill M, Crytzer J, Krause WJ, Eber SL, Hamra FK, Forte LR: Comparison of effects of uroguanylin, guanylin, and *Escherichia coli* heat-stable enterotoxin STa in mouse intestine and kidney: Evidence that uroguanylin is an intestinal natriuretic hormone. J Invest Med 45: 276–282, 1997
62. Ieda H, Naruse S, Kitagawa M, Ishiguro H, Hayakawa T: Effects of guanylin and uroguanylin on rat jejunal fluid and electrolyte transport: Comparison with heat-stable enterotoxin. Reg Pep 79: 165–171, 1999
63. Hamra FK, Eber SL, Chin DT, Currie MG, Forte LR: Regulation of intestinal uroguanylin/guanylin receptor-mediated responses by mucosal acidity. Proc Natl Acad Sci USA 94: 2705–2710, 1997
64. Crane MR, Hugues M, O'Hanley PD, Waldman SA: Identification of two affinity states of low affinity receptors for *Escherichia coli* heat-stable enterotoxin: Correlation of occupation of lower affinity state with guanylate cyclase activation. Mol Pharmacol 41: 1073–1080, 1992
65. Deshmane SP, Carrithers SL, Parkinson SJ, Crupper SS, Robertson DC, Waldman SA: Rat guanylyl cyclase C expressed in COS-7 cells exhibits multiple affinities for *Escherichia coli* heat-stable enterotoxin. Biochemistry 34: 9095–9102, 1995
66. Bakre MM, Ghanekar Y, Visweswariah SS: Homologous desensitization of the human guanylate cyclase C receptor. Cell-specific regulation of catalytic activity. Eur J Biochem 267: 179–187, 2000
67. Bakre MM, Visweswariah SS: Dual regulation of heat-stable enterotoxin-mediated cGMP accumulation in T84 cells by receptor desensitization and increased phosphodiesterase activity. FEBS Lett 408: 345–349, 1997
68. Tesmer JJ, Sunahara RK, Johnson RA, Gosselin G, Gilman AG, Sprang SR: Two-metal-ion catalysis in adenylyl cyclase. Science 285: 756–760, 1999
69. De Jonge HR, Bot AGM, Vaandrager AB: Mechanism of action of *E. coli* heat-stable enterotoxin. Zentralbl Bakteriol Mikrobiol Hyg Abt 1(suppl. 15): 335–340, 1986
70. elDeib MM, Parker CD, White AA: Regulation of intestinal mucosa guanylate cyclase by hemin, heme and protoporphyrin IX. Biochim Biophys Acta 928: 83–91, 1987
71. Parkinson SJ, Waldman SA: An intracellular adenine nucleotide binding site inhibits guanyly cyclase C by a guanine nucleotide-dependent mechanism. Biochemistry 35: 3213–3221, 1996
72. Parkinson SJ, Alekseev AE, Gomez LA, Wagner F, Terzic A, Waldman SA: Interruption of *Escherichia coli* heat-stable enterotoxin-induced guanylyl cyclase signaling and associated chloride current in human intestinal cells by 2-chloroadenosine. J Biol Chem 272: 754–758, 1997
73. Zhang W, Mannan I, Schulz S, Parkinson SJ, Alekseev AE, Gomez LA, Terzic A, Waldman SA: Interruption of transmembrane signaling as a novel antisecretory strategy to treat enterotoxigenic diarrhea. FASEB J 13: 913–922, 1999
74. Greenberg RN, Dunn JA, Guerrant RL: Reduction of the secretory response to *Escherichia coli* heat-stable enterotoxin by thiol and disulfide compounds. Infect Immun 41: 174–180, 1983
75. ElDeib MM, Dove CR, Parker CD, Veum TL, Zinn GM, White AA: Reversal of the biological activity reducing of *Escherichia coli* heat-stable enterotoxin by disulfide-agents. Infect Immun 51: 24–30, 1986
76. Cohen MB, Mann EA, Lau C, Henning SJ, Giannella RA: A gradient in expression of the *Escherichia coli* heat-stable enterotoxin receptor exists along the villus-to-crypt axis of rat small intestine. Biochem Biophys Res Commun 186: 483–490, 1992
77. Li Z, Goy MF: Peptide-regulated guanylate cyclase pathways in rat colon: *In situ* localization of GCA, GCC, and guanylin mRNA. Am J Physiol 265: G394–G402, 1993
78. Lewis LG, Witte DP, Laney DW, Currie MG, Cohen MB: Guanylin mRNA is expressed in villous enterocytes of the rat small intestine and superficial epithelia of the rat colon. Biochem Biophys Res Commun 196: 553–560, 1993
79. Vaandrager AB, De Jonge HR: Effect of cGMP on intestinal transport. Adv Pharmacol 26: 253–283, 1994
80. Krause WJ, Cullingford GL, Freeman RH, Eber SL, Richardson KC, Fok KF, Currie MG, Forte LR: Distribution of heat-stable enterotoxin/guanylin receptors in the intestinal tract of man and other mammals. J Anat 184: 407–417, 1994
81. Cohen MB, Witte DP, Hawkins JA, Currie MG: Immunohistochemical localization of guanylin in the rat small intestine and colon. Biochem Biophys Res Commun 209: 803–808, 1995
82. London RM, Krause WJ, Fan X, Eber SL, Forte LR: Signal transduction pathways via guanylin and uroguanylin in stomach and intestine. Am J Physiol 273: G93–G105, 1997
83. Qian X, Prabhakar S, Nandi A, Visweswariah SS, Goy MF: Expression of GC-C, a receptor-guanylate cyclase, and its endogenous ligands uroguanylin and guanylin along the rostrocaudal axis of the intestine. Endocrinology 141: 3210–24, 2000
84. Carrithers SL, Barber MT, Biswas S, Parkinson SJ, Park PK, Goldstein SD, Waldman SA: Guanylyl cyclase C is a selective marker for metastatic colorectal tumors in human extraintestinal tissues. Proc Natl Acad Sci USA 93: 14827–14832, 1996
85. Schulz S, Chrisman TD, Garbers DL: Cloning and expression of guanylin. Its existence in various mammalian tissues. J Biol Chem 267: 16019–16021, 1992
86. Carrithers SL, Taylor B, Cai WY, Johnson BR, Ott CE, Greenberg RN, Jackson BA: Guanylyl cyclase-C receptor mRNA distribution along the rat nephron. Reg Pep 95: 65–74, 2000
87. Cetin Y, Kulaksiz H, Redecker P, Bargsten G, Adermann K, Grube D, Forssmann WG: Bronchiolar nonciliated secretory (Clara) cells: Source of guanylin in the mammalian lung. Proc Natl Acad Sci USA 92: 5925–5959, 1995
88. Rambotti MG, Giambanco I, Spreca A: Ultracytochemical detection of guanylate cyclase C activity in alimentary tract and associated glands of the rat. Influence of pH, ATP and the ions Mg^{2+} and Mn^{2+}. Histochem J 32: 231–238, 2000
89. John M, Wiedenmann B, Kruhoffer M, Adermann K, Ankorina-Stark I, Schlatter E, Ahnert-Hilger G, Forssmann WG, Kuhn M: Guanylin stimulates regulated secretion from human neuroendocrine pancreatic cells. Gastroenterology 114: 791–797, 1998
90. D'Este L, Kulaksiz H, Rausch U, Vaccaro R, Wenger T, Tokunaga Y, Renda TG, Cetin Y: Expression of guanylin in 'pars tuberalis-specific cells' and gonadotrophs of rat adenohypophysis. Proc Natl Acad Sci USA 97: 1131–1136, 2000

91. Reinecke M, David I, Loffing-Cueni D, Ablinger P, Cetin Y, Kuhn M, Forssmann: Localization, expression, and characterization of guanylin in the rat adrenal medulla. Histochem Cell Biol 106: 367–374, 1996
92. Laney DW Jr, Bezerra JA, Kosiba JL, Degen SJ, Cohen MB: Upregulation of *Escherichia coli* heat-stable enterotoxin receptor in regenerating rat liver. Am J Physiol 266: G899–G906, 1994
93. Scheving LA, Russell WE: Guanylyl cyclase C is up-regulated by nonparenchymal cells and hepatocytes in regenerating rat liver. Cancer Res 56: 5186–5191, 1996
94. Fan X, Wang Y, London RM, Eber SL, Krause WJ, Freeman RH, Forte LR: Signaling pathways for guanylin and uroguanylin in the digestive, renal, central nervous, reproductive, and lymphoid systems. Endocrinology 138: 4636–4648, 1997
95. Forte LR, London RM, Freeman RH, Krause WJ: Guanylin peptides: Renal actions mediated by cyclic GMP. Am J Physiol 278: F180–F191, 2000
96. Fawcus K, Gorton VJ, Lucas ML, McEwan GT: Stimulation of three distinct guanylate cyclases induces mucosal surface alkalinisation in rat small intestine *in vitro*. Comp Biochem Physiol A Physiol 118: 291–295, 1997
97. Joo NS, London RM, Kim HD, Forte LR, Clarke LL: Regulation of intestinal Cl and HCO_3 secretion by uroguanylin. Am J Physiol 274: G633–G644, 1998
98. Guba M, Kuhn M, Forssmann WG, Classen M, Gregor M, Seidler U: Guanylin strongly stimulates rat duodenal HCO_3 secretion: Proposed mechanism and comparison with other secretagogues. Gastroenterology 111: 1558–1568, 1996
99. Shailubhai K, Yu HH, Karunanandaa K, Wang JY, Eber SL, Wang Y, Joo NS, Kim HD, Miedema BW, Abbas SZ, Boddupalli SS, Currie MG, Forte LR: Uroguanylin treatment suppresses polyp formation in the Apc(Min/+) mouse and induces apoptosis in human colon adenocarcinoma cells via cyclic GMP. Cancer Res 60: 5151–5157, 2000
100. Li Z, Knowles JW, Goyeau D, Prabhakar S, Short DB, Perkins AG, Goy MF: Low salt intake down-regulates the guanylin signaling pathway in rat distal colon. Gastroenterology 111: 1714–1721, 1996
101. Moro F, Levenez F, Nemoz-Gaillard E, Pellissier S, Plaisancie P, Cuber JC: Release of guanylin immunoreactivity from the isolated vascularly perfused rat colon. Endocrinology 141: 2594–2599, 2000
102. Martin S, Adermann K, Forssmann WG, Kuhn M: Regulated, side-directed secretion of proguanylin from isolated rat colonic mucosa. Endocrinology 140: 5022–5029, 1999
103. Scheving LA, Jin WH: Circadian regulation of uroguanylin and guanylin in the rat intestine. Am J Physiol 277: C1177–C1183, 1999
104. Chao AC, de Sauvage FJ, Dong YJ, Wagner JA, Goeddel DV, Gardner P: Activation of intestinal CFTR Cl- channel by heat-stable enterotoxin and guanylin via cAMP-dependent protein kinase. EMBO J 13: 1065–1072, 1994
105. Tien XY, Brasitus TA, Kaetzel MA, Dedman JR, Nelson DJ: Activation of the cystic fibrosis transmembrane conductance regulator by cGMP in the human colonic cancer cell line, Caco-2. J Biol Chem 269: 51–54, 1994
106. Cuthbert AW, Hickman ME, MacVinish LJ, Evans MJ, Colledge WH, Ratcliff R, Seale PW, Humphrey PP: Chloride secretion in response to guanylin in colonic epithelial from normal and transgenic cystic fibrosis mice. Br J Pharmacol 112: 31–36, 1994
107. Lohmann SM, Vaandrager AB, Smolenski A, Walter U, De Jonge HR. Distinct and specific functions of cGMP-dependent protein kinases. Trends Biochem Sci 22: 307–312, 1997
108. Seidler U, Blumenstein I, Kretz A, Viellard-Baron D, Rossmann H, Colledge WH, Evans M, Ratcliff R, Gregor M: A functional CFTR protein is required for mouse intestinal cAMP-, cGMP- and Ca^{2+}-dependent HCO_3^- secretion. J Physiol (Lond) 505: 411–423, 1997
109. Hoogerwerf WA, Tsao SC, Devuyst O, Levine SA, Yun CH, Yip JW, Cohen ME, Wilson PD, Lazenby AJ, Tse CM, Donowitz M. NHE2 and NHE3 are human and rabbit intestinal brush-border proteins. Am J Physiol 270: G29–G41, 1996
110. Schultheis PJ, Clarke LL, Meneton P, Miller ML, Soleimani M, Gawenis LR, Riddle TM, Duffy JJ, Doetschman T, Wang T, Giebisch G, Aronson PS, Lorenz JN, Shull G: Renal and intestinal absorptive defects in mice lacking the NHE3 Na^+/H^+ exchanger. Nat Genet 19: 282–285, 1998
111. Kere J, Lohi H, Hoglund P: Congenital chloride diarrhea. Am J Physiol 276: G7–G13, 1999
112. Yun CH, Oh S, Zizak M, Steplock D, Tsao S, Tse CM, Weinman EJ, Donowitz M: cAMP-mediated inhibition of the epithelial brush border Na^+/H^+ exchanger, NHE3, requires an associated regulatory protein. Proc Natl Acad Sci USA 94: 3010–3015, 1997
113. Kurashima K, Yu FH, Cabado AG, Szabo EZ, Grinstein S, Orlowski J: Identification of sites required for down-regulation of Na^+/H^+ exchanger NHE3 activity by cAMP-dependent protein kinase phosphorylation-dependent and -independent mechanisms. J Biol Chem 272: 28672–28679, 1997
114. Vaandrager AB, Bot AG, De Jonge HR: Guanosine 3′,5′-cyclic monophosphate-dependent protein kinase II mediates heat-stable enterotoxin-provoked chloride secretion in rat intestine. Gastroenterology 112: 437–443, 1997
115. Pfeifer A, Aszodi A, Seidler U, Ruth P, Hofmann F, Fassler R: Intestinal secretory defects and dwarfism in mice lacking cGMP-dependent protein kinase II. Science 274: 2082–2086, 1996
116. Vaandrager AB, Bot AG, Ruth P, Pfeifer A, Hofmann F, De Jonge HR: Differential role of cyclic GMP-dependent protein kinase II in ion transport in murine small intestine and colon. Gastroenterology 118: 108–114, 2000
117. Vaandrager AB, Smolenski A, Tilly BC, Houtsmuller AB, Ehlert EM, Bot AG, Edixhoven M, Boomaars WE, Lohmann SM, de Jonge HR: Membrane targeting of cGMP-dependent protein kinase is required for cystic fibrosis transmembrane conductance regulator Cl^- channel activation. Proc Natl Acad Sci USA 95: 1466–1471, 1998
118. Kleizen B, Braakman I, de Jonge HR: Regulated trafficking of the CFTR chloride channel. Eur J Cell Biol 79: 544–556, 2000
119. Korschen HG, Beyermann M, Muller F, Heck M, Vantler M, Koch KW, Kellner R, Wolfrum U, Bode C, Hofmann KP, Kaupp UB: Interaction of glutamic-acid-rich proteins with the cGMP signalling pathway in rod photoreceptors. Nature 400: 761–766, 1999
120. Forte LR, Thorne PK, Eber SL, Krause WJ, Freeman RH, Francis SH, Corbin JD: Stimulation of intestinal Cl^- transport by heat-stable enterotoxin: Activation of cAMP-dependent protein kinase by cGMP. Am J Physiol 263: C607–C6115, 1992

Molecular and Cellular Biochemistry **230**: 85–96, 2002.

Ultracytochemistry as a tool for the study of the cellular and subcellular localization of membrane-bound guanylate cyclase (GC) activity. Applicability to both receptor-activated and receptor-independent GC activity

Maria Grazia Rambotti, Antonio Spreca, Ileana Giambanco, Guglielmo Sorci and Rosario Donato

Department of Experimental Medicine and Biochemical Sciences, Section of Anatomy, University of Perugia, Perugia, Italy

Abstract

Membrane-bound guanylate cyclase activity was detected by ultracytochemistry at the electron microscope level in several mammalian tissues. The technique used in these studies allows the detection of active enzyme at the membrane site where it is located. In a few cases, such as normal and regenerating peripheral nerves and placenta, membrane-bound guanylate cyclase could be detected in the absence of stimulators of enzyme activity. However, in the majority of these studies membrane-bound guanylate cyclase was investigated following stimulation with natriuretic peptides, guanylin, or the Ca^{2+} sensor proteins, S100B and S100A1. In general, membrane-bound guanylate cyclase was localized to plasma membranes, in accordance with the functional role of this enzyme. Yet, in secretory cells the enzyme activity was localized on intracellular membranes, suggesting a role of membrane-bound guanylate cyclase in secretory processes. Finally, S100B and S100A1 were found to colocalize with membrane-bound guanylate cyclase on photoreceptor disc membranes and to stimulate enzyme activity at these sites in dark-adapted retinas in a Ca^{2+}-dependent manner. The results of these analyses are discussed in relation to the proposed functional role(s) of this enzyme. (Mol Cell Biochem **230**: 85–96, 2002)

Key words: guanylate cyclase, ultracytochemistry, natriuretic peptides, guanylin, S100 proteins

Introduction

Guanylate cyclase (GC) is a signal transducing enzyme that catalyzes the formation of the second messenger guanosine 3′,5′-monophosphate (cGMP) from GTP. GC has been detected in a soluble form as well as associated with cellular membranes [1–6].

The family of membrane-bound GC encompasses different forms of the enzyme that share a common structural organization. Each form is composed of an extracellular domain linked by a short transmembrane segment to an intracellular domain containing a protein kinase-like domain and the catalytic domain (for reviews see [2, 6]).

Two subfamilies of membrane-bound GC have been identified in vertebrates. Members of one subfamily are activated by peptides that bind to the enzyme extracellular domain. Two isoforms, GC-A and GC-B, are activated by atrial natriuretic peptide (ANP), brain natriuretic peptide (BNP) and C-type natriuretic peptide (CNP). GC-A responds to stimulation by either ANP or BNP, although ANP is much more effective than BNP in the activation of GC-A. CNP binds selectively to GC-B, whereas BNP activates GC-B

Address for offprints: R. Donato, Department of Experimental Medicine and Biochemical Sciences, Section of Anatomy, University of Perugia, Via del Giochetto C.P. 81 Succ. 3, 06122 Perugia, Italy although only at high concentrations [2, 7–9]. The GC-C isoform is a receptor for bacterial heat-stable enterotoxins and two peptides, i.e. guanylin and uroguanylin [2, 10].

The other subfamily is represented by members of retinal membrane-bound GC. No extracellular ligands have been described for these enzymes, which consistently do not have a receptor domain and are located on intracellular membranes. Their activity is regulated by Ca^{2+}-binding proteins of EF-hand type at low or high concentrations of cytosolic Ca^{2+}. These proteins interact with the GC intracellular domain (for reviews see [5, 11, 12]).

Biochemical studies have thoroughly characterized the activities of membrane-bound GCs in various tissues and in mammalian retina [2, 11, 12]. While biochemical analyses are essential to understand how enzyme activities are being regulated, it is likewise important to determine where these enzyme activities are distributed in a tissue or cell type. Immunocytochemistry and/or *in situ* hybridization methods give information regarding the presence in a tissue/cell type of an enzyme or the precursor of that enzyme independently of its functional state, but neither methods offer information regarding its function. Ultracytochemical methods may represent a valid means to localize an enzyme within a cell and to analyze its functional state in the same time. Thus, by this method one can 'see' an active enzyme at the cell site where it is located. One limitation of this technique stands on its relatively low resolution, i.e. absence of reaction product does not necessarily mean absence of the enzyme or inactive enzyme. This is to say that data from ultracytochemical investigations should be complemented by biochemical and immunocytochemical results. Also, for an enzyme activity to be detected by this technique, the enzyme under investigation should resist tissue fixation. Ultracytochemical methods use the enzyme own activity to localize the enzyme under study, therefore they can also provide information about the ability of the enzyme to operate. In fact, with these methods, one can detect not only the cell site of an enzyme activity in the presence of known stimulators, but also changes of the distribution (actually, the functional state) of enzyme activity under different experimental conditions. Clearly, biochemical investigation is required to develop pertinent ultracytochemical methods for the detection of enzyme activity.

In the case of membrane-bound GC activity, tissue samples undergo a relatively light fixation, then they are washed and incubated in a reaction medium containing low concentrations of lead nitrate and guanylyl-imidodiphosphate sodium salt (Gpp(NH)p), a non-hydrolyzable analog of GTP, as substrate, in the absence or presence of an appropriate stimulator of membrane-bound GC activity. This method allows the detection of a precipitate of enzymatically released imidodiphosphate and lead ions (reaction product) at the site of membrane-bound GC activity. In general, control experiments include incubation of tissue in the absence of the substrate or enzyme inactivation by heat. Then samples are processed by conventional procedures for electron microscope (EM) observation.

Ultracytochemical localization of membrane-bound GC activity

'Basal' guanylate cyclase activity

The cellular and subcellular distribution of membrane-bound GC activity was analyzed on tissues of various organs and on isolated cells under different experimental conditions. The early method we used for the ultracytochemical localization of membrane-bound GC activity was based on previous ultracytochemical procedures aimed at the detection of membrane-bound GC activity at the EM level [13, 14]. This method proved useful for the detection of basal membrane-bound GC activity, i.e. GC activity in the absence of stimulators of the enzyme. One limitation of this method was that low levels of enzyme activity could not be evidenced.

Peripheral nerves

The cellular and subcellular distribution of membrane-bound GC activity was studied in crushed peripheral nerves of rat during degeneration and regeneration [15]. In unlesioned nerves enzyme activity was detected on plasma membranes of Schwann cells and myelinated and non-myelinated fibres (Fig. 1a). At 24 and 48 h after the crush, membrane-bound GC activity was no longer detectable, except on occasional unlesioned nerve fibres. From the 7th to the 28th postlesion day, GC activity could be detected on regenerating fibres, displaying the same localization as in unlesioned nerves. Under identical experimental conditions the pattern of adenylate cyclase (AC) activity distribution appeared to be exactly the opposite [15]. Thus, AC activity was detected in early stages of degeneration of nerve fibres and in periaxonal cells (i.e. transformed Schwann cells, macrophages, polymorphonuclear leukocytes, fibroblasts and elongated cells), and low or no AC activity was observed in unlesioned or regenerated nerves. These data point to different regulatory roles of cGMP and cAMP in nerve degeneration and regeneration, with AC activity playing an important role in cells engaged in the formation of new periaxonal tissue, and membrane-bound GC being mostly involved in the regulation of activities in normal and regenerating axons and differentiated Schwann cells.

Human placenta

Ultracytochemical analyses were conducted on early as well as term human placenta (16). In samples of placental villi

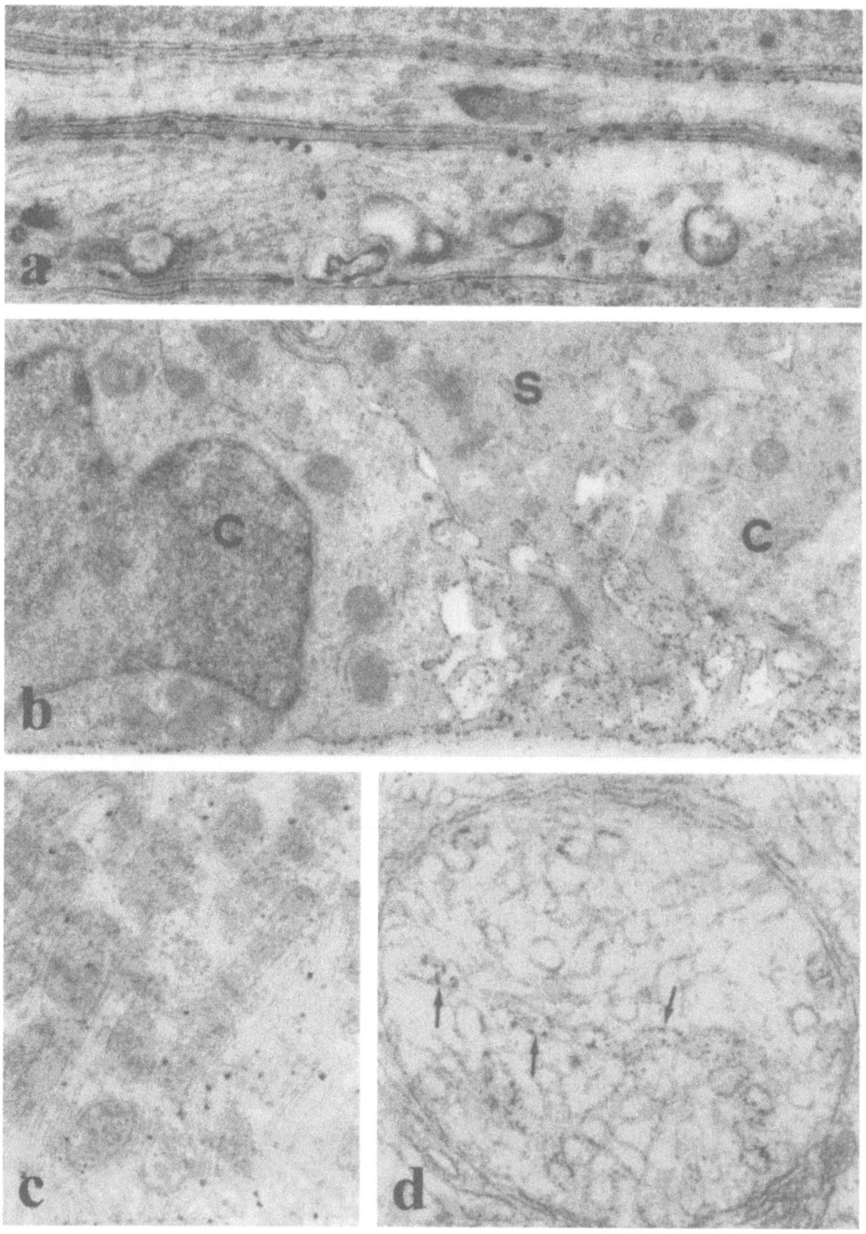

Fig. 1. (a) Ultracytochemical localization of basal membrane-bound GC activity in unmyelinated axons of rat sciatic nerve. GC reaction product is in the form of dots located beneath axonal plasma membranes. (b) Ultracytochemical localization of basal membrane-bound GC activity in human placenta (11th week of pregnancy). GC reaction product is seen on Langhans cytotrophoblast (C) plasma membrane and basal plasma membrane. The syncytiotrophoblast (S) membrane shows a very low level of GC reaction product. (c) Ultracytochemical localization of ANP-stimulated membrane-bound GC activity in adult rat kidney. GC reaction product is abundant in the microvilli plasma membrane (not shown) and the basal plasma membrane of rat proximal convoluted tubule. (d) Ultracytochemical localization of BNP-stimulated membrane-bound GC activity in adult rat adrenal cortex. GC reaction product is found on mitochondrial cristae (arrows) in zona glomerulosa cells. (a) ×33,000 (From Rambotti *et al.*, Glia 1: 260–274, 1988), (b) ×25,000 (From Spreca *et al.*, Cell Mol Biol 35: 449–455, 1989); (c) ×18,000 (From Rambotti *et al.*, Histochem J 22: 469–474, 1990); (d) ×90,000 (From Rambotti *et al.*, Histochem J 25: 384–391, 1993).

taken from the 6th to 12th week of pregnancy, the GC enzyme activity was prevalently found on the plasma membrane of non-dividing and mitotic cells of Langhans cytotrophoblast (Fig. 1b). Only in term placenta could membrane-bound GC activity be detected on the syncytiotrophoblast microvillous border. GC reaction product was also observed on the plasma membrane of endothelial cells. As these analyses were conducted in the absence of GC activity stimulators (e.g. natriuretic peptides), our results have to be referred to basal membrane-bound GC activity; hence, absence of GC reaction product might simply indicate low levels of activity, as pointed out above. With this limitation in mind, our data suggest that cells of Langhans cytotrophoblast display a higher GC activity early in pregnancy, whereas syncytiotrophoblast cells do so on later stages. Whereas the functional meaning of this shift in membrane-bound GC activity is obscure, the presence of GC activity in endothelial cells might be put in relation to the role of GC in the regulation of fetoplacental hemodynamics [17].

Stimulatable membrane-bound guanylate cyclase activity

After the discovery of natriuretic peptides and guanylin, we developed methods for the ultracytochemical detection of the membrane-bound GC activity that is activated by peptide binding to the enzyme extracellular domain. Thus, the cellular localization of membrane-bound GC activity was studied after incubation of samples with ANP, BNP, or CNP. Recently, membrane-bound GC activity stimulated by guanylin (GC-C) was localized ultracytochemically using a technique similar to that previously utilized for the ultracytochemical detection of membrane-bound GC stimulated by natriuretic peptides [18]. GC-C activity can be modulated by adenine nucleotides, Mg^{2+} and Mn^{2+} [19–22]. The activation of GC-C is also markedly influenced by the extracellular pH. Guanylin is a potent activator of GC-C activity at pH ≥ 8.0 [23]. In our hands, best conditions for the ultracytochemical detection of guanylin-stimulated GC activity were pH 8.0 and presence of Mg^{2+} and ATP [18].

Kidney

Membrane-bound GC activity was detected ultracytochemically in rat kidney after ANP stimulation [24]. In the renal corpuscle enzyme activity was localized on podocyte plasma membrane. Other segments of the nephron displaying ultracytochemical evidence of enzyme activity were the proximal convoluted tubule, the thick ascending limbs of the loop of Henle and the collecting tubule (Fig. 1c), whereas cells of the distal tubule did not display ANP-stimulated GC activity. In rat kidney, CNP-stimulated membrane-bound GC could be detected on plasma membranes of podocytes and proximal and distal convoluted tubule cells. By contrast, BNP-stimulated GC activity appeared to be restricted to apical membranes of the proximal convoluted tubule (unpublished data). These data point to different extent of expression and/or activity of the various isoforms of membrane-bound GC activity in the kidney. Guanylin-stimulated GC-C activity was detected on podocyte plasma membrane and proximal convoluted tubules, with remarkable differences between normal and diabetic rats [25]. In general, diabetic rats displayed a larger distribution of guanylin-stimulated GC-C activity than the normal counterpart. These results are summarized in Table 1.

Early studies indicated that the natriuretic effect of ANP was due to a direct effect of the peptide via inhibition of Na^+ reabsorption in various segments of the nephron [26–30]. BNP and CNP were reported to increase urinary fluid and electrolyte excretion as well [31]. Recent studies have demonstrated that the natriuretic effects of ANP depend on regulation of Na^+ entry through amiloride-sensitive cation channels and by modulation of active pumping of Na^+ by Na^+-K^+-ATPase in tubular epithelial cells [32]; that ANP and NO cooperate in regulation of the function of kidney glomeruli [33]; and that ANP decreases Cl^- reabsorption in the thick ascending limb [34]. A strong activation of membrane-bound GC stimulated by CNP in glomeruli, limbs of the loop of Henle and collecting duct was detected in relation to the hydration status, suggesting an additional and novel mechanism for regulating intravascular volume [35]. Also, stimulation of cGMP production in target cells of the nephron by guanylin and uroguanylin elicts diuresis, natriuresis and kaliuresis, particularly when dietary salt levels exceed the body's physiological requirements [36]. Thus, a possible synergism between ANP, guanylin and uroguanylin in natriuretic effects has been proposed. Together, these data indicate that renal structures are a main site of action of natriuretic peptides, which likely operate in a concerted manner with other factors known to affect renal function. Moreover, ANP and BNP were proposed to have a role in the

Table 1. Cellular and subcellular distribution of membrane-bound GC activity in the kidney following activation by peptides

Structure	ANP	BNP	CNP	Guanylin	
				n.r.*	d.r.**
Renal corpuscle					
Endothelial cells	–	–	–	–	–
Podocytes	+	–	+	+	–
Proximal tubule					
Apical membrane	+	+	+	+	+
Basal membrane	+	–	–	–	–
Loop of Henle					
Apical membrane	–	–	–	–	+
Basal membrane	+	–	–	–	+
Distal tubule					
Apical membrane	–	–	–	–	+
Basal membrane	–	–	+	–	+
Collecting tubule					
Apical membrane	+	–	–	–	–
Basal membrane	+	–	–	–	+

*n.r. – normal rat; **d.r. – diabetic rat.

control of renal cell growth in an autocrine manner *in vitro* [37], suggesting the possibility that the two peptides might have a trophic role during development and/or reparative processes.

Further work is required to dissect the functional role of GC-C in kidneys of diabetic rats.

Platelets

In rabbit platelets, ANP-stimulated membrane-bound GC activity was detected on the plasma membrane and membranes of the open canalicular system [38]. No GC activity was detected in unstimulated platelets. Interestingly, AC activity was localized to the same sites as was GC activity [38]. These results suggest that ANP might play a role in the regulation of platelet aggregation and that membrane-bound GC and AC might act in a concerted manner to regulate platelet functions. For instance, the two enzymes might modulate Ca^{2+} entry in human platelets and reorganization of actin-based cytoskeleton [39].

Lung

The distribution of the GC activity stimulated by ANP was studied on sections of rat intrapulmonary bronchioles and alveoli with special reference to their epithelial layers. GC-A activity was detected on mucus secreting goblet cells, and on non-ciliated cells such as Clara cells and great alveolar cells [40]. The presence of GC-A on plasma membranes of great alveolar cells was confirmed by other investigators and ANP was shown to promote accumulation of cGMP in these cells [41, 42]. In the great alveolar cells ANP was reported to inhibit surfactant secretion and the activity of amiloride-sensitive Na^+-channel activity. As Clara cells were proposed to play a role in the reabsorption of Na^+ and water from the airway surface [43], ANP-stimulated GC activity might be implicated in the regulation of water and Na^+ reabsorption in these cells.

Adrenal gland

In rat adrenal gland the distribution of BNP-stimulated GC activity was similar to that of ANP-stimulated GC activity [44]. In the adrenal cortex, enzyme activity was prevalently localized on cells of the zona glomerulosa and zona fascicolata, whereas the zona reticularis showed little membrane-bound GC reaction product. Natriuretic peptide-stimulated GC activity was detected almost exclusively on intracellular membranes and mitochondrial cristae (Fig. 1d). This latter localization might be related to the reported role of ANP in steroidogenesis [45, 46]. The presence of natriuretic peptide-stimulated GC activity in zona glomerulosa might be related to the suppressant effect of ANP on aldosterone secretion [47]. Although reports have challenged the possibility that GC activation by ANP may have any role in suppression of aldosterone secretion [48], others have observed that ANP binding to its receptor on a human adrenocortical cell line causes activation of ANP receptor-coupled GC and concomitant suppression of angiotensin II-stimulated aldosterone secretion [49], and still others have shown that mRNAs for GC-coupled ANP, GC-A and GC-B are down-regulated during pregnancy by oestrogen and progesterone, thus explaining the relative increase in aldosterone secretion under these conditions, with recovery of normal levels of these mRNAs a few days *post-partum* [50]. In the adrenal medulla, the GC reaction product was only detected on adrenaline-containing cells, again on intracellular membranes [44]. Incidentally, GC reaction product could also be observed on membranes of intra-axonal vesicles in preganglionic nerve fibers [44]. Natriuretic peptide-stimulated GC activity in adrenal medulla has been proposed to be important for catecholamine synthesis [51].

Neurohypophysis

The ultracytochemical localization of membrane-bound GC was investigated in the rat neurohypophysis after activation of samples with ANP or BNP [52]. The two hormones caused activation of the enzyme activity at partially distinct sites. The ANP-stimulated GC was localized both to pituicytes (Fig. 2a) and nerve fibres and endings, whereas the BNP-stimulated GC was present exclusively in nerve fibres and endings. These data suggest that both natriuretic peptides may have a role in the regulation of neuroendrocrine processes.

Nasal mucosa

ANP-stimulated GC was localized to the apical portion of the olfactory epithelium, with both olfactory vesicles and supporting cells displaying enzyme activity (Fig. 2b) [53]. CNP-stimulated GC activity was detected on secretory cells of Bowman's glands, on elongated cells in close relation with the Bowman's glands, and on myoepithelial cells. The BNP-stimulated GC activity was restricted to Bowman's gland cells.

The presence of ANP and its receptor in olfactory mucosa was assessed in previous studies. ANP was localized in secretory cells of Bowman's glands and in some cells of the epithelial layer [54], and ANP and membrane-bound GC were also detected in the olfactory bulb [55]. The presence of membrane-bound GC activity at these sites suggests that the ANP, CNP and/or guanylin might be active in promoting gland secretion and in inhibiting reabsorption of water and Na^+ at the mucosa surface in order to prevent dehydration of olfactory mucosa, and/or in the olfactory function.

Currently, we are studying the ultracytochemical localization of CNP-stimulated GC and GC-C stimulated by guanylin in normal human respiratory mucosa and in respiratory mucosa of nasal polyps. Preliminary results [56] indicate that normal and pathological nasal mucosa contain both enzyme activities in both respiratory epithelium (Fig. 2c)

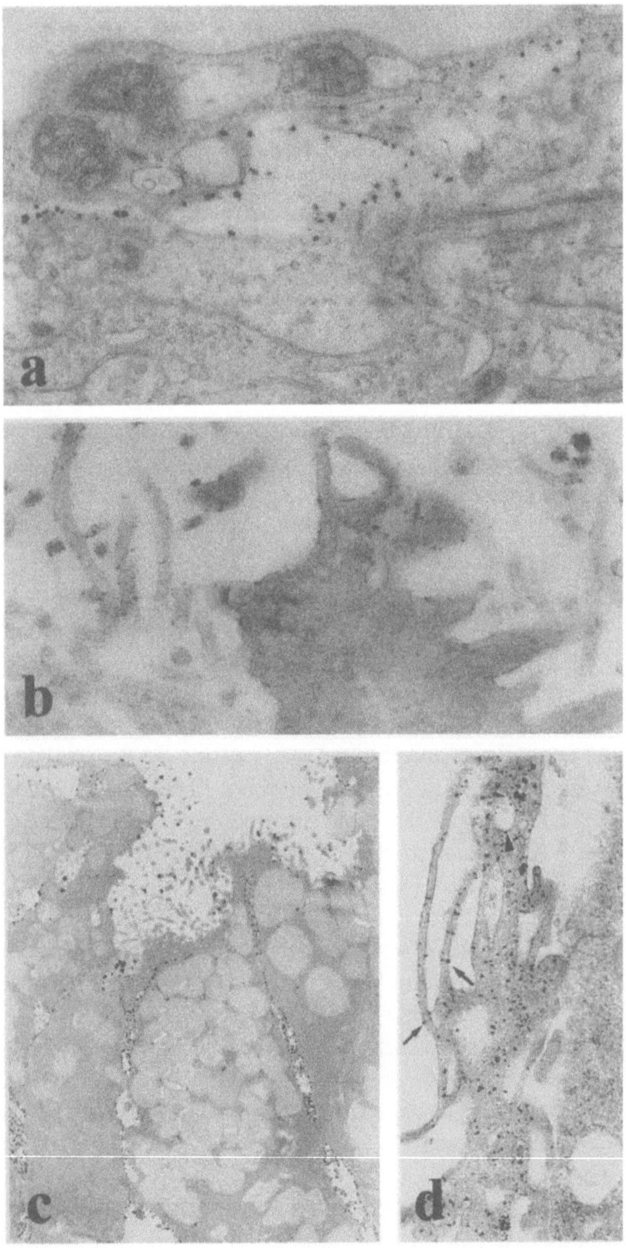

Fig. 2. (a) Ultracytochemical localization of ANP-stimulated membrane-bound GC activity in adult neurohypophysis. GC reaction product is found on pituicyte plasma membrane. (b) Ultracytochemical localization of ANP-stimulated membrane-bound GC activity in olfactory mucosa. GC reaction product is found on plasma membrane of olfactory vesicles and microvilli of supporting cells. (c) Ultracytochemical localization of guanylin-stimulated membrane-bound GC activity in nasal mucosa. GC reaction product is found on the plasma membrane of gland cells. (d) Ultracytochemical localization of CNP-stimulated membrane-bound GC activity in C6 rat glioma cells. GC reaction product is found on plasma membranes (arrows) and membranes (arrowhead) within protoplasmic extensions of dibutyryl cyclic AMP-treated C6 cells. (a) ×27,500 (Rambotti *et al.*, Brain Res 644: 52–58, 1994); (b) ×32,000 (From Spreca and Rambotti, Histochem J 26: 778–786, 1994); (c) ×9,000; (d) ×24,500 (From Sorci *et al.*, Brain Res 683: 51–58, 1995).

and in gland cells. In the respiratory epithelium, the distribution of the enzyme activity shows a gradient of density with the highest amount in the upper portion of the epithelium. CNP- and guanylin-stimulated GCs are present in all cell types, although GC-C seems to be less active in ciliated cells. However, CNP-stimulated GC caused a stronger activation of enzyme activity in glands in polyp mucosa. These data suggest complementary activities of CNP and guanylin in the regulation of mucociliary transport and gland secretion.

CNP was reported to regulate ciliary beating in airway epithelia by activating membrane-bound GC, implying a role of this peptide and GC activity in the regulation of mucociliary clearance [57, 58]. Moreover, CNP stimulated epithelial Cl^- transport by regulating cystic fibrosis transmembrane conductance regulator (CFTR) activity [59]. Also, guanylin and GC-C have been found to be expressed in airway epithelia, where they might be of major importance in the regulation of Cl^- channel activity and transepithelial fluid transport [60].

C6 glioma cells

Activity of membrane-bound GC stimulated with ANP, BNP or CNP was investigated in proliferating and differentiating C6 glioma cells [61]. All peptides were activators of membrane-bound GC, as also demonstrated by biochemical determination of the cGMP produced [61]. In proliferating, subconfluent rat C6 cells natriuretic peptide-stimulated GC activity was localized to membranes within protoplasmic processes, whereas it was detected within both protoplasmic processes and cell bodies in confluent cells and in cell bodies only in post-confluent cells. It should be noted that post-confluent cells were nearly devoid of processes. On the other hand, incubation of confluent cells in the presence of dibutyryl cyclic AMP (db-cAMP) or retinoic acid (RA) caused C6 cells to proliferate at a reduced rate and to extend protoplasmic processes. Under these conditions, the localization of enzyme activity to membranes of protoplasmic processes (Fig. 2d) was restored. These data suggest that the state of growth of C6 cells influences the extent of membrane-bound GC activation by natriuretic peptides in different cell compartments, and that intracellular membranes bearing GC activity tend to concentrate within protoplasmic processes as these cells acquire a more differentiated state. Whether natriuretic peptide-regulated GC activity has a causal role in astrocytes differentiation or the apparent shift of subcellular compartmentation of the enzyme is an epiphenomenon of astrocyte differentiation remains to be determined. Yet, the situation observed with db-cAMP- and RA-treated C6 glioma cells recalls what was observed in regenerating peripheral nerves where 'basal' membrane-bound GC activity was exclusively detected in the protoplasmic processes of Schwann cells surrounding regenerating axons [15].

Alimentary tract

The ultracytochemical localization of membrane-bound GC in the rat gastrointestinal tract was studied after stimulation by ANP, BNP, CNP [62]. ANP and CNP stimulated the enzyme activity on plasma membranes (Fig. 3a), whereas BNP was a much less potent activator of the enzyme. ANP- and CNP-stimulated GC activity was detected in different cell types or in different sites of the same cell, irrespective of the portion of the gastroenteric tract investigated. Thus, enzyme activity was detected on surface epithelium, gland cells, endocrine cells, undifferentiated epithelial cells of the crypts in small and large intestine, and smooth muscle cells. ANP and CNP caused a stronger activation of the enzyme in the stomach and small intestine than in the colon and rectum.

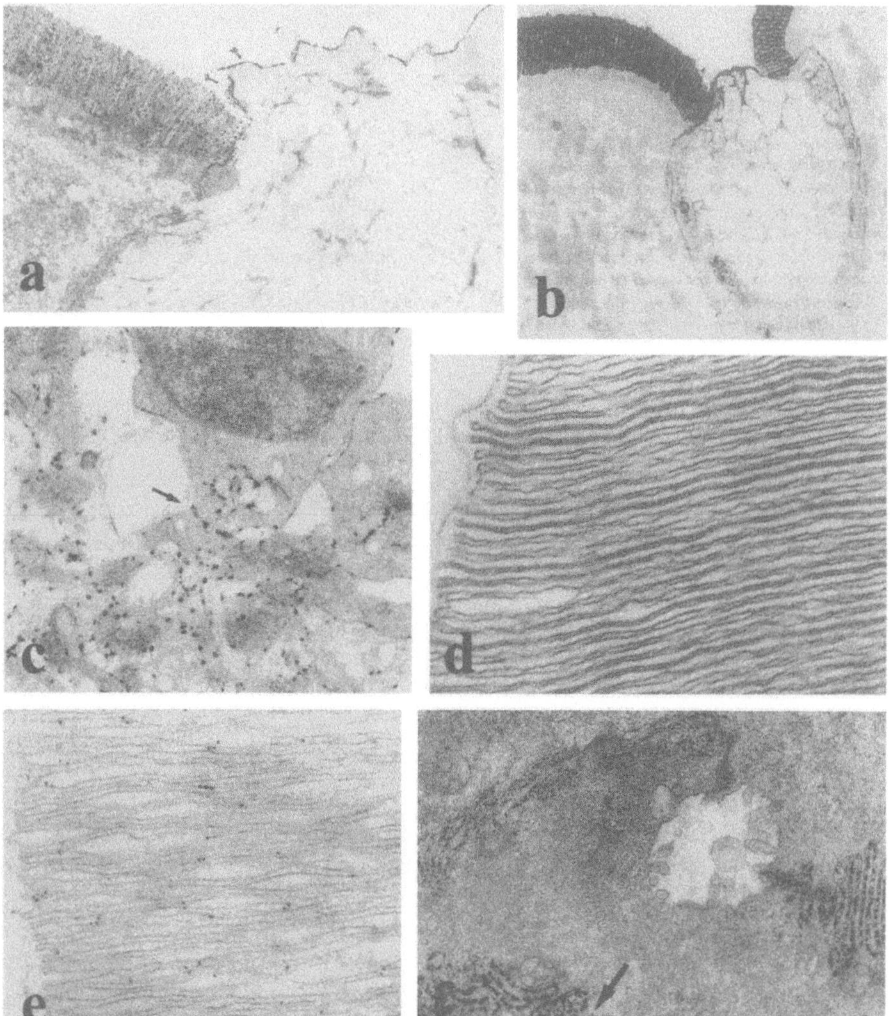

Fig. 3. (a) Ultracytochemical localization of ANP-stimulated membrane-bound GC activity in rat jejunum surface epithelium. GC reaction product is found on apical plasma membranes of mature enterocytes and mucous cells. (b) Ultracytochemical localization of guanylin-stimulated membrane-bound GC activity in rat duodenum surface epithelium. GC reaction product is found on apical plasma membranes of mature enterocytes and mucous cells. (c) Ultracytochemical localization of CNP-stimulated membrane-bound GC activity in bovine retina. Central ending of rod (arrow) and network of neuronal processes show GC reaction product on plasma membranes in the outer plexiform layer. (d) Ultracytochemical localization of S100B-stimulated membrane-bound GC activity in bovine retina. GC reaction product is seen on photoreceptor disc membranes in dark-adapted retina following exposure to S100B in the presence 10 μM free Ca^{2+}. (e) Immunocytochemical localization of S100B in photoreceptors of bovine retina. S100B immunogold particles are detected on disc membranes. (f) Ultracytochemical localization of ANP-stimulated membrane-bound GC activity in human eccrine sweat glands. GC reaction product is seen on lateral and basal (arrow) membranes of clear cells. (a) ×13,000 (From Rambotti *et al.*, Histochem J 29: 117–126, 1997); (b) ×15,000 (From Rambotti *et al.*, Histochem J 29: 117–126, 1997); (c) ×15,000 (From Spreca *et al.*, Histochem J 31: 477–483, 1999); (d) ×100,000 (From Rambotti *et al.*, Neuroscience 92: 1089–1101, 1999); (e) ×88,000 (From Rambotti *et al.*, Neuroscience 92: 1089–1101, 1999); (f) ×21,000 (From Spreca *et al.*, Histochem J 32: 725–731, 2000).

Table 2. Cellular and subcellular distribution of membrane-bound GC activity in the alimentary tract following activation by peptides

Structure	ANP	CNP	BNP
Stomach			
Body glands			
Parietal cells	+	+	+
Chief cells	–	+	–
Cardiac glands	+	–	–
Pyloric glands	+	–	–
Surface epithelium	+	–	–
Muscular layers	+	–	–
Duodenum			
Crypts			
Undifferentiated cells	+	+	–
Endocrine cells	–	+	–
Surface epithelium	+	+	+
Muscular layers – nerve fibres	–	+	–
Jejunum-ileum			
Crypts			
Undifferentiared cells	+	+	–
Surface epithelium	++/–	+	–
Muscular layers – nerve fibres	–	+	–
Colon			
Crypts			
Undifferentiated cells	+	+	–
Surface epithelium	+/–	–	–
Muscular layers	+	–	–
Rectum			
Crypts			
Undifferentiated cells	+	–	–

Table 3. Distribution of membrane-bound GC activity in the alimentary tract following activation by guanylin under various experimental conditions

Structure	Presence of ATP			Absence of ATP		
	8.0*	7.4*	5.0*	8.0*	7.4*	5.0*
Mg^{2-}						
Stomach						
Surface epithelium	++	++	–	–	–	·
Glands	–	–	–	–	–	–
Duodenum						
Surface epithelium	+++	++	–	++	+	–
Crypts	-	–	–	–	–	–
Jejunum – Ileum						
Surface epithelium	++	+	–	+	+	–
Crypts	–	–	–	–	–	–
Colon						
Surface epithelium	++	+	–	–	–	–
Liver	+++	+	–	++	+	–
Exocrine pancreas	+++	+	–	+	+	–
Parotid gland	++	–	–	–	–	–
Submandibolar gland	–	–	–	–	–	–
Mn^{2+}						
Duodenum						
Surface epithelium	++	+	–	+	+	–
Crypts	–	–	–	–	–	–

*Incubations were performed at the indicated pH values in the presence of either Mg^{2+} or Mn^{2+}.

However, ANP-stimulated GC activity was higher in the stomach, whereas CNP-stimulated GC activity was higher in the small intestine. These results are summarized in Table 2.

We also studied the ultracytochemical localization of guanylin-stimulated GC-C activity in the rat alimentary tract and associated glands under various experimental conditions [18]. The highest enzyme activity was detected in samples incubated at pH 8.0 in the presence of guanylin, Mg^{2+} and ATP. GC-C activity was detected on the apical membrane of epithelial cells in the stomach and intestine (Fig. 3b). In the intestine, enzyme activity was higher in the duodenum than in jejunum-ileum and colon. In the small intestine, the activity was more evident in the upper portion than in the basal portion of the villus. The mucus cells of the surface epithelium were also positive for GC-C. In the liver, GC-C activity was detected on membranes lining bile canaliculi. In exocrine pancreas and parotid gland enzyme activity was detected on apical membranes of acinar cells. Centroacinar cells also were positive. These results are summarized in Table 3.

ANP has been found in stomach, and small and large intestine raising the possibility that ANP may act as a local mediator of gastrointestinal function [63]. Previous studies demonstrated that ANP might act as an intestinal secretagogue, given its ability to raise cGMP levels and to enhance electrolyte secretion and accumulation of luminal fluid throughout the intestine tract [64–68]. Also, BNP and CNP were reported to decrease jejunal fluid and electrolyte absorption [31]. A role of ANP in relaxation-contraction of muscular layers at rat jejunum was also described [69]. On the other hand, activation of GC-C in target cells of the gastrointestinal tract was shown to cause increases in both chloride and bicarbonate secretion [23, 70].

Increased cGMP production was observed to activate CFTR via a protein kinase intermediary, and activation of CFTR in turn resulted in increased chloride secretion, decreased Na^+ and HCO_3^- absorption and possibly bicarbonate secretion [71–74].

Retina

The ultracytochemical localization of ANP- and CNP-stimulated GC was studied in light- and dark-adapted retinas. Both peptides stimulated membrane-bound GC activities in light-adapted, but not dark-adapted retinas [75]. The ultracytochemical localization of membrane-bound GC activity was similar irrespective of the natriuretic peptide used. Enzyme activities were detected on plasma membranes of photoreceptor cell body, bipolar, horizontal and ganglion cells, on interneuronal connections in the plexiform layers, and on fibres in the nerve fibres' layer (Fig. 3c). Whereas, the pigmented epithelium presented enzyme activity on both lateral and basal plasma membrane independently of the light- or dark-adapted state, and Müller cells and photoreceptor outer segments did not display natriuretic peptide-stimulated GC reaction product irrespective of the experimental conditions [75]. Previous work has shown that ANP receptor mRNA is expressed in the retina [76]. While our data suggest that plasma membranes of neuronal cell bodies and axons in the retina are the site of action of natriuretic peptides, interpretation of these results is difficult also in consideration that GC activity was only evidenced in light-adapted retinas. The subcellular localization GC activity described here points to natriuretic peptide binding to the GC receptor domain, as expected, and this event should be independent of retina light- or dark-adaptation. Thus, transduction of this interaction into activation of the GC catalytic domain could depend on additional factors acting from inside retinal neurons and being activated by so far unknown light-dependent events. The specificity of natriuretic peptide effect on these neurons can be inferred by the observation that natriuretic peptides activate plasma membrane GC activity in pigmented epithelium irrespective of the light- or dark-adaptation of the retina, whereas Müller cells do not appear to respond to any of the peptides tested, as mentioned above. To our knowledge, this is the first demonstration that engagement of the GC receptor domain by natriuretic peptides can be dissociated from activation of the GC catalytic domain. This issue deserves further investigation.

Ultracytochemistry was also employed for the detection of the activity of membrane-bound GC in the retina in presence of S100B and S100A1, two members of a family of Ca^{2+}-binding proteins of the EF-hand type implicated in the Ca^{2+} signal transduction within cells [77–79]. Both proteins proved to be activators of purified membrane-bound GC activity *in vitro* by binding to the GC cytoplasmic domain [80–83]. In accordance with these observations, experiments in which dark- or light- adapted, fixed bovine retinas were exposed to S100B or S100A1 in the presence or absence of Ca^{2+} without prior permeabilization of tissue with Triton X-100 resulted in no visualization of membrane-bound GC activity [84]. By contrast, in retinas that had been briefly exposed to Triton-X100 before fixation, S100B and S100A1 stimulated enzyme activity associated with photoreceptor disc membranes in dark-adapted, but not light adapted retina in a Ca^{2+}-dependent manner, although with different Ca^{2+} requirements [84]. Whereas S100A1-stimulated GC activity could be observed only at high Ca^{2+} concentrations (1 mM), S100B-stimulated GC activity was observed even at 10 μM free Ca^{2+} (Fig. 3d). These results suggested that the two S100 proteins needed to enter photoreceptors to act on disc membrane-associated GC, i.e. that effects of S100B and/or S100A1 on retinal GC activity required these proteins come in contact with the enzyme, in accordance with biochemical data indicating that each S100 protein binds to the cytoplasmic domain of GC [80–83]. By immunocytochemistry, S100B and S100A1 were localized to disc membranes in photoreceptors (Fig. 3e), i.e. at the same site where GC is found [84]. These observations make it reasonable to hypothesize that endogenous S100B and, possibly, S100A1 might have a physiological role in the Ca^{2+}-dependent regulation of disc membrane-associated GC activity. As activation of photoreceptor outer segment membrane-bound GC has an important role in phototransduction, serving to increase the range of intensities to which the rod can respond [11, 12], S100B and, possibly, S100A1 might play a role in dark-adaptation of photoreceptors in vertebrates or, alternatively, S100A1 and S100B may transduce a different kind of signal in photoreceptors. S100B and S100A1 also stimulated GC activity in photoreceptor cell bodies and Müller cells, but their effect appeared independent of the light- or dark-adapted state of the retina and was observed at relatively high Ca^{2+} concentrations [84]. The functional meaning of these data remains to be investigated. However, the present data suggest that Müller cells express a retinal calcium-dependent guanylate cyclase that is potentially activated by S100B and S100A1, while they do not express natriuretic peptide-sensitive GC activity [75]. In this regard, Müller cells differ from Schwann cells and cultured astrocytes.

Sweat and sebaceous glands

Guanylin-stimulated GC-C and ANP-stimulated GC activity were studied in human sweat glands [85]. In both eccrine and apocrine glands the peptides stimulated enzyme activity in secretory and myoepithelial cells (Fig. 3f). In eccrine glands, cells of the excretory duct also presented enzyme activity.

In sebaceous glands, ANP-stimulated GC was undetectable in undifferentiated cells and detectable in differentiating and fully differentiated cells [86].

The functional role of ANP- and guanylin-stimulated GC activities in clear cells and in cells of the duct of the apocrine eccrine glands is not known at present. ANP and guanylin could regulate water and electrolyte interchange across the plasma membranes of these cells. ANP was proposed to regulate Na^+ entry through amiloride-sensitive cation channels and by modulation of the active pumping of Na^+ by Na^+-K^+-ATPase [32], and to potently inhibit Na^+-K^+-Cl^- co-transport [87].

In our study the activity of membrane-bound GC stimulated by ANP was detected in membranes where also Na^+-K^+-ATPase, Na^+-K^+-Cl^- co-transporter and amiloride-sensitive channels are present. Effects of guanylin on activation of GC-C and production of cGMP can be related to its control on Cl^- channels by CFTR, as demonstrated in the intestine [71–74]. CFTR is abundant in epithelial cells of sweat glands [88], and the sweat formation is altered in CF patients with mutations of the cystic fibrosis (CF) gene [89]. In dark cells of eccrine glands and secretory cells of apocrine glands, both containing secretory vesicles, the activity of membrane-bound GC stimulated by ANP or guanylin was localized on membranes which participate directly in the secretory mechanism [85], suggesting a role of membrane-bound GC in the secretory function.

Conclusions

Ultracytochemical analyses of different isoforms of membrane-bound GC activity in various organs and in cultured cells reveal that the enzyme activity is mainly located on plasma membranes. The presence of GC on apical and/or baso-lateral plasma membranes in polarized cells correlates with the suggested role of this enzyme in a number of activities, as outlined above. In exocrine and endocrine cells, platelets and retinal Müller cells GC activity also was detected on intracellular membranes such as the agranular endoplasmic reticulum, Golgi complex and membranes of vesicles of secretion, which participate in secretory processes, and in some nerve endings the enzyme activity was detected on membranes of vesicles containing neurosecretory granules. The mechanism whereby natriuretic peptides activate GC on intracellular membranes remains to be elucidated.

In the last few years, we investigated the ultracytochemical localization of ANP-, BNP-, CNP- and guanylin-activated GC in various organs, as morphological correlates of this activity were lacking. When the ultracytochemical localization of enzyme activity of more isoforms was investigated in the same tissue or organ, we obtained a range of results. The activities of two or more isoforms of peptide-stimulated GC were found to have a similar distribution in a tissue, or to be localized in different cell types or at different sites of the same cell. In the same tissue or organ, the various peptides caused different extents of GC activation, and in some cases a single peptide appeared not to activate the enzyme sufficiently. The coexistence of receptors coupled to GC at the same site or at different sites of the same cell suggests that different peptides might have complementary activities.

The ultracytochemical method we used throughout allowed to detect variations of the state of activation of membrane-bound GC activity in tissues and cells under different functional conditions. Particularly, changes in extent of enzyme activity were observed during tissue growth and cellular proliferation as well as in retina depending on the light- or dark-adaptation. Also, in pathologic conditions differences in the extent of enzyme activity were observed in regenerating peripheral nerves, in diabetic kidneys and in mucosa of nasal polyps compared with normal structures. All these issues deserve more thorough biochemical and morphological investigation.

The functional meaning of some localizations of membrane-bound GC activity can be inferred from the enzyme subcellular localization and sensitivity to activators. Membrane-bound GC activity was localized prevalently in cells where the enzyme is involved in regulating water and electrolytes homeostasis and in promoting secretion. Our data suggest that the enzyme might be involved in the regulation of regeneration, cell growth and proliferation. In the retina, experimental data support the possibility that S100B- and S100A1-stimulated membrane-bound GC activity is related to visual mechanisms, whereas the functional roles of other GC isoforms that are particularly active in retinal neurons and are sensitive to natriuretic peptides are unknown thus far.

In conclusion, besides immunocytochemical and in situ hybridization methods, ultracytochemical studies can provide additional clues regarding the function of membrane-bound GC.

Acknowledgements

This work was supported in part by Perugia University funds to M.G.R. and European Commission (Contract No. BIO4CT960083) funds to R.D.

References

1. Tremblay J, Gerzer R, Hamet P: Cyclic GMP in cell function. Adv Sec Mess Phosphoprot Res 22: 319 -383, 1988
2. Drewett JG, Garbers DL: The family of guanylyl cyclase receptors and their ligands. Endocrine Rev 15: 135–162, 1994
3. Dizhoor AM: Regulation of cGMP synthesis in photoreceptors: Role in signal transduction and congenital diseases of the retina. Cell Signal 12: 711–719, 2000
4. Lucas KA, Pitari GM, Kazerounian S, Ruiz-Stewart I, Park J, Schulz S, Chepenik KP, Waldman SA: Guanylyl cyclases and signaling by cyclic GMP. Pharmacol Rev 52: 375–414, 2000
5. Palczewski K, Polans AS, Baehr W, Ames JB: Ca^{2+}-binding proteins in the retina: structure, function, and the etiology of human visual diseases. Bioessays 22: 337–350, 2000
6. Potter LR, Hunter T: Guanylyl cyclase-linked natriuretic peptide receptors: Structure and regulation. J Biol Chem 276: 6057–6060, 2001
7. Chang M, Lowe DG, Lewis M, Hellmiss R, Chen E, Goeddel DV: Differential activation by atrial and brain natriuretic peptides of two different receptor guanylate cyclases. Nature 341: 68–72, 1989
8. Schulz S, Sing S, Tunn DJ, Chin H, Garbers DL: The primary structure of a plasma membrane guanylate cyclase demonstrates diversity within this new receptor family. Cell 58: 1155–1162, 1989
9. Koller K, Lowe DG, Bennet GL, Minamino N, Kangawa K, Matsuo H, Goeddel DV: Selective activation of the B natriuretic peptide receptor by C-type natriuretic peptide (CNP). Science 252: 120–123, 1991
10. Forte LR, London RM, Krause WJ, Freeman RH: Mechanisms of guanylin action via cyclic GMP in the kidney. Annu Rev Physiol 62: 673–695, 2000
11. Polans A, Baeher W, Palczewski K: Turned on by Ca^{2+}! The physiology and pathology of Ca^{2+}-binding proteins in the retina. Trends Neurosci 19: 547–554, 1996

12. Pugh EN Jr, Duda T, Sitaramayya A, Sharma RK: Photoreceptor guanylate cyclases: A review. Biosci Rep 17: 429–473, 1997
13. Kang JH, Sahai A, Criss WE, West WL: Ultracytochemical localization of estrogen-stimulated guanylate cyclase in rat uterus. J Histochem Cytochem 30: 331–342, 1982.
14. Schulze W, Krause EG: Cytochemical demonstration of guanylate cyclase activity in cardiac muscle. Histochemistry 77: 243–254, 1983
15. Rambotti MG, Spreca A, Rende M, Donato R: Ultracytochemical localization of adenylate cyclase and guanylate cyclase in crushed peripheral nerves. Glia 1: 260–274, 1988
16. Spreca A, Rambotti MG, Saccardi C, Ferrarese R, Pomili G, Angelini A: Ultracytochemical localization of guanylate cyclase in early human placenta. Cell Mol Biol 35: 449–455, 1989
17. McQueen J, Kingdom JC, Whittle MJ, Connell JM: Characterization of atrial natriuretic peptide receptors in human fetoplacental vasculature. Am J Physiol 264: H798–H804, 1993
18. Rambotti MG, Giambanco I, Spreca A: Ultracytochemical detection of guanylate cyclase C activity in alimentary tract and associated glands of the rat. Influence of pH, ATP and the ions Mg^{2+} and Mn^{2+}. Histochem J 32: 231–238, 2000
19. Katwa LC, Parker CD, Dubyng JK, White AA: Nucleotide regulation of heat-stable enterotoxin receptor binding and of guanylate cyclase activation. Biochem J 283: 331–342, 1992
20. Khare S, Wilson D, Wali RK, Tien X-Y, Bissonnette M, Niedziele SM, Bolt JG, Sitrin MD, Brasitus TA: Guanylin activates rat colonic particulate guanylate cyclase. Biochem Biophys Res Commun 203: 1432–1437, 1994
21. Parkinson SJ, Carrithers SL, Waldman SA: Opposing adenine nucleotide-dependent pathways regulate guanylyl cyclase C in rat intestine. J Biol Chem 269: 22683–22690, 1994
22. Deshmane SP, Parkinson SJ, Crupper SS, Robertson DC, Schulz S, Waldman SA: Cytoplasmic domains mediate the ligand-induced affinity shift of guanylate cyclase C. Biochemistry 36: 12921–12929, 1997
23. Fan X, Hamra FK, London RM, Eber SL, Krause WJ, Freeman RH, Smith CE, Currie MG, Forte LR: Structure and activity of uroguanylin and guanylin from the intestine and urine of rats. Am J Physiol 273: E957–E964, 1997
24. Rambotti MG, Saccardi C, Spreca A: Evidence for particulate guanylate cyclase in rat kidney after stimulation by atrial natriuretic factor. An ultracytochemical study. Histochem J 22: 469–474, 1990
25. Rambotti MG, Spreca A, Mariucci G, Tantucci M, Ambrosini MV: Guanylate cyclase-C in nephrons of normal and diabetics rats. It J Anat Embriol 104(suppl. 2): 201, 1999
26. Sonnenberg H, Cupples WA, De Bold AJ, Veress AT: Intrarenal localization of the natriuretic effect of cardiac atrial extract. Can J Physiol Pharmacol 60: 1149–1152, 1982
27. Ishikawa S, Saito T, Okada K, Kuzuya T, Kangawa K, Matsuo H: Atrial natriuretic factor increases cyclic GMP and inhibits cyclic AMP in rat renal papillary collecting tubule cells in culture. Biochem Biophys Res Commun 130: 1147–1153, 1985
28. Tremblay J, Gerzer R, Vinay P, Pang SC, Beliveau R, Hamet P: The increase of cGMP by atrial natriuretic factor correlates with the distribution of particulate guanylate cyclase. FEBS Lett 181: 17–22, 1985
29. Koseki C, Hayashi Y, Torikai S, Furuya M, Ohnuma N, Imai M: Localization of binding sites for rat atrial natriuretic polypeptide in the rat kidney. Am J Physiol 250: F210–F216, 1986
30. Hammond TG, Yusufi ANK, Knox FG, Dousa TP: Administration of atrial natriuretic factor inhibits sodium-coupled transport in proximal tubules. J Clin Invest 75: 1983–1989, 1985
31. Morita H, Hagiike M, Horiba T, Miyake K, Ohyama H, Yamanouchi H, Hosomi H, Kangawa K, Minamino N, Matsuo H: Effects of brain natriuretic peptide and C-type natriuretic peptide infusion on urine flow and jejunal absorption in anesthetized dogs. Jpn J Physiol 42: 349–353, 1992
32. Scavone C, Scanlon C, McKee M, Nathanson JA: Atrial natriuretic peptide modulates sodium and potassium-activated adenosine triphosphate through a mechanism involving cyclic GMP and cyclic GMP-dependent protein kinase. J Pharmacol Exp Ther 272: 1036–1043, 1995
33. Stepinski J, Wendt U, Lewko B, Angielski S: Co-operation between particulate and soluble guanylyl cyclase systems in the rat renal glomeruli. J Physiol Pharmacol 51: 497–511, 2000
34. Bailly C: Effect of luminal atrial natriuretic peptide on chloride reabsorption in mouse cortical thick ascending limb: Inhibition by endothelin. J Am Soc Nephrol 11: 1791–1797, 2000
35. Dean AD, Vehaskari VM, Ritter D, Greenwald JE: Distribution and regulation of guanylyl cyclase type B in the rat nephron. Am J Physiol 270: F311–F318, 1996
36. Forte LR, London RM, Freeman RH, Krause WJ: Guanylin peptides: Renal actions mediated by cyclic GMP. Am J Physiol 278: F180–F191, 2000
37. Mistry S, Lussert B, Stewart K, Hawksworth GM, Struthers A, McLay JS: The expression and secretion of atrial natriuretic factor and brain natriuretic peptide by rat proximal tubular cells. Biochem Pharmacol 59: 783–790, 2000
38. Spreca A, Rambotti MG, Donato R: Particulate guanylate cyclase and adenylate cyclase activities after activation with various agents in rabbit platelets. An ultracytochemical study. Histochem J 23: 143–148, 1991
39. Rosado JA, Sage SO: The actin cytoskeleton in store-mediated calcium entry. J Physiol 526: 221–229, 2000
40. Rambotti MG, Spreca A: Ultrastructural demonstration of guanylate cyclase in rat lung. Cell Mol Biol 37: 455–462, 1991
41. Panchenko MP, Joyce-Brady M, Starikova MG, Oakes SM, Adachi R, Brody JS, Dickey BF: Atrial natriuretic peptide modulates alveolar type 2 cell adenylyl and guanylyl cyclases and inhibits surfactant secretion. Biochim Biophys Acta 1403: 115–125, 1998
42. Tharaux PL, Dussaule JC, Coulettes, Clerici C: Evidence for functional ANP receptors in cultured alveolar type II cells. Am J Physiol 274: L244–L251, 1998
43. Van Scott MR, Davis CW, Boucher RC: Na^+ and Cl^- transport across rabbit nonciliated bronchiolar epithelial (Clara) cells. Am J Physiol 256: C893–C901, 1989
44. Rambotti MG, Ercolani P, Spreca A: Ultracytochemical localization of particulate guanylate cyclase in rat adrenal gland exposed to stimulation by porcine brain natriuretic peptide. Histochem J 25: 384–391, 1993
45. Sharma RK, Marala RB, Duda TM: Purification and characterization of the 180-kDa membrane guanylate cyclase containing atrial natriuretic factor receptor from rat adrenal gland and its regulation by protein kinase C. Steroids 53: 437–460, 1989
46. Khurana ML, Pandey KN: Receptor-mediated stimulatory effect of atrial natriuretic factor, brain natriuretic peptide, and C-type natriuretic peptide on testosterone production in purified mouse Leydig cells: Activation of cholesterol side-chain cleavage enzyme. Endocrinology 133: 2141–2149, 1993
47. Brenner BM, Bollermann BJ, Gunning ME, Zeidel MI: Diverse biological activities of atrial natriuretic peptide. Physiol Rev 70: 655–699, 1990
48. Hanke CJ, Drewett JG, Myers CR, Campbell WB: Nitric oxide inhibits aldosterone synthesis by a guanylyl cyclase-independent effect. Endocrinology 139: 4052–4060, 1998
49. Bodart V, Rainey WE, Fournier A, Ong H, De Lean A: The H295R human adrenocortical cell line contains functional atrial natriuretic peptide receptors that inhibits aldosterone biosynthesis. Mol Cell Endocrinol 118: 134–137, 1996
50. Vaillancourt P, Omer S, Palfree R, Varma DR, Mulay S: Down-regulation of adrenal atrial natriuretic peptide receptors and proteins by pregnancy in rats. J Endocrinol 155: 523–530, 1997

51. Tsutsui M, Yanagihara N, Minami K, Kobayashi H, Nakashima Y, Kuroiwa A, Izumi F: C-type natriuretic peptide stimulates catecholamine synthesis through the accumulation of cyclic GMP in cultured bovine adrenal medullary cells. J Pharmacol Exp Ther 268: 584–589, 1994
52. Rambotti MG, Mughetti D, Spreca A: Detection of particulate guanylate cyclase in rat neurohypophysis after stimulation with ANF and BNP: An ultracytochemical study. Brain Res 644: 52–58, 1994
53. Spreca A, Rambotti MG: Ultracytochemical localization of particulate guanylate cyclase after stimulation with natriuretic peptides in lamb olfactory mucosa. Histochem J 26: 778–786, 1994
54. Gutkowska J, Marcinkiewicz M: Does the olfactory system mediate water-and mineral-regulating mechanisms? Evidence of immunoreactive atrial natriuretic factor within olfactory mucosa. J Neurochem 53: 1830–1836, 1989
55. Gutkowska J: Natriuretic peptides in the rat olfactory system. Rev Bras Biol 56: 79–88, 1996
56. Rambotti MG, Altissimi G, Spreca A: Guanylate and adenylate cyclases in normal and pathologic human nasal mucosa. It J Anat Embriol 105(suppl. 1): 88, 2000
57. Geary CA, Goy MF, Boucher RC: Synthesis and vectorial export of cGMP in airway epithelium: Expression of soluble and CNP-specific guanylate cyclases. Am J Physiol 265: L598–L605, 1993
58. Geary CA, Davis CW, Paradiso AM, Boucher RC: Role of CNP in human airways: cGMP-Mediated stimulation of ciliary beat frequency. Am J Physiol 268: L1021–L1028, 1995
59. Kelley TJ, Calvin UC, Mitchell LD: *In vivo* activation of CFTR-dependent chloride transport in murine airways epithelium by CNP. Am J physiol 273: L1065–L1072, 1997
60. Zhang ZH, Jow F, Numann R, Hinson J: The airways-epithelium: A novel site of action by guanylin. Biochem Biophys Res Commun 244: 50–56, 1998
61. Sorci G, Spreca A, Donato R, Rambotti MG: Detection of membrane-bound guanylate cyclase activity in rat C6 glioma cells at different growth states following activation by natriuretic peptides. Brain Res 683: 51–58, 1995
62. Rambotti MG, Giambanco I, Spreca A: Detection of guanylate cyclases A and B stimulated by natriuretic peptides in gastrointestinal tract of rat. Histochem J 29: 117–126, 1997
63. Gower WR Jr, Dietz JR, Vesely DL, Finley CL, Skolnick KA, Fabri PJ, Cooper DR, Chalfant CE: Atrial natriuretic peptide gene expression in the rat gastrointestinal tract. Biochem Biophys Res Commun 202: 562–570, 1994
64. Moriarty KJ, Higgs NB, Lees M, Tonge A, Wardle TD, Warhurst G: Influence of atrial natriuretic peptide on mammalian large intestine. Gastroenterology 98: 647–653, 1990
65. Catto-Smith AG, Hardin JA, Patrick MK, O'Loughlin EV, Gall DG: The effect of atrial natriuretic peptide on intestinal electrolyte transport. Reg Pep 36: 29–44, 1991
66. Guarino A, Tarallo L, Canani RB, Nocerino A, Iafusco M, Raimondi F, Rubino A: Comparative effects of atrial natriuretic peptide and *E. coli* heat-stable toxin on rat intestinal transport. J Pediatr Gastroenterol Nutr 15: 105–111, 1992
67. Vesely DL: Peptides from the N-terminus of the atrial natriuretic factor prohormone enhance guanylate cyclase activity and increase cyclic GMP levels in a wide variety of tissues. Mol Cell Biochem 109: 43–50, 1992
68. Li Z, Goy MF: Peptide-regulated guanylate cyclase, pathways in rat colon: *In situ* localization of GCA, GCC, and guanylin mRNA Am J Physiol 265: G394–G402, 1993
69. Scott RB, Maric M: The effect of atrial natriuretic peptide on small intestine contractility and transit. Peptides 12: 799–803, 1991
70. London RM, Krause WJ, Fan X, Eber SL, Forte LR: Signal transduction pathways via guanylin and uroguanylin in stomach and intestine. Am J Physiol 273: G93–G105, 1997
71. Forte LR, Thorne PK, Eber SL, Krause WJ, Freeman RR, Francis SH, Corbin JD: Stimulation of intestinal Cl^- transport by heat-stable enterotoxin activation of cAMP-dependent protein kinase by cGMP. Am J Physiol 263: C607–C615, 1992
72. Markert T, Vaandrager AB, Gambaryan S, Pohler D, Hauser C, Walter U, DeJong HR, Jarchau T, Lohmann SM: Endogenous expression of type II cGMP-dependent protein kinase mRNA and protein in rat intestine. J Clin Invest 96: 822–830, 1995
73. Vaandrager AB, Bot AGM, DeJong HR: Guanosine 3′,5′-cyclic monophosphate-dependent protein kinase II mediates heat-stable enterotoxin-provoked chloride secretion in rat intestine. Gastroenterology 112: 437–443, 1997
74. Guba M, Kuhn M, Forssmann W-G, Classen M, Gregor M, Seidler U: Guanylin strongly stimulates rat duodenal HCO_3^- secretion: Proposed mechanism and comparison with other secretagogues. Gastroenterology 111: 1558–1568, 1996
75. Spreca A, Giambanco I, Rambotti MG: Ultracytochemical study of guanylates cyclases A and B in light- and dark-adapted retinas. Histochem J 31: 477–483, 1999
76. Pardhasaradhi K, Kutty RK, Gentleman S, Krishna G: Expression of mRNA for atrial natriuretic peptide receptor guanylate cyclase (ANPRA) in human retina. Cell Mol Neurobiol 14: 1–7, 1994
77. Schäfer BW, Heizmann CW: The S100 family of EF-hand calcium-binding proteins: Functions and pathology. Trends Biochem Sci 21: 134–140, 1996
78. Donato R: Functional roles of S100 proteins, calcium-binding proteins of the EF-hand type. Biochim Biophys Acta 1450: 191–231, 1999
79. Donato R: S100: A multigenic family of calcium-modulated proteins of the EF-hand type with intracellular and extracellular functional roles. Int J Biochem Cell Biol 33: 637–668, 2001
80. Margulis A, Pazdnyakov N, Sitaramayya A: Activation of bovine photoreceptor guanylate cyclase by S100 proteins. Biochem Biophys Res Commun 218: 243–247, 1996
81. Duda T, Goraczniak RM, Sharma RK: Molecular characterization of S100A1-S100B protein in retina and its activation mechanism of bovine photoreceptor guanylate cyclase. Biochemistry 35: 6263–6266, 1996
82. Pozdnyakoz N, Goraczniak R, Margulis A, Duda T, Sharma RK, Yoshida A, Sitaramayya A: Structural and functional characterization of retinal calcium-dependent guanylate cyclase activator protein (CD-GCAP): Identity with S100β protein. Biochemistry 36: 14159–14166, 1997
83. Pozdnyakov N, Margulis A, Sitaramayya A: Identification of effector binding sites of S100β: Studies with guanylate cyclase and p80, a retinal phosphoprotein. Biochemistry 37: 10701–10708, 1998
84. Rambotti MG, Giambanco I, Spreca A, Donato R: S100B and S100A1 proteins in bovine retina: Their calcium-dependent stimulation of a membrane-bound guanylate cyclase activity as investigated by ultracytochemistry. Neuroscience 92: 1089–1101, 1999
85. Spreca A, Simonetti S, Rambotti MG: Atrial natriuretic peptide and guanylin-activated guanylate cyclase isoforms in human sweat glands. Histochem J 32: 725–731, 2000
86. Spreca A, Giambanco I, Rambotti MG: Guanylate cyclase-A in sebaceous glands. It J Anat Embriol 103: 168, 1998
87. O'Donnell, ME: Regulation of Na-K-Cl cotransport in endothelial cells by atrial natriuretic factor. Am J Physiol 257: C36–C44, 1989
88. Crawford I, Malone PC, Zeitlin PL, Guggino WB, Hyde SC, Turley H, Gatter KC, Harris A, Higgins CF: Immunocytochemical localization of the cystic fibrosis gene product CFTR. Proc Natl Acad Sci USA 1991 88: 9262–9266, 1991
89. Widdicombe JH: Yet another role for the cystic fibrosis transmembrane conductance regulator. Am J Resp Cell Mol Biol 22: 11–14, 2000

Molecular and Cellular Biochemistry **230**: 97–106, 2002.

Photoreceptor specific guanylate cyclases in vertebrate phototransduction

Karl-Wilhelm Koch,[1] Teresa Duda[2] and Rameshwar K. Sharma[2]

[1]*Institut für Biologische Informationsverarbeitung 1, Forschungszentrum Jülich, Jülich, Germany;* [2]*The Unit of Regulatory and Molecular Biology, Department of Cell Biology and Ophthalmology, SOM and NJMS, University of Medicine and Dentistry of New Jersey, Stratford, NJ, USA*

Abstract

Two membrane bound guanylate cyclases are expressed in vertebrate photoreceptor cells. They serve a key function in photoreceptor physiology as they synthesize the intracellular transmitter of photoexcitation guanosine 3′,5′-cyclic monophosphate (cGMP). Both cyclases named ROS-GC1 and ROS-GC2 form a subclass of membrane bound cyclases and differ in many aspects from hormone peptide receptor guanylate cyclases. One unique feature is their regulation by three small Ca^{2+}-binding proteins called GCAPs. These regulatory proteins sense changes in the cytoplasmic Ca^{2+}-concentration [Ca^{2+}] during illumination and activate ROS-GCs when the [Ca^{2+}] decreases below the value in a dark adapted cell of 500–600 nM. Recent work has identified the target regions of GCAP-1 in ROS-GC1. In addition to GCAPs several other proteins including aktin, tubulin, a glutamic-acid-rich protein and a GTPase accelerating protein (RGS9) were found to interact with ROS-GC1 and probably form a multiprotein complex. (Mol Cell Biochem **230**: 97–106, 2002)

Key words: phototransduction, membrane bound guanylate cyclases, guanylate cyclase-activating proteins, Ca^{2+}-signaling

Phototransduction

Phototransduction is a biochemical process by which the rods and cones generate electrical signals in response to captured photons (reviewed in [1–5]). This process occurs in the rod (or cone) outer segments (ROS). The photon causes hyperpolarization in the plasma membrane of the photoreceptor cell through an enzymatic cascade, which involves interaction between the photopigment rhodopsin, a heterotrimeric G-protein called transducin and a cGMP specific phosphodiesterase (PDE). In the dark-adapted state the cytoplasmic free cGMP concentration in a photoreceptor cell is approximately 5 µM, which is enough to keep a fraction of cyclic nucleotide-gated (CNG) channels in the plasma membrane open. Light-induced hydrolysis of cGMP leads to the closure of CNG-channels and subsequently to a change in membrane potential. The CNG-channel in rod and cone cells is also permeable to Ca^{2+}-ions. About 10% of the current flowing into the cell in the dark state is carried by Ca^{2+}-ions. A continuous operation of a $Na^+/Ca^{2+},K^+$-exchanger removes the intracellular Ca^{2+}, a mechanism that keeps the intracellular free [Ca^{2+}] at a resting level of 500–600 nM. Closure of the CNG-channels leads to a decrease of [Ca^{2+}] to a level below 100 nM, because the extrusion of Ca^{2+} by the exchanger is maintained after a light pulse. The change in intracellular [Ca^{2+}] is a trigger for several Ca^{2+}-binding proteins to modulate the function of their target proteins [6–10]. One key event in this Ca^{2+}-mediated feedback system is the control of a membrane bound guanylate cyclase (ROS-GC) by guanylate cyclase-activating proteins (GCAPs). ROS-GC activity increases when intracellular [Ca^{2+}] drops. This mechanism replenishes the exhausted cGMP pool and works in concert with several inactivation steps to restore the dark state of the cell. In addition, Regulation of ROS-GC activity is among several different molecular mechanisms that regulate the light sensitivity of a photoreceptor cell (light adaptation) [6–10].

A simplified scheme representing the main steps of phototransduction is shown in Fig. 1. A detailed description of the underlying biochemical steps in photoreceptor excitation and

Address for offprints: K.-W. Koch, Institut für Biologische Informationsverarbeitung 1, Forschungszentrum Jülich, Postfach 1913, D-52425 Jülich, Germany (E-mail: k.w.koch@fz-juelich.de)

adaptation has been covered in many reviews [1–10] and is not described here. This review will be focused on the physiological role and the molecular properties of the membrane-bound family of photoreceptor specific guanylate cyclases. These cyclases are called 'ROS-GCs' because of their expression in ROS, the site of phototransduction. The details of the important historical developments in the field of phototransduction from the year 1970 to 1985 covering the discovery of cGMP and Ca^{2+} as the internal messengers of phototransduction, and between the years 1993 and 1994 defining the precise molecular identity of the first member of ROS-GC – directly in ROS – have been reviewed earlier [1–5].

Molecular properties of photoreceptor specific membrane GCs (mGCs)

Cloning of retinal GC genes

Electrophysiological recordings on intact and truncated rod cells revealed the importance of GC activity in photoreceptor physiology [11–13] and for summary see ref. [4]. These studies showed an intriguing feature of this cyclase, which was that its activity is Ca^{2+}-inhibited, and the cyclase is membrane bound. Four subtypes of membrane guanylate cyclases have been cloned form the retina. Two are peptide receptors and two are not. The peptide receptors, ANF-RGC [14] and CNP-RGC [15], are not regulated by the intracellular Ca^{2+} spikes, hence, these retinal guanylate cyclases have no role in phototransduction. The non-peptide receptors, ROS-GC1 and ROS-GC2 [16, 17], reside in photoreceptors and they are Ca^{2+}-inhibited with a pattern that duplicates the native cyclase activity under conditions of phototransduction. Thus, they are regulated by the intracellular Ca^{2+} spikes in ROS to meet the criterion of being linked with phototransduction. ROS-GC1 and ROS-GC2 refer to the bovine forms, their counter parts in human and rat have also been cloned and they have been referred to respectively as retGC1 and retGC2, and GC-E and GC-F [18–20]. To keep the nomenclature simple, irrespective of their species-specificity, in this review these Ca^{2+}-modulated membrane guanylate cyclases are referred as ROS-GCs. ROS-GCs constitute a subfamily of membrane guanylate cyclases that is Ca^{2+}-modulated.

The nucleotide and the amino acid sequence of the cDNA corresponds to an open reading frame of 1110 and 1103 amino acids for ROS-GC1 and ROS-GC2, respectively [16, 17]. Both sequences contain an N-terminal leader sequence of 56 and 50 amino acids, which is not present in the mature protein. The processed proteins have calculated molecular masses of 114 and 118 kDa. The predicted domain structure (Fig. 2) consists of an extracellular domain (ECD) or intradiskal domain (IDD) in GC molecules that reside in the disk membrane, a transmembrane segment (TM), a juxtamembrane part (JMD), a kinase homology domain (KHD), a dimerization domain (DD), a cytoplasmic catalytic domain (CCD) and a C-terminal extended domain (CTE). The JMD was originally assigned as part of the KHD [16], but sequence comparison with other membrane bound GCs show that this 115 amino acid long domain is highly conserved between ROS-GC1 and ROS-GC2, but does not show significant homology (~ 7%) with peptide receptor GCs [21]. There is one striking structural feature which sets the bovine ROS-GC2 apart from its human and rat counter parts and from ROS-GC1: It contains a signature sequence segment of five amino acids – QKSSW – at its C-terminal end which is unique to itself [17]. Together with GC-D [22–24], a cyclase that is expressed in olfactory neuroepithelium these two ROS-GCs form a subclass of membrane bound cyclases according to the following criteria: (a) ROS-GC1 and ROS-GC2 are structurally closer to GC-D, than to the peptide hormone receptors: ANF-RGC and CNP-RGC; (b) ROS-GC1, ROS-GC2 and GC-D are not activated by the peptide hormones; (c) Both ROS-GCs and GC-D reside in the sensory neurons; (d) ROS-GC1 and ROS-GC2 are regulated by the intracellularly-generated Ca^{2+} spikes. This Ca^{2+}-sensitive feature has not yet been reported for GC-D.

Purification of ROS-GC from retinal sources

Cloning of photoreceptor GCs was preceded by the purification of a membrane bound GC from bovine and amphibian rod outer segments [25–27]. This enzyme showed two important biochemical features: (1) It was unresponsive to the peptide hormone ANF [27]; (2) it did not respond directly to the free Ca^{2+} concentrations [25]. These observations indicated that the cyclase was not an ANF-RGC, and probably did not belong to the hormone peptide receptor family of the membrane guanylate cyclases, at that time the only recognized family of membrane bound guanylate cyclases. Based on an earlier clue that the photoreceptor guanylate cyclase responded to Ca^{2+} signals through a Ca^{2+}-binding protein [28], exhibition of the second feature was consistent with the idea that the purified form of the cyclase required the presence of a missing Ca^{2+}-binding protein for its activity. These two became the hallmark features for the future cloning of the native ROS-GC.

Estimated molecular masses of the purified proteins (110–115 kDa) are consistent with the predicted masses from the deduced amino acid sequence of the cloned ROS-GCs. Fragments of GC purified from bovine ROS were obtained after limited proteolysis and were used for direct amino acid sequencing [16, 27–29]. All sequences match the sequence of the cloned form of bovine ROS-GC1 [16, 27–29]. These findings showed that the native purified enzyme from ROS and the cloned enzyme from retina were identical. An important aspect of the sequencing studies was that it experimentally validated the position of the N-terminus amino acid of the mature protein. The studies also showed that native ROS-GC1 is glycosylated at its N-terminus [29]. To date, of all the membrane GCs cloned, ROS-GC1 remains the only one cloned on the basis of its protein sequence. Its structure serves as the standard for the identity of other cloned members of the guanylate cyclase family. It aided in the correction of the structure of a previously cloned human form of retinal guanylate cyclase [18], and settled, at the time, an important mistaken belief that the human form was a distinct subtype to that of the sequenced native bovine form [16]. The mistaken belief was based on the premise that a fragment of the native ROS-GC possessed a sequence that was not present in the human retinal form.

In contrast to ROS-GC1, the sequence of ROS-GC2 derived from its cDNA indicates that it is not N-glycosylated [17, 19, 20]. So far, ROS-GC2 has not been purified from retinal sources. Since the original purification of ROS-GC1 from ROS was achieved by testing enzyme activities and subsequent enrichment of cyclase activity [25, 26], this protein either represents the more prominent or the more active enzyme in ROS.

Cellular localization

Transcripts of ROS-GC1 and ROS-GC2 were detected by in *situ hybridization* in retinal sections of Rhesus monkeys. Specific labeling was restricted to the outer nuclear layer and inner segments, which indicates a photoreceptor specific expression of both GCs [18, 19]. The photoreceptor specific expression of two membrane bound ROS-GCs raises several questions: Does a rod/cone specific expression pattern exist, or are both ROS-GCs expressed in the same cell type and do they contribute to the maintenance of cytoplasmic [cGMP]? If so, are ROS-GCs differently regulated for example, by GCAPs? Is there a different expression during development?

Several localization studies with anti-peptide antibodies specific for ROS-GC1 were performed using light microscope immunocytochemistry [30–33]. By this method the presence of ROS-GC1 was detected in the outer segments of several species including human, monkey, rat, cat and chicken. Additional labeling was observed in the plexiform layers [33]. Staining of cones was more intense in human, monkey, cat and chicken retina and could reflect a higher concentration of ROS-GC1 in cones than in rods. Alternatively, epitope masking of ROS-GC1 in rods could account for this difference. Subcellular distribution of both GCs was investigated by electron microscopy and immunogold labeling of ROS [31, 34]. Both GCs appear to be localized in the same rod outer segment and are distributed over the marginal region of disk and plasma membranes [34]. However, these studies could not give a conclusive answer as to, whether the expression of one GC form is restricted to cones (e.g. ROS-GC1) and whether the other form is more abundant in rods (e.g. ROS-GC2).

Disruption of the ROS-GC1 gene in mice has led to the complete disappearance of cones within 5 weeks after birth, whereas rods appeared normal after several months [35]. Electroretinogram (ERG) recordings show an almost complete loss of cone responses and a reduced amplitude of the rod response in null mice. Single cell recordings on rods show a normal dark current, but significant changes in the flash responses [35]. For example, there is larger amplitude for a given photoexcitation and accelerated recovery kinetics. While these results are consistent with the view that ROS-GC1 is the predominant (if not the only) GC in cones, it remains unclear, what expression level of ROS-GC1 is normal in rods. Furthermore, it is not clear whether compensatory effects by the activity of ROS-GC2 could account for the normal dark current in null mice. On the other hand, the interpretation that ROS-GC1 is not expressed in rods is inconsistent with other findings: (a) Only ROS-GC1 has been biochemically purified from rod rich (bovine) retinae (s. above); (b) human retinal diseases that affect rod and cone vision (Leber´s congenital amaurosis and CORD6) are only linked to the ROS-GC1 gene (for biochemical consequences of these disease forms see article by Duda and Koch, this issue); (c) all reports on immunostaining of retinal sections show labeling of rods by ROS-GC1 specific antibodies.

We therefore conclude that ROS-GC1 is the prominent GC in rods and cones. Expression of ROS-GC2 is probably restricted to rods. The exact quantitative amount has to be examined. Loss of ROS-GC1 function in ROS-GC1 null mice can be partially compensated by ROS-GC2 activity.

Regulation of photoreceptor GC activity

Modulation by GCAPs and free [Ca^{2+}]

The unique feature of photoreceptor specific GCs is their negative regulation by free Ca^{2+} through a subfamily of neuronal Ca^{2+}-binding proteins named guanylate cyclase-activating proteins (GCAPs). Koch and Stryer published the first evidence for the existence of GCAPs in 1988 [28]. These authors observed that a soluble extract of ROS proteins could activate a membrane bound guanylate cyclase in a Ca^{2+}-dependent fashion, the activity increased when the free [Ca^{2+}] decreased. The effect was highly cooperative (Hill coefficient n = 3.9) and half-maximal at 90 nM free [Ca^{2+}]. Thus, the regulation of membrane bound guanylate cyclase occurred in the physiologically relevant range of free [Ca^{2+}] in a photoreceptor cell. A few years later, two Ca^{2+}-binding proteins, named GCAP-1 and GCAP-2 have been purified from retinal sources and cloned from retinal cDNA libraries of several vertebrate species [36–41]. In addition, a third form (GCAP-3) has recently been identified in mammalian retina [42] and one additional form called guanylate cyclase-inhibitory protein (GCIP) in an amphibian retina [43]. GCAPs contain three functional Ca^{2+}-binding EF-hand motifs. A fourth EF-hand is probably nonfunctional. Native as well as heterologously expressed recombinant forms of GCAP-1 and GCAP-2 activate guanylate cyclase in washed ROS membranes in a Ca^{2+}-dependent way [36, 39–41]. When heterologously expressed GCs were assayed in the presence of GCAPs, all three GCAPs at low free [Ca^{2+}] activate ROS-GC1 [42, 44]. GCAP-2 and GCAP-3 only activate ROS-GC2 [42, 44]. At high free [Ca^{2+}] GCAP-2 and GCAP-3 inhibit ROS-GCs [42, 45, 46].

The activation of GCs at saturating GCAP-1 and GCAP-2 is half-maximal between 50 and 290 nM free [Ca^{2+}] (see ref. in review [4]). These values of free [Ca^{2+}] span the physiological range of free [Ca^{2+}] after illumination. Gray-Keller and Detwiler [47] measured the free [Ca^{2+}] in a vertebrate rod cell at different background light intensities by dialysis of the fluorescent Ca^{2+}-indicator indo-dextran into a detached lizard ROS. Cytoplasmic [Ca^{2+}] reaches a steady state level of 252 nM when rods were illuminated with subsaturating light for 60 sec and free cytoplasmic [Ca^{2+}] decreases to 50 nM after a saturating light pulse. At present it is not clear whether the variability of measured EC_{50} values of GC activation simply reflect slight differences of experimental protocols of different laboratories or whether they reflect an additional physiological control step. Half-maximal activation of ROS-GCs could be shifted to higher or lower free [Ca^{2+}] according to different background light intensities. Such an additional regulatory mechanism of ROS-GC activity would increase the dynamic response capability of GCs and could significantly attribute to the light sensitivity of the cell.

Target regions of GCAPs in ROS-GC1

Recent work to identify the binding and/or regulatory regions of GCAPs has been mainly focused on the modulation of ROS-GC1 by GCAP-1. Initial work has excluded any participation of the ECD of ROS-GC1 in the regulation by GCAPs [48, 49]. Instead, the KHD has been identified as a necessary regulatory domain for modulation by GCAP-1 [48]. Very recently, two groups – Lange *et al.* [21] and Sokal *et al.* [50] – have narrowed down the possible target regions of GCAP-1 to stretches of 15–20 amino acids in ROS-GC1. Although both groups used similar approaches, screening of an overlapping peptide library and testing of specifically constructed ROS-GC1 mutants, the identified regions were located in different parts of ROS-GC1. Lange *et al.* [21] identified two regions in the juxtamembrane domain (^{445}M-^{456}L; ^{503}L-^{522}I), former region representing the transducer and latter the binding domain (Fig. 2). In this study the screening of the peptide library was restricted. It included the ROS-GC1 domain between amino acids ^{437}R-^{552}G. The library was screened with peptides overlapping by 10 out of 12 amino acids. This region was chosen because in a prior study, deletion- and hybrid-analysis of ROS-GC1 had revealed that this domain was pivotal in the GCAP modulation of the cyclase [48]. GCAP1 regulatory regions identified in the study by Lange et al. are in line in explaining a rod dystrophy, LCA1, where the patients are born blind or soon become blind after birth [51, 52]. In these patients there is a point mutation in ROS-GC1 gene between the region ^{503}L-^{522}I. Biochemical studies on heterologous ROS-GC1 mutants show that this LCA mutant has completely lost its sensitivity to modulation by GCAP-1 [53].

Sokal *et al.* [50] identified a region in the catalytic domain, ^{968}G-^{985}G, which corresponds to ^{912}G-^{929}G in the mature protein (Fig. 2). In this study the authors screened the whole cytoplasmic domain with peptides 15 amino acids in length with 8 amino acids overlapping. The region, however, shows a 2–3 fold higher affinity for GCAP2 than for GCAP1. Thus, the identified region in this study may represent the GCAP2 site in ROS-GC1. This interpretation is consistent with the results of earlier studies, which, through deletion and hybrid analyses, provided abundant evidence that the GCAP2-modulated domain in ROS-GC1 resides at its C-terminus [44, 54]. It is then possible that the primary interaction site of GCAP1 in ROS-GC1 is between amino acids ^{503}L-^{522}I and the secondary site is between ^{912}G-^{929}G (mature protein).

Modulation by ATP

ATP is obligatory for the transduction machinery of both natriuretic factor receptor guanylate cyclases (NF-RGCs), ANF-RGC and CNP-RGC (reviewed in [55]). Operational selectivity of the machinery resides in the receptor component of the cyclase, which exists in the extracellular domain of the cyclase. This is where the hormone binds and generates the natriuretic factor signal. When the hormone binds at the extracellular domain, subsequent transduction steps occur in the intracellular region of the cyclase. ATP regulates these steps by directly binding to a domain, which is juxtaposed to the transmembrane region of the cyclase. A glycine-rich seven amino acid sequence motif – GXGXXXG – plays a critical role in these steps. This sequence motif resides within a 40 amino acid domain after the transmembrane domain of the cyclase. In a recent study a three-dimensional atomic model of this ATP-regulated domain (ARM) has been constructed and experimentally validated [56]. The model shows that the glycine-rich sequence motif is critical for the geometry of the ATP-binding pocket. The motif residues, however, do not directly interact with ATP. The striking feature of the model is that the middle glycine of the motif, GXGSXXG, controls the orientation of the adjacent serine residue. Through its phosphorylation, this residue has been suggested to play an important role in ANF signaling [57]. The model shows that ATP drives ANF signaling first by its direct binding to ARM and then activating the catalytic module where one of the steps appears to be phosphorylation of the cyclase.

ATP is not obligatory for the GCAP1-mediated signaling of ROS-GC1. At its physiological concentration of 0.2 mM or higher it inhibits ROS-GC1 [58, 59], yet at lower concentrations (0.1–0.2 mM) it causes a 2-fold increase of GC activity [41, 60, 61]. There is about 20% homology between the ARM domain of ANF-RGC and the corresponding ROS-GC1 domain. Importantly, the GXGXXXG motif and the residues T513, K535, E551 critical in the geometry of the ATP binding pocket in the ARM domain of ANF-RGC [56] are missing in the corresponding ROS-GC1 domain. Thus, the ATP regulatory mechanisms involved in NF-RGC and ROS-GC are different. ATP stimulation of ROS-GC is enhanced at low free [Ca^{2+}] and saturating levels of GCAPs. In addition to this stimulating effect, ATP stabilizes ROS-GCs and has a protective effect that is not mediated by phosphorylation [62]. A kinase activity is presumably located in the KHD of ROS-GC1 [63]. ROS-GC1 undergoes substoichiometrical autophosphorylation and can also phosphorylate exogenous kinase substrates as myelin basic protein [63], but endogeneous substrates in ROS are unknown so far. In addition, protein kinase A and protein kinase C activities can modulate GC activities, but the underlying concrete reaction pathway has not been deciphered [64, 65].

Catalytic mechanism of cGMP synthesis

Membrane bound GCs have a single cyclase catalytic domain (CCD) that shares about 30% sequence homology to the corresponding domains in soluble guanylate cyclases and adenylate cyclases [16–20]. Membrane GCs function as homodimers, they convert GTP into cGMP and posses two active sites per homodimer [66]. Magnesium or manganese are essential cofactors and form with GTP a metal-β-γ-bidentate complex (the divalent metal ion is coordinated to the β- and γ-phosphate of GTP). The cyclization reaction proceeds by a single nucleophilic displacement via an inversion of the configuration at the α-phosphor atom [67]. In this base catalyzed mechanism the 3′-OH group of the ribose attacks the α-phosphor atom. In the transition state a pentacoordinated phosphor atom with bipyramidal geometry is formed [67] and probably stabilized by an Arg residue at position 976 (unprocessed human ROS-GC1) [68]. The nucleotide binding pocket has been revealed by site-directed mutagenesis based on homology modeling of adenylate and guanylate cyclases [69]. Nucleotide specificity can be converted to that of an adenylate cyclase by only two amino acid substitutions at E925K and C995D (unprocessed human ROS-GC1).

The formation of a dimer-dimer interface seems important for an efficient catalysis [66, 68–70]. The hypothetical dimerization domain that is located between the KHD and CCD is critical for the correct dimerization. Mutations in the DD in the ROS-GC1 gene are linked to a cone-rod dystrophy (CORD6) [71, 72] and cause a severe distortion of dimer formation and GC activity [73–75].

Photoreceptor GC is part of a multi-protein complex

Early fractionation studies of photoreceptor GC have indicated that the enzyme is associated with Triton X-100 insoluble cytoskeletal proteins [25, 76, 77]. In more recent studies aktin and tubulin have been identified as molecular components that interact with ROS-GC1 (Fig. 3) [32, 78], in the case of tubulin a complex of GCAP-1, tubulin and ROS-GC1 was co-immunoprecipitated [78]. In addition, ROS-GC1 was identified as a major component in a multi-protein complex that is assembled in rod, not cone photoreceptors by glutamic-acid-rich proteins (GARPs) [79]. GARPs exist in two soluble forms and as a large cytoplasmic domain of the β-subunit of the CNG-channel in rods [80]. They serve as multivalent adapter proteins to assemble key components of the cGMP signaling cascade (ROS-GC, PDE, rod CNG-channel) at the rim region of the disks [79]. They also bind a retina-specific ATP-binding cassette transporter (formerly named rim protein) [81–83]. It has been hypothesized that this protein complex is dynamically organized and may control cGMP turnover [79]. The GTPase accelerating protein RGS9 has recently been proposed to act as a mediator between cGMP hydrolysis and synthesis [84]. GTPase accelerating proteins constitute a family of so called RGS proteins (regulators of G protein signaling). RGS9 accelerates the GTPase activity of the photoreceptor G protein transducin [85]. Seno *et al.* [84] reported that RGS9 interacts directly with ROS-GC (Fig. 3) and inhibits cyclase activity (for a more detailed description see review by Yamazaki in this issue).

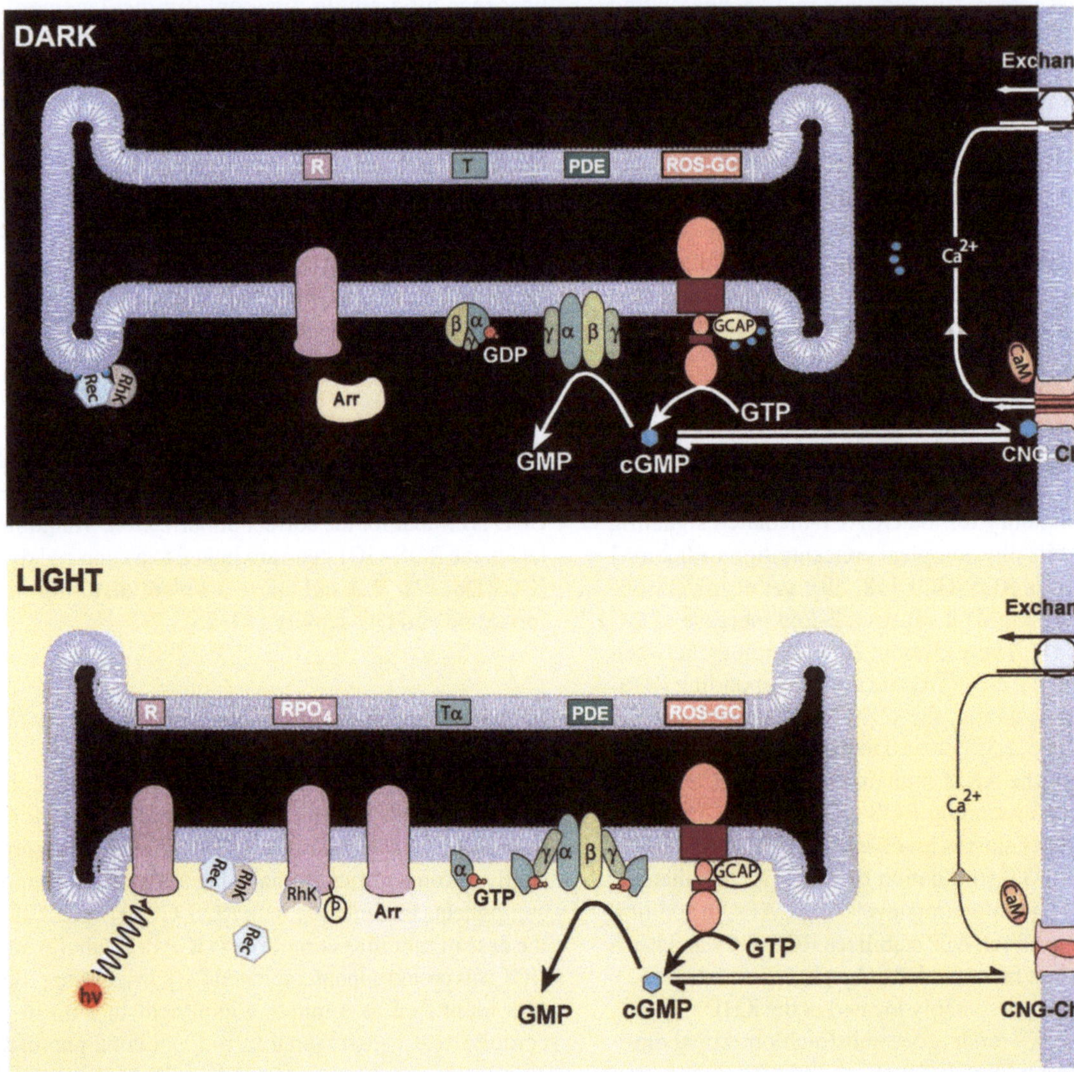

Figure 1

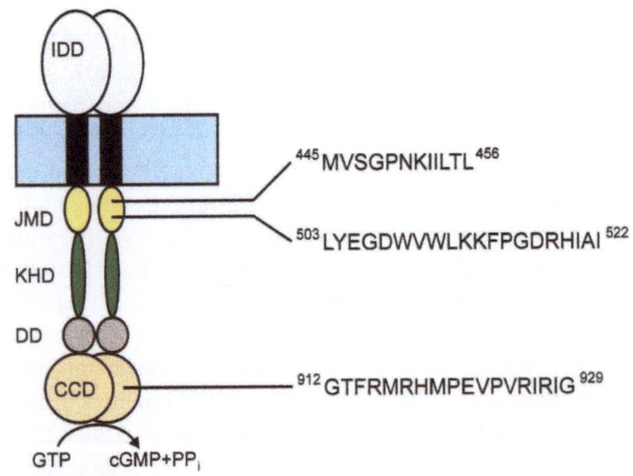

Figure 2

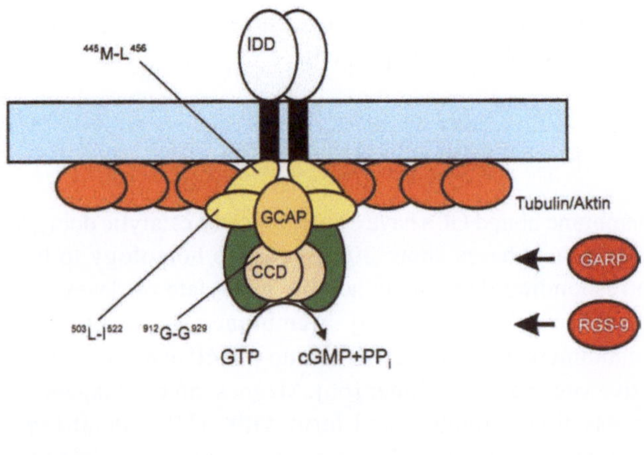

Figure 3

Concluding remarks and future directions

In the past decade molecular recognition of the ROS-GC component in the phototransduction machinery has been a milestone discovery. The discovery made it possible to study and define the process of phototransduction in precise molecular terms. It revealed an intriguing mechanism of Ca^{2+} signaling in ROS where Ca^{2+} increments inhibit the cyclase and cause hyper-polarization. ROS-GC exists in two forms, ROS-GC1 and ROS-GC2; and it senses Ca^{2+} via two GCAPs: GCAP-1 and GCAP-2. At present, we have no knowledge of the relative physiological relevance of these two ROS-GCs and two GCAPs in vertebrate phototransduction. One important future goal is to address this issue. Which of these signal transduction components are involved and how they regulate phototransduction in rods?

To date most of the phototransduction studies have been centered on rods; although physiological information is available on cones [86] we lack data on biochemical properties of cones. Our view on vertebrate phototransduction will be incomplete until we have this information. To obtain this information, we need a pure cone preparation model. The second goal will be to establish such a model.

ROS-GCs are part of a larger protein complex (Fig. 3) that involves a set of different proteins (GCAPs, S100 proteins, cytoskeletal proteins, RGS9 and GARP). It is not known what parameters control the composition of this complex and how they influence the dynamics of its assembly and disassembly. The third goal will be to relate these protein-protein interactions to photoreceptor physiology.

An unresolved issue is the biological function of the intradiskal domain (IDD) that corresponds to the extracellular domain (ECD) of hormone peptide receptor GCs. Although it is well established that ROS-GCs are not regulated by hormone peptides, the IDD might bind to an unidentified small molecule in the lumen of the disks and provide an additional regulatory control on Ca^{2+} signaling of ROS-GC. This will form the fourth goal of the studies.

In the fifth goal, two critical challenges lie in the determination of high-resolution three-dimensional structures of ROS-GC. In particular, a co-crystal of ROS-GC with its main regulators (GCAP-1 and GCAP-2) will provide the necessary information to understand the mechanistic details of the Ca^{2+} switch steps that are crucial for photoreceptor physiology.

In a final sixth goal, several recent findings indicate a physiological role of ROS-GC1 beyond phototransduction: for example in pinealocytes of the pineal gland [87] and mitral cells of the olfactory bulb [88]. There may be other systems yet to be found where ROS-GCs or their homologues are part of signaling pathways. The goal will be to identify those signaling systems.

←

Fig. 1. Scheme of phototransduction in rod outer segments. In the dark (upper part) a high concentration of cytoplasmic cGMP keeps a fraction of CNG-channels in the plasma membrane open. Ca^{2+}-ions enter the cell via the CNG-channel and are extruded via the $Na^+/K^+,Ca^{2+}$-exchanger. Synthesis and hydrolysis of cGMP by ROS-GC and PDE occur at a low rate. The heterotrimeric G protein transducin (T) is in its GDP-bound state and non-active. The Ca^{2+}-binding proteins calmodulin (CaM) recoverin (Rec) and GCAP bind to their target proteins, the CNG-channel, rhodopsin kinase (RhK) and ROS-GC, respectively. Absorption of light (lower part) by the visual pigment rhodopsin leads to the activation of the transduction cascade: the GTP-bound α-subunit of transducin activates PDE that rapidly hydrolyzes cGMP. Subsequently the CNG-channels close and the Ca^{2+}-concentration decreases. The change in cytoplasmic $[Ca^{2+}]$ is sensed by Ca^{2+}-binding proteins: CaM dissociates from the CNG-channel which leads to an increase in cGMP sensitivity of the channel, inhibition of rhodopsin kinase by recoverin is terminated and rhodopsin can be phosphorylated. GCAPs activate ROS-GC and synthesis of cGMP increases. Arrestin (Arr) binds to phosphorylated rhodopsin and interferes with the binding and further activation of transducin. Enhancement of cGMP synthesis and termination of the cascade leads to reopening of CNG-channels.

Fig. 2. Schematic representation of ROS-GC1 as a dimer. Different parts of the molecule are the intradiskal domain (IDD in grey), the transmembrane domain in black (blue is a phospholipid membrane), the juxtamembrane domain (JMD in yellow), the kinase homolgy domain (KHD in green), the dimerization domain (DD in grey) and the cyclase catalytic domain (CCD in ochre). Target regions of GCAP are indicated by the corresponding amino acid sequences of the mature protein: the transducer unit ^{445}M-^{456}L and the binding region ^{503}L-^{522}I in the JMD (Lange *et al.* [21]) and a second binding region in the CCD (^{912}G-^{929}G) identified by Sokal *et al.* [50].

Fig. 3. Protein complex at the surface of the disk membrane (blue) that involves tubulin/aktin (orange), GCAP-1 (yellow), GARP, RGS9 (red) and ROS-GC (domains are similar as in Fig. 2). GCAP-1 interacts with at least two binding sites in ROS-GC1 located in the JMD and the CCD.

Acknowledgements

The studies in the investigators laboratories covered in this review are supported by awards from the Deutsche Forschungsgemeinschaft (K.-W.K.), and USPHS award EY 10828 (RKS); by the affiliated support of the Research to Prevent Blindness Inc., New York, USA and the Lions Eye Research Foundation, New Jersey, USA.

References

1. Kaupp UB, Koch K-W: Role of cGMP and Ca^{2+} in vertebrate photoreceptor excitation and adaptation. Annu Rev Physiol. 54: 153–175, 1992
2. Pugh EN Jr, Lamb TD: Amplification and kinetics of the activation steps in phototransduction. Biochim Biophys Acta 1141: 111–149, 1993
3. Yau K-W: Phototransduction mechanism in retinal rods and cones. Invest Ophthalmol Vis Sci 35: 9–32, 1994

4. Pugh EN Jr, Duda T, Sitaramayya A, Sharma RK: Photoreceptor guanylate cyclases: A review. Biosci Rep 17: 429–473, 1997
5. Molday RS: Photoreceptor membrane proteins, phototransduction, and retinal degenerative diseases. Invest Ophthalmol Vis Sci 39: 2493–2513, 1998
6. Koch K-W: Control of photoreceptor proteins by Ca^{2+}. Cell Calcium 18: 314–321, 1995
7. Koutalos Y, Yau K-W: Regulation of sensitivity in vertebrate rod photoreceptors by calcium. Trends Neurosci 19: 73–81, 1996
8. Pugh EN Jr, Nikonov S, Lamb TD: Molecular mechanisms of vertebrate photoreceptor light adaptation. Curr Opin Neurobiol 9: 410–418, 1999
9. Palczewski K, Polans AS, Baehr W, Ames JB: Ca^{2+}-binding proteins in the retina: Structure, function, and the etiology of human visual diseases. BioEssays 22: 337–350, 2000
10. Fain GL, Matthews HR, Cornwall MC, Koutalos Y: Adaptation in vertebrate photoreceptors. Physiol Rev 81: 117–151, 2001
11. Hodgkin AL, Nunn BJ: Control of light-sensitive current in salamander rods. J Physiol 403: 439–471, 1988
12. Kawamura S, Murakami M: Regulation of cGMP levels by guanylate cyclase in truncated frog rod outer segments. J Gen Physiol 94: 649–668, 1989
13. Koutalos Y, Nakatani K, Tamura T, Yau K-W: Characterization of guanylate cyclase activity in single retinal rod outer segments. J Gen Physiol 106: 863–890, 1995
14. Kutty RK, Fletcher RT, Chader GJ, Krishna G: Expression of guanylate cyclase-A mRNA in the rat retina: Detection using polymerase chain reaction. Biochem Biophys Res Commun 182: 851–857, 1992
15. Duda T, Goraczniak RM, Sitaramayya A, Sharma RK: Cloning and expression of an ATP-regulated human retina C-type natriuretic factor receptor guanylate cyclase. Biochemistry 32: 1391–1395, 1993
16. Goraczniak RM, Duda T, Sitaramayya A, Sharma RK: Structural and functional characterization of the rod outer segment membrane guanylate cyclase. Biochem J 302: 455–461, 1994
17. Goraczniak R, Duda T, Sharma RK: Structural and functional characterization of a second subfamily member of the calcium-modulated bovine rod outer segment membrane guanylate cyclase, ROS-GC2. Biochem Biophys Res Commun 234: 666–670, 1997
18. Shyjan AW, de Sauvage FJ, Gillett NA, Goeddel DV, Lowe DG: Molecular cloning of a retina-specific membrane guanylyl cyclase. Neuron 9: 727–737, 1992
19. Lowe DG, Dizhoor AM, Liu K, Gu Q, Spencer M, Laura R, Lu L, Hurley JB: Cloning and expression of a second photoreceptor-specific membrane retina guanylyl cyclase (RetGC), RetGC-2. Proc Natl Acad Sci USA 92: 5535–5539, 1995
20. Yang R-B, Foster DC, Garbers DL, Fülle H-J: Two membrane forms of guanylyl cyclase found in the eye. Proc Natl Acad Sci USA 92: 602–606, 1995
21. Lange C, Duda T, Beyermann M, Sharma RK, Koch K-W: Regions in vertebrate photoreceptor guanylyl cyclase ROS-GC1 involved in Ca^{2+}-dependent regulation by guanylyl cyclase-activating protein GCAP-1. FEBS Lett 460: 27–31, 1999
22. Fülle H-J, Vassar R, Foster DC, Yang R-B, Axel R, Garbers DL: A receptor guanylyl cyclase expressed specifically in olfactory sensory neurons. Proc Natl Acad Sci USA 92: 3571–3575, 1995
23. Juilfs DM, Fülle H-J, Zhao AZ, Houslay MD, Garbers DL, Beavo JA: A subset of olfactory neurons that selectively express cGMP-stimulated phosphodiesterase (PDE2) and guanylyl cyclase-D define a unique olfactory signal transduction pathway. Proc Natl Acad Sci USA 94: 3388–3395, 1997
24. Meyer MR, Angele A, Kremmer E, Kaupp UB, Müller F: A cGMP-signaling pathway in a subset of olfactory sensory neurons. Proc Natl Acad Sci USA 97: 10595–10600, 2000
25. Koch K-W: Purification and identification of photoreceptor guanylate cyclase. J Biol Chem 266: 8634–8637, 1991
26. Hayashi F, Yamazaki A: Polymorphism in purified guanylate cyclase from vertebrate rod photoreceptors. Proc Natl Acad Sci USA 88: 4746–4750, 1991
27. Margulis A, Goraczniak RM, Duda T, Sharma RK, Sitaramayya A: Structural and biochemical identity of retinal rod outer segment membrane guanylate cyclase. Biochem Biophys Res Commun 194: 855–861, 1993
28. Koch K-W, Stryer L: Highly cooperative feedback control of retinal rod guanylate cyclase by calcium ions. Nature 334: 64–66, 1988
29. Koch K-W, Stecher P, Kellner R: Bovine retinal rod guanyl cyclase represents a new N-glycosylated subtype of membrane-bound guanyl cyclases. Eur J Biochem 222: 589–595, 1994
30. Dizhoor AM, Lowe DG, Olshevskaya EV, Laura RP, Hurley JB: The human photoreceptor membrane guanylyl cyclase, RetGC, is present in outer segments and is regulated by calcium and a soluble activator. Neuron 12: 1345–1352, 1994
31. Liu X, Seno K, Nishizawa Y, Hayashi F, Yamazaki A, Matsumoto H, Wakabayashi T, Usukura J: Ultrastructural localization of retinal guanylate cyclase in human and monkey retinas. Exp Eye Res 59: 761–768, 1994
32. Hallett MA, Delaat JL, Arikawa K, Schlamp CL, Kong F, Williams DS: Distribution of guanylate cyclase within photoreceptor outer segments. J Cell Sci 109: 1803–1812, 1996
33. Cooper N, Liu L, Yoshida A, Pozdnyakov N, Margulis A, Sitaramayya A: The bovine rod outer segment guanylate cyclase, ROS-GC, is present in both outer segment and synaptic layers of the retina. J Mol Neurosci 6: 211–222, 1995
34. Yang R-B, Garbers DL: Two eye guanylyl cyclases are expressed in the same photoreceptor cells and form homomers in preference to heteromers. J Biol Chem 272: 13738–13742, 1997
35. Yang H, Wanner IB, Roper SD, Chaudhari N: An optimized method for *in situ* hybridization with signal amplification that allows the detection of rare mRNAs. J Histochem Cytochem 47: 431–445, 1999
36. Gorczyca WA, Gray-Keller MP, Detwiler PB, Palczewski K: Purification and physiological evaluation of a guanylate cyclase activating protein from retinal rods. Proc Natl Acad Sci USA 91: 4014–4018, 1994
37. Palczewski K, Subbaraya I, Gorczyca WA, Helekar BS, Ruiz CC, Ohguro H, Huang J, Zhao X, Crabb JW, Johnson RS, Walsh KA, Gray-Keller MP, Detwiler PB, Baehr W: Molecular cloning and characterization of retinal photoreceptor guanylyl cyclase-activating protein. Neuron 13: 395–404, 1994
38. Subbaraya I, Ruiz CC, Helekar BS, Zhao X, Gorczyca WA, Pettenati MJ, Rao PN, Palczewski K, Baehr W: Molecular characterization of human and mouse photoreceptor guanylate cyclase-activating protein (GCAP) and chromosomal localization of the human gene. J Biol Chem 269: 31080–31089, 1994
39. Dizhoor AM, Olshevskaya EV, Henzel WJ, Wong SC, Stults JT, Ankoudinova I, Hurley JB: Cloning, sequencing, and expression of a 24-kDa Ca^{2+}-binding protein activating photoreceptor guanylyl cyclase. J Biol Chem 270: 25200–25206, 1995
40. Gorczyca WA, Polans AS, Surgucheva IG, Subbaraya I, Baehr W, Palczewski K: Guanylyl cyclase activating protein. J Biol Chem 270: 22029–22036, 1995
41. Frins S, Bönigk W, Müller F, Kellner R, Koch K-W: Functional characterization of a guanylyl cyclase-activating protein from vertebrate rods. J Biol Chem 271: 8022–8027, 1996
42. Haeseleer F, Sokal I, Li N, Pettenati M, Rao N, Bronson D, Wechter R, Baehr W, Palczewski K: Molecular characterization of a third member of the guanylyl cyclase-activating protein subfamily. J Biol Chem 274: 6526–6535, 1999

43. Li N, Fariss RN, Zhang K, Otto-Bruc A, Haeseleer F, Bronson D, Qin N, Yamazaki A, Subbaraya I, Milam AH, Palczewski K, Baehr W: Guanylate-cyclase-inhibitory protein is a frog retinal Ca^{2+}-binding protein related to mammalian guanylate-cyclase-activating proteins. Eur J Biochem 252: 591–599, 1998
44. Goraczniak RM, Duda T, Sharma RK: Calcium modulated signaling site in type 2 rod outer segment membrane guanylate cyclase (ROS-GC2). Biochem Biophys Res Commun 245: 447–453, 1998
45. Dizhoor AM, Hurley JB: Inactivation of EF-hands makes GCAP-2 (p24) a constitutive activator of photoreceptor guanylyl cyclase by preventing a Ca^{2+}-induced 'activator-to-inhibitor' transition. J Biol Chem 271: 19346–19350, 1996
46. Olshevskaya EV, Hughes RE, Hurley JB, Dizhoor AM: Calcium binding, but not a calcium-myristoyl switch, controls the ability of guanylyl cyclase-activating protein GCAP-2 to regulate photoreceptor guanylyl cyclase. J Biol Chem 272: 14327–14333, 1997
47. Gray-Keller MP, Detwiler PB: The calcium feedback signal in the phototransduction cascade of vertebrate rods. Neuron 13: 849–861, 1994
48. Duda T, Goraczniak R, Surgucheva I, Rudnicka-Nawrot M, Gorczyca WA, Palczewski K, Sitaramayya A, Baehr W, Sharma RK: Calcium modulation of bovine photoreceptor guanylate cyclase. Biochemistry 35: 8478–8482, 1996
49. Laura RP, Dizhoor AM, Hurley JB: The membrane guanylyl cyclase, retinal guanylyl cyclase-1, is activated through its intracellular domain. J Biol Chem 271: 11646–11651, 1996
50. Sokal I, Haeseleer F, Arendt A, Adman ET, Hargrave PA, Palczewski K: Identification of a guanylyl cyclase-activating protein-binding site within the catalytic domain of retinal guanylyl cyclase 1. Biochemistry 38: 1387–1393, 1999
51. Perrault I, Rozet JM, Calvas P, Gerber S, Camuzat A, Dollfus H, Châtelin S, Souied E, Ghazi I, Leowski C, Bonnemaison M, Le Paslier D, Frézal J, Dufier J-L, Pittler S, Munnich A, Kaplan J: Retinal-specific guanylate cyclase gene mutations in Leber's congenital amaurosis. Nature Genet 14: 461–464, 1996
52. Perrault I, Rozet J-M, Gerber S, Ghazi I, Leowski C, Ducroq D, Souied E, Dufier J-L, Munnich A, Kaplan J: Leber congenital amaurosis. Mol Genet Metab 68: 200–218, 1999
53. Duda T, Venkataraman V, Goraczniak R, Lange C, Koch K-W, Sharma RK: Functional consequences of a rod outer segment membrane guanylate cyclase (ROS-GC1) gene mutation linked with Leber's congenital amaurosis. Biochemistry 38: 509–515, 1999
54. Krishnan A, Goraczniak RM, Duda T, Sharma RK: Third calcium-modulated rod outer segment membrane guanylate cyclase transduction mechanism. Mol Cell Biochem 178: 251–259, 1998
55. Sharma RK, Duda T, Goraczniak R, Sitaramayya A: Membrane guanylate cyclase signal transduction system. Ind J Biochem Biophys 34: 40–49, 1997
56. Duda T, Yadav P, Jankowska A, Venkataraman V, Sharma RK: Three dimensional atomic model and experimental validation for the ATP-regulated module (ARM) of the atrial natriuretic factor receptor guanylate cyclase. Mol Cell Biochem 217: 165–172, 2001
57. Potter LR, Hunter T: Phosphorylation of the kinase homology domain is essential for activation of the A-type natriuretic peptide receptor. Mol Cell Biol 18: 2164–2172, 1998
58. Sitaramayya A, Marala RB, Hakki S, Sharma RK: Interactions of nucleotide analogues with rod outer segment guanylate cyclase. Biochemistry 30: 6742–6747, 1991
59. Duda T, Goraczniak R, Sharma RK: Distinct inhibitory ATP-regulated modulatory domain (ARM_i) in membrane guanylate cyclases. Biochem J 319: 279–283, 1996
60. Lambrecht H-G, Koch K-W: Phosphorylation of recoverin, the calcium-sensitive activator of photoreceptor guanylyl cyclase. FEBS Lett 294: 207–209, 1991
61. Gorczyca WA, Van Hooser JP, Palczewski K: Nucleotide inhibitors and activators of retinal guanylyl cyclase. Biochemistry 33: 3217–3222, 1994
62. Tucker CL, Laura RP, Hurley JB: Domain-specific stabilization of photoreceptor membrane guanylyl cyclase by adenine nucleotides and guanylyl cyclase activating proteins (GCAPs). Biochemistry 36: 11995–12000, 1997
63. Aparicio JG, Applebury ML: The photoreceptor guanylate cylcase is an autophosphorylating protein kinase. J Biol Chem 271: 27083–27089, 1996
64. Wolbring G, Schnetkamp PPM: Modulation of the calcium sensitivity of bovine retinal rod outer segment guanylyl cyclase by sodium ions and protein kinase A. Biochemistry 35: 11013–11018, 1996
65. Wolbring G, Schnetkamp PPM: Activation by PKC of the Ca^{2+}-sensitive guanylyl cyclase in bovine retinal rod outer segments measured with an optical assay. Biochemistry 34: 4689–4695, 1995
66. Hurley JH: The adenylyl and guanylyl cyclase superfamily. Curr Opin Struct Biol 8: 770–777, 1998
67. Koch K-W, Eckstein F, Stryer L: Stereochemical course of the reaction catalyzed by guanylate cyclase from bovine retinal rod outer segments. J Biol Chem 265: 9659–9663, 1990
68. Liu Y, Ruoho AE, Rao VD, Hurley JH: Catalytic mechanism of the adenylyl and guanylyl cyclases: Modeling and mutational analysis. Proc Natl Acad Sci USA 94: 13414–13419, 1997
69. Tucker CL, Hurley JH, Miller TR, Hurley JB: Two amino acid substitutions convert a guanylyl cyclase, RetGC-1, into an adenylyl cyclase. Proc Natl Acad Sci USA 95: 5993–5997, 1998
70. Yu H, Olshevskaya E, Duda T, Seno K, Hayashi F, Sharma RK, Dizhoor AM, Yamazaki A: Activation of retinal guanylyl cyclase-1 by Ca^{2+}-binding proteins involves its dimerization. J Biol Chem 274: 15547–15555, 1999
71. Kelsell RE, Gregory-Evans K, Payne AM, Perrault I, Kaplan J, Yang R-B, Garbers DL, Bird AC, Moore AT, Hunt DM: Mutations in the retinal guanylate cyclase (RETGC-1) gene in dominant cone-rod dystrophy. Hum Mol Genet 7: 1179–1184, 1998
72. Gregory-Evans K, Kelsell RE, Gregory-Evans CY, Downes SM, Fitzke FW, Holder GE, Simunovic M, Mollon JD, Taylor R, Hunt DM, Bird AC, Moore AT: Autosomal dominant cone-rod retinal dystrophy (CORD6) from heterozygous mutation of *GUCY23D*, which encodes retinal guanylate cyclase. Ophthalmology 107: 55–61, 2000
73. Duda T, Krishnan A, Venkataraman V, Lange C, Koch K-W, Sharma RK: Mutations in the rod outer segment membrane guanylate cyclase in a cone-rod dystrophy cause defects in calcium signaling. Biochemistry 38: 13912–13919, 1999
74. Tucker CL, Woodcock SC, Kelsell RE, Ramamurthy V, Hunt DM, Hurley JB: Biochemical analysis of a dimerization domain mutation in RetGC-1 associated with dominant cone-rod dystrophy. Proc Natl Acad Sci USA 96: 9039–9044, 1999
75. Duda T, Venkataraman V, Jankowska A, Lange C, Koch K-W, Sharma RK: Impairment of the rod outer segment membrane guanylate cyclase dimerization in A cone-rod dystrophy results in defective calcium signaling. Biochemistry 39: 12522–12533, 2000
76. Fleischman D, Denisevich M: Guanylate cyclase of isolated bovine retinal rod axonemes. Biochemistry 18: 5060–5066, 1979
77. Hakki S, Sitaramayya A: Guanylate cyclase from bovine rod outer segments: Solubilization, partial purification, and regulation by inorganic pyrophosphate. Biochemistry 29: 1088–1094, 1990
78. Schrem A, Lange C, Beyermann M, Koch K-W: Identification of a domain in guanylyl cyclase-activating protein 1 that interacts with a complex of guanylyl cyclase and tubulin in photoreceptors. J Biol Chem 274: 6244–6249, 1999

79. Körschen HG, Beyermann M, Müller F, Heck M, Vantler M, Koch K-W, Kellner R, Wolfrum U, Bode C, Hofmann KP, Kaupp UB: Interaction of glutamic-acid-rich proteins with the cGMP signalling pathway in rod photoreceptors. Nature 400: 761–766, 1999
80. Körschen HG, Illing M, Seifert R, Sesti F, Williams A, Gotzes S, Colville C, Müller F, Dosé A, Godde M, Molday L, Kaupp UB, Molday RS: A 240 kDa protein represents the complete β subunit of the cyclic nucleotide-gated channel from rod photoreceptor. Neuron 15: 627-636, 1995
81. Allikmets R, Singh N, Sun H, Shroyer NF, Hutchinson A, Chidambaram A, Gerrard B, Baird L, Stauffer D, Pfeiffer A, Rattner A, Smallwood P, Li Y, Anderson KL, Lewis RA, Nathans J, Leppert M, Dean M, Lupski JR: A photoreceptor cell-specific ATP-binding transporter gene (ABCR) is mutated in recessive Stargardt macular dystrophy. Nature Genet 15: 236–246, 1997
82. Illing M, Molday LL, Molday RS: The 220-kDa rim protein of retinal rod outer segment is a member of the ABC transporter superfamily. J Biol Chem 272: 10303–10310, 1997
83. Sun H, Molday RS, Nathans J: Retinal stimulates ATP hydrolysis by purified and reconstituted ABCR, the photoreceptor-specific ATP-binding cassette transporter responsible for Stargardt Disease. J Biol Chem 274: 8269–8281, 1999
84. Seno K, Kishigami A, Ihara S, Maeda T, Bondarenko VA, Nishizawa Y, Usukura J, Yamazaki A, Hayashi F: A possible role of RGS9 in phototransduction. J Biol Chem 273: 22169–22172, 1998
85. He W, Cowan CW, Wensel TG: RGS9, a GTPase accelerator for phototransduction. Neuron 20: 95–102, 1998
86. Miller JL, Picones A, Korenbrot JI: Differences in transduction between rod and cone photoreceptors: an exploration of the role of calcium homeostasis. Curr Opin Neurobiol 4: 488–495, 1994
87. Venkataraman V, Nagele R, Duda T, Sharma RK: Rod outer segment membrane guanylate cyclase type 1-linked stimulatory and inhibitory calcium signaling systems in the pineal gland: Biochemical, molecular and immunohistochemical evidence. Biochemistry 39: 6042–6052, 2000
88. Duda T, Venkataraman V, Krishnan A, Nagele RG, Sharma RK: Negatively calcium-modulated membrane guanylate cyclase (ROS-GC1) signaling systems in the rat olfactory bulb. Biochemistry 40: 2001 (in press)

Molecular and Cellular Biochemistry **230:** 107–116, 2002.

Calcium-modulated membrane guanylate cyclase in synaptic transmission?

Teresa Duda[1] and Karl-Wilhelm Koch[2]

[1]*The Unit of Regulatory and Molecular Biology, Departments of Cell Biology and Ophthalmology, SOM and NJMS University of Medicine and Dentistry of New Jersey, Stratford, NJ, USA;* [2]*Institut fur Biologische Informationsverarbeitung 1, Jülich, Germany*

Abstract

Rod outer segment guanylate cyclase 1 (ROS-GC1) is a pivotal enzyme for vertebrate phototransduction and the systematically growing evidence point to its connection with processes other than phototransduction within and outside the retina. ROS-GC1 activity is regulated by Ca^{2+} in two opposite modes. This regulation is indirect and occurs through Ca^{2+}-binding proteins. At nanomolar Ca^{2+} concentrations, ROS-GC1 is activated by GCAPs and at micromolar Ca^{2+}-concentrations, by S100β and neurocalcin. The former mode operates in phototransduction and the latter was proposed to play a role in synaptic activity. The last possibility was supported by findings of ROS-GC1 expression not only in various retinal layers other than photoreceptor outer segments but also outside the retina, in pineal gland and olfactory bulb. If ROS-GC1 indeed is to play a role in neurotransmission its expression must be colocalized with its Ca^{2+}-dependent regulators and with possible targets of an increased cyclic GMP concentration, cyclic nucleotide-gated channels or cyclic GMP-dependent protein kinase, in synaptic regions. In this review these aspects of ROS-GC1 expression in retina, pineal gland and olfactory bulb are discussed. (Mol Cell Biochem **230**: 107–116, 2002)

Key words: ROS-GC1, calcium, retina, pineal gland, olfactory bulb

Introduction

The presence and functional significance of membrane guanylate cyclase in photoreceptor outer segments is well established [1–3]. The cyclase exists there in two forms, ROS-GC1 and ROS-GC2. Both these cyclases were cloned and expressed in heterologous systems. The properties of the recombinant cyclases mimic the properties of native cyclase [1–3]. They are stimulated by calcium binding proteins, GCAP1 and/or GCAP2 in low free [Ca^{2+}] of 10–50 nM, which is similar to the free [Ca^{2+}] that exists in outer segments after illumination [2, 4–15]. The GCAP-dependent stimulation is inhibited by increasing [Ca^{2+}] and the half maximal inhibition (IC_{50}) occurs at 90–200 nM Ca^{2+} [2]. Thus, both cyclases are unequivocally linked with phototransduction. They replenish the cyclic GMP pool lowered after illumination and their activities are controlled by a Ca^{2+}-feedback loop that is a key step in light adaptation [4]. The detailed characterization of ROS-GC1 and ROS-GC2 in view of the current model of phototransduction is presented elsewhere [Koch *et al.*, this issue].

Another line of research showed that the expression of ROS-GC1 is not restricted to outer segments. Immunocytochemical studies have revealed that in addition to outer segments ROS-GC1 is also expressed in other retinal neurons [16, 17]. Especially intriguing was the finding of ROS-GC1 prominent immunoreactivity in both retinal synaptic regions, outer and inner plexiform layers [16, 17]. The presence of ROS-GC1 in these regions indicated that the protein might also be related to other than phototransduction neurosensory processes. In accordance with these findings were *in vitro* experiments that showed regulation of ROS-GC1 by S100β and neurocalcin at elevated free [Ca^{2+}] of 2–30 μM [18–20]. More recently, the presence of ROS-GC1 in the pineal gland and in the olfactory bulb was demonstrated [21, 22].

Address for offprints: T. Duda, The Unit of Regulatory and Molecular Biology, Departments of Cell Biology and Ophthalmology, University of Medicine and Dentistry of New Jersey, Stratford, New Jersey 08084, USA (E-mail: dudatm@umdnj.edu)

So far no physiological role for a membrane bound guanylate cyclase in retinal synaptic regions, in the pineal gland and in the olfactory bulb has been established, but it was hypothesized that the Ca^{2+}-dependent regulation of ROS-GC1 activity might play a role in synaptic transmission [23]. In this review we discuss the criteria that need to be fulfilled in order to establish such a role. In particular we focus on the following questions: Is there evidence of localization of ROS-GC1 in synaptic regions? Does the Ca^{2+}-dependent regulation of ROS-GC1 by Ca^{2+}-binding proteins fit to a possible function in synapses? Are ROS-GC1 and its regulators colocalized? Are possible targets of an increased cyclic GMP concentration (e.g. cyclic nucleotide-gated channels, cyclic GMP-dependent protein kinase) colocalized with ROS-GC1 in synaptic regions? Does a role of ROS-GC1 fit into the current concept of synaptic transmission? We will discuss these points along three biological systems, the vertebrate retina, pineal gland and olfactory bulb.

ROS-GC1 in retinal cells

Signal flow in the retina

Vertebrate photoreceptors (rods and cones) are highly differentiated retinal neurons that generate electrical signals in response to photons captured by the photosensitive membrane protein, rhodopsin in rods, opsin in cones [24]. The cascade of biochemical processes involved in generation of these signals is termed as phototransduction. The information about a state of illumination is conveyed from photoreceptor outer segments (the site of phototransduction) to ganglion cells and further to the brain. There is no, however, direct communication between photoreceptors and ganglion cells [25, 26]. The information is transmitted through a set of vertically and horizontally arranged retinal neurons (Fig. 1) which combine the signals arriving from various photoreceptors and bland them in such a way that the signals received by ganglion cells reflect the pattern of illumination received by the retina.

Photoreceptor cells that are in (*on*) the center of ganglion cell's receptive field directly synapse with bipolar cells that in turn directly synaps with ganglion cells. This represents the direct or vertical flow of information. Signals from photoreceptors that are *off* the center of the ganglion cell's receptive field are transferred to bipolar and ganglion cells indirectly through horizontal and amacrine cells. This represents the horizontal or lateral flow of information (Fig. 1). Information from rod photoreceptors is transmitted exclusively through the *on* type bipolar cells but both types, *on* and *off* of bipolar cells, transmit the information from cone photoreceptors. The majority of the synaptic contacts in the retina occur within two distinct layers, which are termed the outer and inner plexiform layers. The outer plexiform layer is made up by the processes of photoreceptor, horizontal and bipolar cells and the inner plexiform layer, by the processes of bipolar, amacrine and ganglion cells. Cells of the direct (vertical) pathway use generally glutamate as their neurotransmitter [27–33]; cells of the indirect (horizontal) pathway use gamma aminobutyric acid (GABA) and glycine [29, 34–38]. Some retinal cells also form gap junctions [39–46].

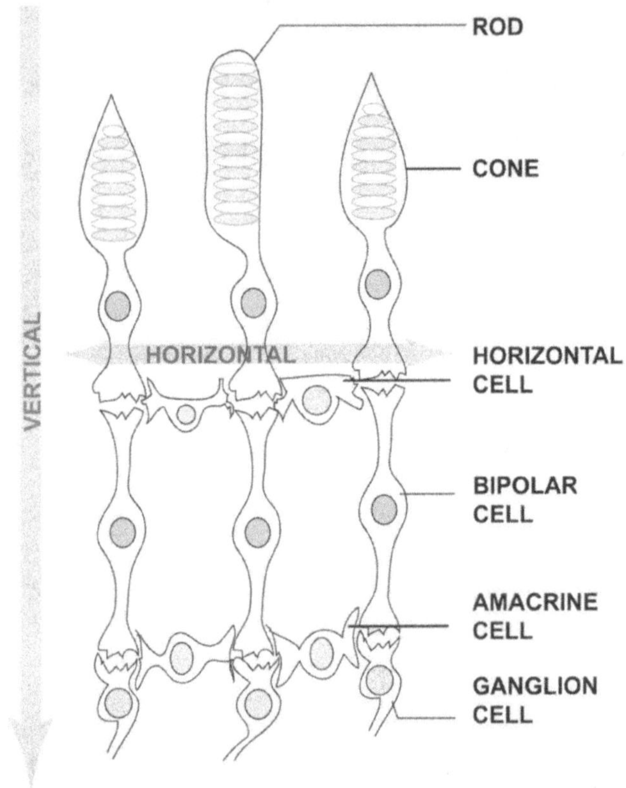

Fig. 1. Schematic representation of the arrangement of cells in the retina and the vertical and horizontal flow of information among these cells.

ROS-GC1 and its Ca^{2+}-dependent modulators colocalize outside outer segments

Components of cyclic GMP signaling cascades were identified in several retinal cells outside rod and cone outer segments. These include cone synaptic terminals [47], on-bipolar cells [48, 49], ganglion cells [50, 51] and Müller (glial) cells [52]. Immunolabelling of retinae from monkey, cat, rat and chicken showed the expression of ROS-GC1 in the outer and inner plexiform layers that correspond to the synaptic regions [16, 17] (Fig. 2). These exciting findings that expanded the possible role of ROS-GC1 beyond the boundaries of photo-

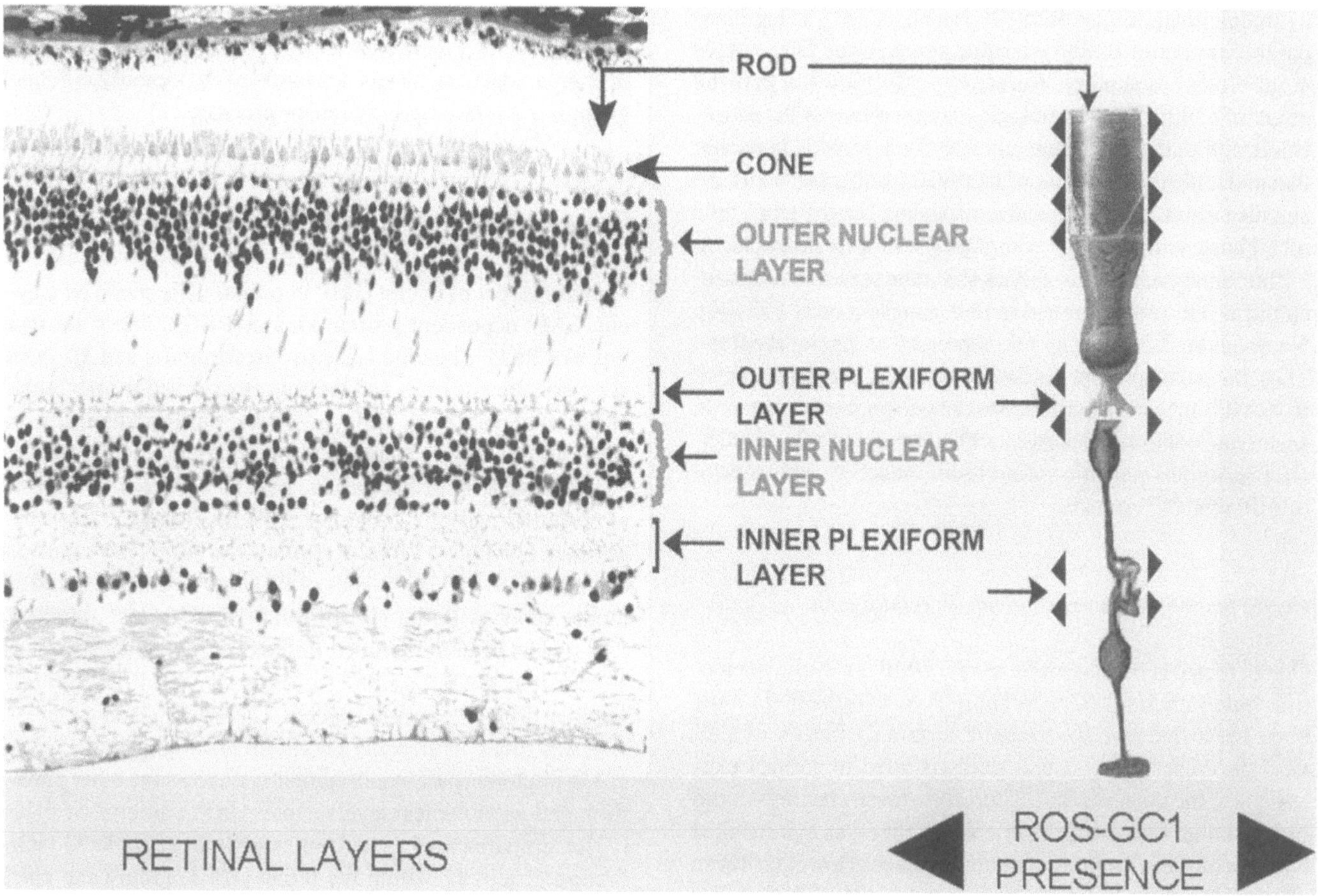

Fig. 2. Cross-section through the bovine retina showing the retinal layers (left panel), and the illustration of the sites of ROS-GC1 expression (right panel). The left panel which shows the cross-section of the retina is a paraffin section stained with hematoxylin and eosin. The right panel is a graphic representation. The two different photoreceptor cell types and the two nuclear and plexiform layers are indicated.

transduction, posed questions about regulation of the cyclase outside outer segments. Until then the only known ROS-GC1 regulators were GCAPs, which stimulated the cyclase activity at low nanomolar Ca^{2+} concentrations [4–15]; the synaptic activity, however, requires Ca^{2+} concentrations in micromolar range [53–55]. This functional puzzle was solved with the purification of a retinal protein that was able to stimulate ROS-GC1 activity in micromolar Ca^{2+} concentrations [56]. The protein was termed calcium dependent GCAP, CD-GCAP [56]. Sequencing of CD-GCAP revealed its identity with S100β protein [57]. When purified S100β was tested for its effect on ROS-GC1 activity, at micromolar Ca^{2+} concentrations the dose-dependent stimulation of the cyclase activity was observed with either native (washed ROS membranes) or recombinant (expressed in COS cells) ROS-GC1 [18, 19]. The half maximal cyclase stimulation by S100β was at [Ca^{2+}] of 2–30 μM. Thus, ROS-GC1 can operate in a range of free [Ca^{2+}] existing in synaptic terminals. It was then postulated that the S100β-mediated activation of ROS-GC1 has a role in retinal synaptic layers [16, 18, 19, 56]. This hypothesis was validated through experiments in which S100β-dependent increase of cyclic GMP synthesis was shown in retinal cells [58]. Fixed and permeabilized rat retinae were incubated with S100 proteins and various free [Ca^{2+}] (10–1000 μM) and the localization of membrane guanylate cyclase activity was determined by ultracytochemistry [58]. The results showed that S100β stimulated cyclic GMP synthesis by membrane guanylate cyclase in Müller cell processes and photoreceptor cells outside outer segments. Under these conditions some membrane guanylate cyclase activity was also observed in outer segments of dark-adapted retinae. From these results the authors concluded that Ca^{2+}- and S100β-dependent regulation of membrane guanylate cyclase activity plays a role in retinal processes other than phototransduction [58].

Recently, a Ca^{2+} binding protein belonging to a family of neuronal calcium sensor proteins [59–61] named neurocalcin was found to stimulate ROS-GC1 activity in the presence of micromolar [Ca^{2+}] [20]. The stimulation is dose-dependent and Ca^{2+}-dependent [20]. The Ca^{2+} concentration at which the half-maximal stimulation occurs is ~20 μM [20]. Although neurocalcin stimulates ROS-GC1 with an EC_{50} value comparable to that of S100β (for both proteins the EC_{50} values are at ~1 μM) its maximal stimulatory effect is lower than the effect of S100β. The physiological implications of the neurocalcin stimulatory effect are unclear. They predict, however, that under identical conditions ROS-GC1 will sense both Ca^{2+} signals carried by S100β and neurocalcin but the amplitude of response will be higher with S100β than with neurocalcin.

Immunocytochemical studies show the presence of neurocalcin in the retinal amacrine and ganglion cells [62–64]. Neurocalcin, however, is not expressed in photoreceptors. Thus, the paired presence of S100β/ROS-GC1 or neurocalcin/ROS-GC1 in various retinal cells outside photoreceptor outer segments would be consistent with a presumed role of ROS-GC1 in retinal synaptic transmission under diverse control of different Ca^{2+}-sensors.

Cyclic nucleotide gated channels in retinal cells

One of the possible targets for cyclic GMP is a cyclic nucleotide-gated channel (CNG-channel). CNG-channels have been identified in cone synaptic terminals. Influx of Ca^{2+} through these channels was demonstrated to control exocytosis at the synapse [65]. Therefore, these channels could be a principal target for cyclic GMP that was synthesized by a membrane bound guanylate cyclase. However, these channels seem to be under the control of nitric oxide (NO) [66]. Presence of a soluble guanylate cyclase, which is the predominant target of NO and is activated by NO was indeed found in inner segments and synaptic regions of photoreceptor cells [67–69]. The activity of the soluble guanylate cyclase does indirectly depend on free $[Ca^{2+}]$, since the neuronal nitric oxide synthase in retinal cells is Ca^{2+}-calmodulin dependent [67].

Taken together these findings lead to a conceptual problem: two different forms of a guanylate cyclase are localized in the same compartment of the cell. Both are positively regulated by increasing levels of Ca^{2+}, although by different mechanisms. Exocytosis is triggered by Ca^{2+}-influx through voltage -activated Ca^{2+}-channels in both rods and cones. In addition, the influx of Ca^{2+} through CNG-channels in cone inner segments and synaptic terminals also initiates exocytosis. What are two Ca^{2+}-regulated cyclases for? It is conceivable that a small fraction of CNG-channels is constitutively opened by resting levels of cyclic GMP. The membrane bound cyclase would be the source of the resting cyclic GMP level. Diffusion of NO to the synaptic region would then activate the soluble cyclase and increasing levels of cyclic GMP would open a larger fraction of CNG-channels leading to exocytosis. Another speculative but possible explanation could be that the two cyclases operate at different time intervals. Activation of the soluble GC requires Ca^{2+}-dependent synthesis of NO by NO-synthase and diffusion of NO to the target by crossing the cell membranes. This could cause a time delay in comparison to an instant activation of the membrane bound guanylate cyclase by Ca^{2+}-sensor proteins.

Cyclic GMP-dependent protein kinase (PKG)

Another target of cyclic GMP in retinal cells could be a cyclic GMP dependent protein kinase (PKG). There are two types of PKG. The type 1 has two isoforms 1α and 1β; both are cytosolic enzymes and their sequences differ only in the N-terminal parts [70]. The type 2 is membrane bound and its substrate specificity is distinct from the type 1 isoforms [71, 72]. In other than ocular tissues, type 1 PKG has been shown to stimulate various cellular events including neurotransmission, and type 2 PKG, to regulate the activity of chloride channels [70–73]. Both types of PKG have been localized in mouse retina with the 1β isoform being the most abundant [74]. It has been determined through immunocytochemical studies that the three PKG isoforms show similar, but not identical protein expression pattern. The most abundant expression was found in the inner plexiform layer, in ganglion cells and in photoreceptor inner segments, whereas the outer plexiform and inner nuclear layers express lower amount of PKG [73]. Comparison of these studies with those on ROS-GC1 immunoreactivity within the retina [16, 17] allows to conclude that PKG and ROS-GC1 colocalize in outer and inner plexiform layers, photoreceptor inner segments and in ganglion cells. However, as yet, there is no report linking PKG with ROS-GC1, instead, PKG effects in combination with NO and soluble guanylate cyclase have been described [75]. Thus, the possible consequences of ROS-GC1 and PKG colocalization remain to be determined.

Is there any evidence of ROS-GC1 involvement in retinal synaptic activity?

Although present-day knowledge is not sufficient for constructing a complete physiological model of ROS-GC1 role in retinal synaptic transmission, very significant substantiation of this hypothesis came from analyses of the electroretinograms (ERGs) and the *on–off* responses elicited in patients with CORD6 (cone-rod dystrophy, a retinopathy caused by mutations in ROS-GC1 gene) [76]. Electronegative ERG responses and the abnormal *on–off* responses in these patients were linked to dysfunctional synaptic connections between rod and cone photoreceptors and the inner retina. Since the abnormal responses are observed in the early stages of the disease it was suggested that the synaptic dysfunction accompanies the onset of the dystrophy [76].

ROS-GC1 in other tissues

Pineal gland

In invertebrates and lower vertebrates the pineal gland senses and responds to light. This ability is lost in higher vertebrates including mammals although their pineal glands express proteins specific for the process of phototransduction occurring in retinal outer segments [77, 78]. In line with this trail of protein expression, ROS-GC1 transcript was detected in chicken, rat and bovine pineal [21, 79, 80] and the ROS-GC1 cDNA was cloned from bovine pineal gland [21]. In addition, also GCAP1 was cloned from bovine pineal gland and the pineal GCAP1 was identical with the retinal GCAP1 in all aspects [21]. The cloning experiments led to questions: (1) is ROS-GC2 also expressed in pineal; (2) is ROS-GC11 expressed in all pineal cells; and (3) is GCAP1 the only ROS-GC1 modulator expressed in this gland. Detailed biochemical, *in situ* hybridization and immunocytochemical analyses were used to answer these questions. When the bovine pineal membranes were exposed to varying concentrations of GCAP1, GCAP2 and S100β at appropriate [Ca^{2+}] the profiles of cyclase responses to these modulators were identical to that of ROS-GC1 [21]. Absence of ROS-GC2 expression was confirmed by western analysis and RT-PCR [21]. The pattern of cellular expression of ROS-GC1 was determined through *in situ* hybridization and immunocytochemical analyses. These experiments showed that out of two cell types, pinealocytes and glial cells, present in the pineal gland, both ROS-GC1 transcript and the encoded protein exist only in pinealocytes [21]. The issue of ROS-GC1 modulators expressed in pineal gland was resolved through two independent sets of experiments. Initially, by western analyses the presence of GCAP1 and S100β but absence of GCAP2 was determined. The co-presence of GCAP1, S100β and ROS-GC1 was scrutinized by immunocytochemistry. These experiments clearly showed that all pinealocytes express ROS-GC1 with GCAP1 or S100β but not with both of them. S100β expression was found to be restricted to only a small population of pinealocytes, whereas S100β negative cells expressed GCAP1. Thus, it was concluded that the two ROS-GC1 transduction systems, one involving GCAP1 and the other involving S100β are not present in the same pinealocytes.

The physiological significance of these two ROS-GC1 transduction systems in the pineal gland is not known yet. However, because both are present in pinealocytes together with the epinephrine/norepinephrine activated $\alpha_{2A/D}$-adrenergic receptor [21], an attractive speculation was presented: a photosensitive regulation of epinephrine/norepinephrine signaling would cause cellular oscillations of [Ca^{2+}] which in turn triggers light-regulated melatonin secretion [21].

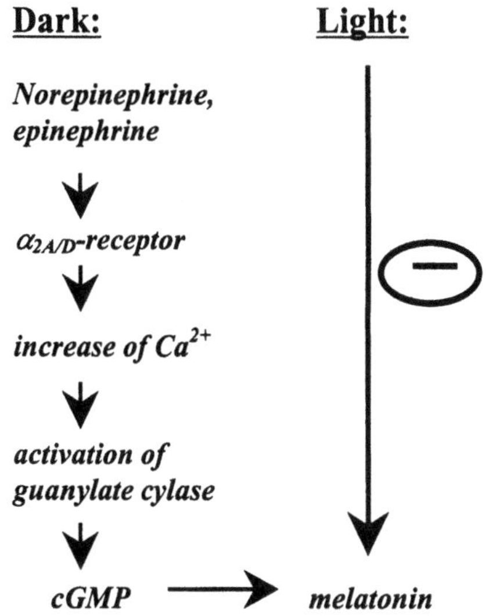

Light inhibits the synthesis of melatonin, and melatonin secretion is regulated by epinephrine/norepinephrine. At midnight, there is a peak of norepinephrine levels in the pineal gland [81]. In the proposed model, the norepinephrine via its $\alpha_{2A/D}$-adrenergic receptor causes an increase in intracellular [Ca^{2+}], which binds to S100β and activates ROS-GC1. Cyclic GMP, thus produced, mediates melatonin synthesis and secretion. In addition, elevation of intracellular [Ca^{2+}] resulting from epinephrine/norepinephrine signaling would cause a dark-like state in pinealocytes, and the removal of the signal, a light-like state. Thus, according to this hypothesis, cycling between dark and light in the pineal gland is mediated by photosensitive oscillations of epinephrine/norepinephrine levels. Interestingly, the nerve fibers that innervate the pineal gland and release epinephrine in a circadian fashion arise from the superior cervical ganglion, which receives input from the retinal ganglion cells through the suprachiasmatic nucleus, paraventricular nucleus, and intermediolateral column [82].

Olfactory bulb

Three neuronal compartments are involved in the perception of smell in mammals. The first is the olfactory epithelium, where the sensory cells for olfaction reside; the second is the olfactory bulb where the axons emanating from sensory cells project, and the third is the olfactory cortex, which is linked with the olfactory tract (Fig. 3). The odor signal is sensed within the olfactory epithelium, and smell perception occurs within the olfactory cortex. Thus, the sensory neurons of the olfactory epithelium are internetted with central cortical neurons via intermediary neurons of the olfactory bulb. A general scheme of olfactory signaling occurring in the sensory cells of the olfactory epithelium has been worked out, but

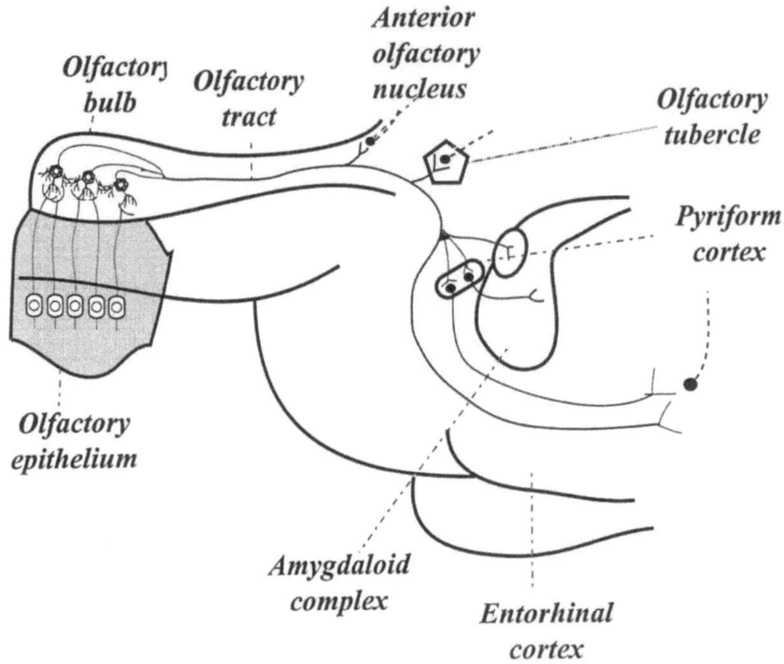

Fig. 3. Schematic diagram of the olfactory system. The major areas involved in the perception of smell are represented. The olfactory epithelium is the site for sensing the odorant molecules. Odorant transduction is initiated at the bipolar sensory neurons of the epithelium. The information is sent to the olfactory bulb, where it is processed. The processed information, via the olfactory tract, is then transmitted to the five major centers indicated.

many molecular details are still missing. It is well established that odorants bind to odorant receptors that belong to the family of G-protein–coupled receptors [83–86]. In mammals binding of an odorant to a receptor results in the activation of a G-protein dependent adenylate cyclase type III. Increase of cyclic AMP leads to an opening of CNG-channels located in the ciliary plasma membrane. Thus the membrane depolarizes after stimulation by an odorant [87–89]. This is opposite to the effect of light on photoreceptor plasma membrane. Recently a membrane guanylate cyclase, GC-D, has been cloned from the olfactory epithelium [90] and determined to exist in the ciliary membrane of a small population of olfactory neurons that project to a subtype of glomeruli (necklace glomeruli) in the olfactory bulb [91,92]. The restricted expression of the cyclase is, however, disputed by another study showing its expression throughout the olfactory epithelium [93]. Two different functions have been suggested for this cyclase. According to one suggestion, the cyclase serves as a receptor of an as yet unidentified odorant [90, 91] another, however, proposes its involvement in amplification of an odorant-induced cyclic AMP response [93]. Even less is known about the neural pathway beginning with the depolarization of the cilia and ending in the olfactory cortex. The axons of olfactory receptor cells terminate in the synaptic areas of the olfactory bulb termed glomeruli. Therein, the information is modulated and relayed via the bulb intermediary neurons [94]. They include mitral, tufted, granule and periglomerular cells. The information is received by the dendrites of these neurons, processed in the soma and transmitted, eventually, via the olfactory tract to the five specified areas of the olfactory cortex: anterior olfactory nucleus, olfactory tubercle, pyriform cortex, amygdaloid complex and entorhinal complex. These areas decode the sensory input and translate the original odorant signal into the perception of smell. There is almost no information about the identity of the biochemical events of this post-odorant transduction – signal processing machinery.

Very recently the presence of ROS-GC1 and its modulator GCAP1 has been established in the olfactory bulb membranes [22]. *In situ* hybridization and immunohistochemical analysis showed that both the cyclase and the modulator coexist within mitral cells. To construct a physiological model of the ROS-GC1/GCAP1 transduction system in mitral cells, the following established features of these cells were considered: (1) The cells receive excitatory signals from the olfactory receptor cells [94–96]; (2) They receive inhibitory signals from the neighboring cells [97, 98]; (3) They undergo autoexcitation [99]; (4) The integration of the excitatory and inhibitory signals occurs in the cell soma; (5) Mitral cells express both voltage-dependent Ca^{2+}-channels and cyclic nucleotide gated channels [100–104]. These considerations were taken into account to propose a model for a ROS-GC1/

GCAP1 signal transduction system that is under control of nanomolar to micromolar changes of intracellular [Ca^{2+}]. After odorant stimulation, the neurosensory cells of the olfactory epithelium send excitatory signals to the mitral cells. These signals are integrated with the dendrodentritic inhibitory signals received from neighboring cells [97, 98]. As a result, the excitatory input is either stopped or enhanced [99]. The site of integration of these signals is the soma of the mitral cells, where the process of autoexcitation occurs. Autoexcitation causes release of glutamate in response to Ca^{2+} influx. A possible scenario is that at low [Ca^{2+}] (<50 nM) within mitral cells ROS-GC1 will be activated by GCAP1. Cyclic GMP formed in response to ROS-GC1 activation will facilitate the regenerative calcium spikes by regulating the activity of Ca^{2+}-permeable channels. Cyclic GMP could influence the intracellular [Ca^{2+}] in two independent ways. It could directly bind to CNG channels causing their opening and Ca^{2+} influx, or might modulate the activity of voltage-activated Ca^{2+} channels, by phosphorylation via PKG.

Conclusion

To consider the possible involvement of ROS-GC1 in synaptic transmission, we posed several questions at the beginning of the chapter concerning the localization of ROS-GC1, of the regulatory Ca^{2+}-binding proteins and of the cyclic GMP targets. As it is evident from the above discussion, ROS-GC1 colocalizes with its Ca^{2+}-sensitive regulators and components of cyclic GMP signaling cascades. Therefore, some criteria are fulfilled to propose a role of ROS-GC1 in synaptic transmission. But to fit completely ROS-GC1 into the current concept in synaptic transmission more questions need to be answered. First, the aspect of the spatial and temporal activation pattern of ROS-GC1 during synaptic excitation and inhibition needs to be determined. Second, whether the two guanylate cyclases, soluble and membrane bound, act together within the same synaptic terminal or have their specific sites remains to be established.

Acknowledgments

This study was supported by the facilities provided by the Unit of Regulatory and Molecular Biology supported by USPHS awards EY 10828 and HL 58151; by the affiliated support of the Research to Prevent Blindness Inc., New York; the Lions Eye Research Foundation, New Jersey, and by the facilities provided by UMDNJ-SOM.

References

1. Dizhoor AM: Regulation of cGMP synthesis in photoreceptors: Role in signal transduction and congenital diseases of the retina. Cell Signal 12: 711–719, 2000
2. Pugh EN Jr, Duda T, Sitaramayya A, Sharma RK: Photoreceptor guanylate cyclases: A review. Biosci Rep 17: 429–473, 1997
3. Koch KW: Biochemical mechanism of light adaptation in vertebrate photoreceptors. Trends Biochem Sci 17: 307–311, 1992
4. Koch KW, Stryer L: Highly cooperative feedback control of retinal rod guanylate cyclase by calcium ions. Nature 334: 64–66, 1988
5. Lambrecht HG, Koch KW: A 26 kd calcium binding protein from bovine rod outer segments as modulator of photoreceptor guanylate cyclase. EMBO J 10: 793–798, 1991
6. Gorczyca WA, Gray-Keller MP, Detwiler PB, Palczewski K: Purification and physiological evaluation of a guanylate cyclase activating protein from retinal rods. Proc Natl Acad Sci U S A 91: 4014–4018, 1994
7. Palczewski K, Subbaraya I, Gorczyca WA, Helekar BS, Ruiz CC, Ohguro H, Huang J, Zhao X, Crabb JW, Johnson RS, Walsh KA, Gray-Keller MP, Detwiler PB, Baehr W: Molecular cloning and characterization of retinal photoreceptor guanylyl cyclase-activating protein. Neuron 13: 395–404, 1994
8. Subbaraya I, Ruiz CC, Helekar BS, Zhao X, Gorczyca WA, Pettenati MJ, Rao PN, Palczewski K, Baehr W: Molecular characterization of human and mouse photoreceptor guanylate cyclase-activating protein (GCAP) and chromosomal localization of the human gene. J Biol Chem 269: 31080–31089, 1994
9. Dizhoor AM, Olshevskaya EV, Henzel WJ, Wong SC, Stults JT, Ankoudinova I, Hurley JB: Cloning, sequencing, and expression of a 24-kDa Ca(2+)-binding protein activating photoreceptor guanylyl cyclase. J Biol Chem 270: 25200–25206, 1995
10. Gorczyca WA, Polans AS, Surgucheva IG, Subbaraya I, Baehr W, Palczewski K: Guanylyl cyclase activating protein. A calcium-sensitive regulator of phototransduction. J Biol Chem 270: 22029–22036, 1995
11. Frins S, Bonigk W, Muller F, Kellner R, Koch KW: Functional characterization of a guanylyl cyclase-activating protein from vertebrate rods. Cloning, heterologous expression, and localization. J Biol Chem 271: 8022–8027, 1996
12. Duda T, Goraczniak R, Surgucheva I, Rudnicka-Nawrot M, Gorczyca WA, Palczewski K, Sitaramayya A, Baehr W, Sharma RK: Calcium modulation of bovine photoreceptor guanylate cyclase. Biochemistry 35: 8478–8482, 1996
13. Goraczniak RM, Duda T, Sharma RK: Calcium modulated signaling site in type 2 rod outer segment membrane guanylate cyclase (ROS-GC2). Biochem Biophys Res Commun 245: 447–453, 1998
14. Krishnan A, Goraczniak RM, Duda T, Sharma RK: Third calcium-modulated rod outer segment membrane guanylate cyclase transduction mechanism. Mol Cell Biochem 178: 251–259, 1998
15. Gray-Keller MP, Detwiler PB: The calcium feedback signal in the phototransduction cascade of vertebrate rods. Neuron 13: 849–861, 1994
16. Cooper N, Liu L, Yoshida A, Pozdnyakov N, Margulis A, Sitaramayya A: The bovine rod outer segment guanylate cyclase, ROS-GC, is present in both outer segment and synaptic layers of the retina. J Mol Neurosci 6: 211–222, 1995
17. Liu X, Seno K, Nishizawa Y, Hayashi F, Yamazaki A, Matsumoto H, Wakabayashi T, Usukura J: Ultrastructural localization of retinal guanylate cyclase in human and monkey retinas. Exp Eye Res 59: 761–768, 1994

18. Margulis A, Pozdnyakov N, Sitaramayya A: Activation of bovine photoreceptor guanylate cyclase by S100 proteins. Biochem Biophys Res Commun 218: 243–247, 1996
19. Duda T, Goraczniak RM, Sharma RK: Molecular characterization of S100A1–S100B protein in retina and its activation mechanism of bovine photoreceptor guanylate cyclase. Biochemistry 35: 6263–6266, 1996
20. Kumar VD, Vijay-Kumar S, Krishnan A, Duda T, Sharma RK: A second calcium regulator of rod outer segment membrane guanylate cyclase, ROS-GC1: Neurocalcin. Biochemistry 38: 12614–112620, 1999
21. Venkataraman V, Nagele R, Duda T, Sharma RK: Rod outer segment membrane guanylate cyclase type 1-linked stimulatory and inhibitory calcium signaling systems in the pineal gland: Biochemical, molecular, and immunohistochemical evidence. Biochemistry 39: 6042–6052, 2000
22. Duda T, Venkataraman V, Krishnan A, Nagele RG, Sharma RK: Negatively calcium-modulated membrane guanylate cyclase signaling system in the rat olfactory bulb. Biochemistry 40: 4654–4662, 2001
23. Pozdnyakov N, Yoshida A, Cooper NG, Margulis A, Duda T, Sharma RK, Sitaramayya A: A novel calcium-dependent activator of retinal rod outer segment membrane guanylate cyclase. Biochemistry 34: 14279–14283, 1995
24. Ebrey T, Koutalos Y: Vertebrate photoreceptors. Prog Retin Eye Res 20: 49–94, 2001
25. Borg-Graham LJ: The computation of directional selectivity in the retina occurs presynaptic to the ganglion cell. Nat Neurosci 4: 176–183, 2001
26. Wu SM, Gao F, Maple BR: Functional architecture of synapses in the inner retina: Segregation of visual signals by stratification of bipolar cell axon terminals. J Neurosci 20: 4462–4470, 2000
27. Brandstatter JH, Hack I: Localization of glutamate receptors at a complex synapse. The mammalian photoreceptor synapse. Cell Tissue Res 303: 14, 2001
28. Vardi N, Duvoisin R, Wu G, Sterling P: Localization of mGluR6 to dendrites of ON bipolar cells in primate retina. J Comp Neurol 423: 402–412, 2000
29. Barnstable CJ: Glutamate and GABA in retinal circuitry. Curr Opin Neurobiol 3: 520–525, 1993
30. Copenhagen DR: Synaptic transmission in the retina. Curr Opin Neurobiol 1: 258–262, 1991
31. Redburn DA: Neurotransmitter systems in the outer plexiform layer of mammalian retina. Neurosci Res Suppl 8: S127–136, 1988
32. Nakanishi S, Masu M, Bessho Y, Nakajima Y, Hayashi Y, Shigemoto R: Molecular diversity of glutamate receptors and their physiological functions. EXS 71: 71–80, 1994
33. Brandstatter JH, Koulen P, Wassle H: Diversity of glutamate receptors in the mammalian retina. Vision Res 38: 1385–1397, 1998
34. Grunert U: Distribution of GABA and glycine receptors on bipolar and ganglion cells in the mammalian retina. Microsc Res Tech 50: 130–140, 2000
35. Duarte CB, Santos PF, Carvalho AP: Corelease of two functionally opposite neurotransmitters by retinal amacrine cells: Experimental evidence and functional significance. J Neurosci Res 58: 475–479, 1999
36. Kalloniatis M, Tomisich G: Amino acid neurochemistry of the vertebrate retina. Prog Retin Eye Res 18: 811–866, 1999
37. Kamermans M, Spekreijse H: The feedback pathway from horizontal cells to cones. A mini review with a look ahead. Vision Res 39: 2449–2468, 1999
38. Wassle H, Koulen P, Brandstatter JH, Fletcher EL, Becker CM: Glycine and GABA receptors in the mammalian retina. Vision Res 38: 1411–1430, 1998
39. Becker D, Bonness VV, Mobbs P: Cell coupling in the retina: Patterns and purpose. Cell Biol Int 22: 781–792, 1998
40. Mills SL, Massey SC: A series of biotinylated tracers distinguishes three types of gap junction in retina. J Neurosci 20: 8629–8636, 2000
41. Sohl G, Guldenagel M, Traub O, Willecke K: Connexin expression in the retina. Brain Res Brain Res Rev 32: 138–145, 2000
42. Zhang DQ, McMahon DG: Direct gating by retinoic acid of retinal electrical synapses. Proc Natl Acad Sci USA 97: 14754–14759, 2000
43. Massey SC, Mills SL: Gap junctions between AII amacrine cells and calbindin-positive bipolar cells in the rabbit retina. Vis Neurosci 16: 1181–1189, 1999
44. Weiler R, He S, Vaney DI: Retinoic acid modulates gap junctional permeability between horizontal cells of the mammalian retina. Eur J Neurosci 11: 3346–3350, 1999
45. Kenyon GT, Marshak DW: Gap junctions with amacrine cells provide a feedback pathway for ganglion cells within the retina. Proc R Soc Lond B Biol Sci 265: 919–925, 1998
46. Jacoby R, Stafford D, Kouyama N, Marshak D: Synaptic inputs to ON parasol ganglion cells in the primate retina. J Neurosci 16: 8041–8056, 1996,
47. Savchenko A, Barnes S, Kramer RH: Cyclic-nucleotide-gated channels mediate synaptic feedback by nitric oxide. Nature 390: 694–698, 1997
48. Leconte L, Barnstable CJ: Impairment of rod cGMP-gated channel alpha-subunit expression leads to photoreceptor and bipolar cell degeneration. Invest Ophthalmol Vis Sci 41: 917–926, 2000
49. Shiells RA, Falk G: A rise in intracellular Ca^{2+} underlies light adaptation in dogfish retinal 'on' bipolar cells. J Physiol 514: 343–350, 1999
50. Ahmad I, Leinders-Zufall T, Kocsis JD, Shepherd GM, Zufall F, Barnstable CJ: Retinal ganglion cells express a cGMP-gated cation conductance activatable by nitric oxide donors. Neuron 12: 155–165, 1994
51. Hirooka K, Kourennyi DE, Barnes S: Calcium channel activation facilitated by nitric oxide in retinal ganglion cells. J Neurophysiol 83: 198–206, 2000
52. Kusaka S, Dabin I, Barnstable CJ, Puro DG: cGMP-mediated effects on the physiology of bovine and human retinal Muller (glial) cells. J Physiol 497: 813–824, 1996
53. Neher E, Augustine GJ: Calcium gradients and buffers in bovine chromaffin cells. J Physiol 450: 273–301, 1992
54. Augustine GJ, Neher E: Neuronal Ca^{2+} signalling takes the local route. Curr Opin Neurobiol 2: 302–307, 1992
55. Augustine GJ, Neher E: Calcium requirements for secretion in bovine chromaffin cells. J Physiol 450: 247–271, 1992
56. Pozdnyakov N, Yoshida A, Cooper NG, Margulis A, Duda T, Sharma RK, Sitaramayya A: A novel calcium-dependent activator of retinal rod outer segment membrane guanylate cyclase. Biochemistry 34: 14279–14283, 1995
57. Pozdnyakov N, Goraczniak R, Margulis A, Duda T, Sharma RK, Yoshida A, Sitaramayya A: Structural and functional characterization of retinal calcium-dependent guanylate cyclase activator protein (CD-GCAP): Identity with S100beta protein. Biochemistry 36: 14159–14166, 1997
58. Rambotti MG, Giambanco I, Spreca A, Donato R: S100B and S100A1 proteins in bovine retina: Their calcium-dependent stimulation of a membrane-bound guanylate cyclase activity as investigated by ultracytochemistry. Neuroscience 92: 1089–1101, 1999
59. Burgoyne RD, Weiss JL: The neuronal calcium sensor family of Ca^{2+}-binding proteins. Biochem J 353: 1–12, 2001
60. Paterlini M, Revilla V, Grant AL, Wisden W: Expression of the neuronal calcium sensor protein family in the rat brain. Neuroscience 99: 205–216, 2000

61. Terasawa M, Nakano A, Kobayashi R, Hidaka H: Neurocalcin: A novel calcium-binding protein from bovine brain. J Biol Chem 267: 9596–9599, 1992
62. De Raad S, Comte M, Nef P, Lenz SE, Gundelfinger ED, Cox JA: Distribution pattern of three neural calcium-binding proteins (NCS-1, VILIP and recoverin) in chicken, bovine and rat retina. Histochem J 27: 524–535, 1995
63. Bastianelli E, Takamatsu K, Okazaki K, Hidaka H, Pochet R: Hippocalcin in rat retina. Comparison with calbindin-D28k, calretinin and neurocalcin. Exp Eye Res 60: 257–266, 1995
64. Nakano A, Terasawa M, Watanabe M, Usuda N, Morita T, Hidaka H: Neurocalcin, a novel calcium binding protein with three EF-hand domains, expressed in retinal amacrine cells and ganglion cells. Biochem Biophys Res Commun 186: 1207–1211, 1992
65. Rieke F, Schwartz EA: A cGMP-gated current can control exocytosis at cone synapses. Neuron 13: 863–873, 1994
66. Savchenko A, Barnes S, Kramer RH: Cyclic-nucleotide-gated channels mediate synaptic feedback by nitric oxide. Nature 390: 694–698, 1997
67. Koch KW, Lambrecht HG, Haberecht M, Redburn D, Schmidt H: Functional coupling of a Ca^{2+}/calmodulin dependent nitric oxide synthase and a soluble guanylyl cyclase in vertebrate photoreceptor cells. EMBO J 13: 3312–3320, 1994
68. Margulis A, Pozdnyakov N, Dang L, Sitaramayya A: Soluble guanylate cyclase and nitric oxide synthase in synaptosomal fractions of bovine retina. Vis Neurosci 15: 867–873, 1998
69. Blute TA, Velasco P, Eldred WD: Functional localization of soluble guanylate cyclase in turtle retina: Modulation of cGMP by nitric oxide donors. Vis Neurosci 15: 485–498, 1998
70. Francis SH, Corbin JD: Progress in understanding the mechanism and function of cyclic GMP-dependent protein kinase. Adv Pharmacol 26: 115–170, 1994
71. Vaandrager AB, Ehlert EM, Jarchau T, Lohmann SM, de Jonge H: N-terminal myristoylation is required for membrane localization of cGMP-dependent protein kinase type II. J Biol Chem 271: 7025–7029, 1996
72. Gamm DM, Francis SH, Angelotti TP, Corbin JD, Uhler MD: The type II isoform of cGMP-dependent protein kinase is dimeric and possesses regulatory and catalytic properties distinct from the type I isoforms. J Biol Chem 270: 27380–27388, 1995
73. Vaandrager AB, Tilly BC, Smolenski A, Schneider-Rasp S, Bot AG, Edixhoven M, Scholte BJ, Jarchau T, Walter U, Lohmann SM, Poller WC, de Jonge HR: cGMP stimulation of cystic fibrosis transmembrane conductance regulator Cl-channels co-expressed with cGMP-dependent protein kinase type II but not type I beta. J Biol Chem 272: 4195–4200, 1997
74. Gamm DM, Barthel LK, Raymond PA, Uhler MD: Localization of cGMP-dependent protein kinase isoforms in mouse eye. Invest Ophthalmol Vis Sci 41: 2766–2773, 2000
75. Wexler EM, Stanton PK, Nawy S: Nitric oxide depresses GABAA receptor function via coactivation of cGMP-dependent kinase and phosphodiesterase. J Neurosci 18: 2342–2349, 1998
76. Gregory-Evans K, Kelsell RE, Gregory-Evans CY, Downes SM, Fitzke FW, Holder GE, Simunovic M, Mollon JD, Taylor R, Hunt DM, Bird AC, Moore AT: Autosomal dominant cone-rod retinal dystrophy (CORD6) from heterozygous mutation of GUCY2D, which encodes retinal guanylate cyclase. Ophthalmology 107: 55–61, 2000
77. Blackshaw S, Snyder SH: Developmental expression pattern of phototransduction components in mammalian pineal implies a light-sensing function. J Neurosci 17: 8074–8082, 1997
78. Schomerus C, Ruth P, Korf HW: Photoreceptor-specific proteins in the mammalian pineal organ: Immunocytochemical data and functional considerations. Acta Neurobiol Exp (Warsz) 54 (suppl): 9–17, 1994
79. Semple-Rowland SL, Larkin P, Bronson JD, Nykamp K, Streit WJ, Baehr W: Characterization of the chicken GCAP gene array and analyses of GCAP1, GCAP2, and GC1 gene expression in normal and rd chicken pineal. Mol Vis 5: 14, 1999
80. Yang RB, Foster DC, Garbers DL, Fulle HJ: Two membrane forms of guanylyl cyclase found in the eye. Proc Natl Acad Sci USA 92: 602–606, 1995
81. Racke K, Krupa H, Schroder H, Vollrath L: *In vitro* synthesis of dopamine and noradrenaline in the isolated rat pineal gland: Day–night variations and effects of electrical stimulation. J Neurochem 53: 354–361, 1989
82. Korf HW, Schomerus C, Stehle JH: The pineal organ, its hormone melatonin, and the photoneuroendocrine system. Adv Anat Embryol Cell Biol 146: 1–100, 1998
83. Buck L, Axel R: A novel multigene family may encode odorant receptors: A molecular basis for odor recognition. Cell 65: 175–187, 1991
84. Shepherd GM: Discrimination of molecular signals by the olfactory receptor neuron. Neuron 13: 771–790, 1994
85. Probst WC, Snyder LA, Schuster DI, Brosius J, Sealfon SC: Sequence alignment of the G-protein coupled receptor superfamily. DNA Cell Biol 11: 1–20, 1992
86. Pilpel Y, Lancet D: The variable and conserved interfaces of modeled olfactory receptor proteins. Protein Sci 8: 969–977, 1999
87. Menini A: Calcium signalling and regulation in olfactory neurons. Curr Opin Neurobiol 9: 419–426, 1999
88. Torre V, Ashmore JF, Lamb TD, Menini A: Transduction and adaptation in sensory receptor cells. J Neurosci Dec 15: 7757–7768, 1995
89. Zufall F, Leinders-Zufall T: The cellular and molecular basis of odor adaptation. Chem Senses 25: 473–481, 2000
90. Fulle HJ, Vassar R, Foster DC, Yang RB, Axel R, Garbers DL: A receptor guanylyl cyclase expressed specifically in olfactory sensory neurons. Proc Natl Acad Sci USA 92: 3571–3575, 1995
91. Juilfs DM, Fulle HJ, Zhao AZ, Houslay MD, Garbers DL, Beavo JA: A subset of olfactory neurons that selectively express cGMP-stimulated phosphodiesterase (PDE2) and guanylyl cyclase-D define a unique olfactory signal transduction pathway. Proc Natl Acad Sci U S A 94: 3388–3395, 1997
92. Meyer MR, Angele A, Kremmer E, Kaupp UB, Muller F: A cGMP-signaling pathway in a subset of olfactory sensory neurons. Proc Natl Acad Sci USA 97: 10595–10600, 2000
93. Moon C, Jaberi P, Otto-Bruc A, Baehr W, Palczewski K, Ronnett GV: Calcium-sensitive particulate guanylyl cyclase as a modulator of cAMP in olfactory receptor neurons. J Neurosci 18: 3195–3205, 1998
94. Mori K, Nagao H, Yoshihara Y: The olfactory bulb: Coding and processing of odor molecule information. Science 286: 711–715, 1999
95. Shepherd GM, Greer CA: In: G.M. Shepherd (ed), The Synaptic Organization of the Brain. Oxford University Press, New York, 1998, pp 159–202
96. Pinching AJ, Powell TP: The neuropil of the glomeruli of the olfactory bulb. J Cell Sci 9: 347–377, 1971
97. Price JL, Powell TP: The mitral and short axon cells of the olfactory bulb. J Cell Sci 7: 631–651, 1970
98. Rall W, Shepherd GM, Reese TS, Brightman MW: Dendrodendritic synaptic pathway for inhibition in the olfactory bulb. Exp Neurol 14: 44–456, 1966
99. Friedman D, Strowbridge BW: Functional role of NMDA autoreceptors in olfactory mitral cells. J Neurophysiol 84: 39–50, 2000
100. Tanaka O, Sakagami H, Kondo H: Localization of mRNAs of voltage-dependent Ca(2+)-channels: Four subtypes of alpha 1- and beta-subunits in developing and mature rat brain. Brain Res Mol Brain Res 30: 1–16, 1995

101. Ludwig A, Flockerzi V, Hofmann F: Regional expression and cellular localization of the alpha1 and beta subunit of high voltage-activated calcium channels in rat brain. Neurosci 17: 1339–1349, 1997
102. Park SH, Suh YS, Kim H, Rhyu IJ, Kim HL: Chromosomal localization and neural distribution of voltage dependent calcium channel beta 3 subunit gene. Mol Cells 7: 200–203, 1997
103. Hobom M, Dai S, Marais E, Lacinova L, Hofmann F, Klugbauer N: Neuronal distribution and functional characterization of the calcium channel alpha2delta-2 subunit. Eur J Neurosci 12: 1217–1226, 2000
104. Strijbos PJLM, Pratt GD, Khan, S, Charles IG and Garthwaite J: Molecular characterization and *in situ* localization of a full-length cyclic nucleotide-gated channel in rat brain. Eur J Neurosci 11: 4463–4467, 1999

Molecular and Cellular Biochemistry **230:** 117–124, 2002.

Calcium-sensitive ROS-GC1 signaling outside of photoreceptors: A common theme

Venkateswar Venkataraman[1] and Robert G. Nagele[2]

[1]*The Unit of Regulatory and Molecular Biology, Departments of Cell Biology and Ophthalmology;*
[2]*Department of Molecular Biology, SOM and NJMS University of Medicine and Dentistry, Stratford, NJ, USA*

Abstract

Until now, ROS-GC1 signal transduction system was thought to be exclusive to photoreceptors in the retina. Two recent reports, however, now show that this is not the case. In one, the ROS-GC1 signal transduction system has been identified and characterized in pinealocyte neurons. This signaling is modulated by norepinephrine. However, the response of the individual pinealocyte neuron to the norepinephrine signal depends on whether the GCAP1-linked (results in hyperpolarization) or S100β-linked (results in depolarization) pathway is operational in the pinealocyte. The GCAP1-linked pathway results in hyperpolarization, while the S100β-linked pathway, in depolarization. The two pathways are mutually exclusive. In the other report, the calcium-modulated ROS-GC1:GCAP1 signaling system has been discovered in mitral cells of the olfactory bulb. These findings raise the possibility that a common theme of calcium-modulated ROS-GC signaling may be utilized in a wide variety of neurosensory cells. This idea is also supported from evolutionary and functional perspectives. (Mol Cell Biochem **230**: 117–124, 2002)

Key words: ROS-GC1, norepinephrine, calcium, pineal gland, olfactory bulb

Introduction

ROS-GC1, a membrane guanylate cyclase (mGC), is a pivotal molecule in phototransduction (reviewed in [1]). It anchors the reciprocal feedback loop involving Ca^{2+} and cyclic GMP, which governs the dark, light and recovery phases in the phototransduction process. Until now, this is the only function that has been ascribed to it, because mutations of ROS-GC1 have been linked to congenital retinal disorders causing blindness [2–8]. An extensive analysis of the biochemical and functional properties of ROS-GC1 has, largely, made it possible to explain the pathology of these disorders based on the known molecular mechanism of ROS-GC1 signaling [2–4, 8]. ROS-GC1 is finely tuned to respond to changes in cellular Ca^{2+} levels. To date, the Ca^{2+}-modulated ROS-GC signaling system has been believed to be unique to photoreceptor cells. There it plays a vital role in the light, dark and recovery cycles, anchoring, via cyclic GMP, the feedback loop linking the cellular calcium levels and the polarization state of the photoreceptor (reviewed in [1]). ROS-GC1 in photoreceptors is Ca^{2+}-inhibited. In non-photoreceptor neurons, ROS-GC1 is both Ca^{2+}-inhibited and Ca^{2+}-stimulated. Thus, ROS-GC1 is a bimodal Ca^{2+} switch (reviewed in [1]). In the inhibitory mode, increasing Ca^{2+} concentrations, with a $K_{1/2}$ near 100 nM, progressively inhibits the cyclase [9–14]. In the stimulatory mode, Ca^{2+} stimulates the cyclase with a $K_{1/2}$ near 1 μM [15–18; reviewed in 20]. These two opposing modes of Ca^{2+}-mediated modulation are conducted by two different classes of Ca^{2+}-binding proteins: guanylate cyclase activating proteins (GCAPs) and calcium-dependent GCAPs (CD-GCAPs). Each GCAP or CD-GCAP modulates ROS-GC1 through its specified binding motif (module), which resides in the intracellular region of the cyclase [11, 12, 17, 18, 20, 21]. Upon binding to its module, it stimulates or inhibits ROS-GC1. The inhibitory action, mediated primarily by GCAP1 (a GCAP), is thought to be specific for phototransduction. The stimulatory action, primarily mediated by S100β (a CD-GCAP), has been proposed to play a role in synaptic activity. Various aspects of the role of ROS-GC1 in phototransduction – the biochemical basis of congenital disorders due to ROS-GC1 mutations, the modulation of ROS-GC1 by GCAP1 and S100β, their physiological

Address for offprints: V. Venkataraman, The Unit of Regulatory and Molecular Biology, Departments of Cell Biology and Ophthalmology, University of Medicine and Dentistry, Stratford, NJ 08084, USA (E-mail: venkatar@umdnj.edu)

implications – have all been reviewed individually and exhaustively in this issue. However, recent reports have provided evidence for an unexpected versatility of the Ca^{2+}-modulated ROS-GC signaling system, which was thought to be unique to photoreceptors. Its detection in the pineal gland and linkage to norepinephrine signaling suggested that ROS-GC1 might be involved in general neuronal signaling. Modulation by Ca^{2+}, the ability to influence the polarity of the cell membrane through cGMP-gated channels, the responsiveness to a neurotransmitter and extremely rapid response times (nanoseconds) makes ROS-GC transduction an attractive candidate for neuronal signaling. This review presents the idea that modulation of ROS-GC1 by Ca^{2+} is a common theme adapted universally by neurosensory cells and that photoreceptors are only one example where this signaling pathway has been recruited. Recent biochemical, functional, molecular and immunocytochemical evidence that have led to this idea are presented. In addition, a model for neuronal ROS-GC1 signaling is proposed and the implications of the widening horizon of ROS-GC1 functions are discussed, including an evolutionary perspective to ROS-GC functions in the neurosensory system.

The ROS-GC1 signal transduction system is present in pinealocyte neurons and is modulated by norepinephrine

On the trail of a norepinephrine-modulated mGC

The seed for the field of ROS-GC1 signaling in neurosensory cells was planted more than two decades ago by two initial observations. (1) An epinephrine-dependent mGC transduction system, not present in the normal adrenal cortex, is present in the rat adrenocortical carcinoma 494 cells; (2) epinephrine causes the activation of the mGC, which is mediated by an α-adrenergic receptor (α-AR) [22, 23]. Given the lack of knowledge on the existence of multiple forms of α-AR and mGCs at that time, the biochemical linkage between the signaling receptor and the transduction component remained unresolved. Subsequent purification, characterization [24, 25], and cloning [26, 27] of the α-AR demonstrated that the receptor was, pharmacologically, a newly defined α_{2D}-AR subtype [27–29]. This α_{2D}-AR receptor subtype is a species variant of the pharmacologically distinct α_{2A}-AR [27, 29–31] and therefore, is referred to as $\alpha_{2D/A}$-AR [32]. These findings lead to the inference that the original carcinoma receptor subtype linked to the mGC is $\alpha_{2D/A}$-AR. The identity of the mGC, however, remained elusive: other than epinephrine-responsiveness, the only other biochemical feature known about the mGC was that it was Ca^{2+}-modulated [33]. Thus, it did not belong to the hormone peptide receptor family whose members are regulated by the peptide hormones (reviewed in [1]). The mystery surrounding the identity of mGC and its linkage with $\alpha_{2D/A}$-AR signaling has now been solved by the discovery of a rod outer segment membrane guanylate cyclase (ROS-GC) subfamily [9, 34, 35]. This subfamily is Ca^{2+}-modulated, and one member of the subfamily – ROS-GC1 – is linked with $\alpha_{2D/A}$-AR signaling in the pinealocyte neurons.

Calcium couples norepinephrine-activated $\alpha_{2D/A}$-AR and ROS-GC1 in pinealocyte neurons

The existence of an $\alpha_{2D/A}$-AR-linked ROS-GC1 signal transduction system was first demonstrated in the pineal gland [36]. This system shares its features with the original rat carcinoma signaling system because: (1) stimulation of the pineal ROS-GC1 is specifically mediated by α_{2D}-AR and involves Ca^{2+}; (2) the signaling is indirect, and; (3) stimulation of the cyclase can be independently mimicked by the pineal soluble fraction in a Ca^{2+}-dependent fashion [36]. Based on these findings, the following sequence of events are envisioned for the signal transduction model: (i) the signaling system is activated by binding of epinephrine to its α_{2D}-AR; (ii) this causes an elevation in levels of intracellular free Ca^{2+}; (iii) Ca^{2+} then binds to a ROS-GC1 modulator; and (iv) the Ca^{2+}-bound modulator alters the catalytic activity of ROS-GC1, influencing the production of cGMP [36].

Two opposing modes of coupling occur in distinct subpopulations of pinealocyte neurons

Detailed biochemical, molecular and immunocytochemical analysis of the pineal gland has shown that the ROS-GC transduction system is present only in the pinealocyte neurons and not in surrounding glial cells [37]. These analyses provide important insights into operational differences between pinealocyte and photoreceptor ROS-GC transduction systems. The pinealocytes express only ROS-GC1 along with $\alpha_{2D/A}$-AR. Unlike photoreceptor cells, they do not express ROS-GC2 [37, 38]. A surprising and highly intriguing finding is that the receptor and the cyclase are always accompanied by only one modulator, S100β or GCAP1, and never both [37]. In the majority of the pinealocytes the modulator is GCAP1 [37]. This indicates that the S100β-dependent Ca^{2+} signaling is restrictive, occurring only in a relatively small number of pinealocytes. The third modulator, GCAP2, expressed in photoreceptors [39], is not found in pinealocytes [37]. This indicates that, at a functional level, there are two types of pinealocytes; one behaving like a 'photoreceptor' and containing GCAP1, and the other behaving like a non-photoreceptor, retinal 'neuron' and containing S100β. Both pinealocyte types contain a common ROS-GC1 and also are linked to a common $\alpha_{2D/A}$-AR [37]. However, it is the identity and nature of the ROS-GC1 modulator in each pinealocyte that dictates the end result of the operation of the ROS-GC system in that pinealocyte. In response to a Ca^{2+} spike, the GCAP1-driven system is inhibited and the S100β-driven system, stimulated. A model illustrating the operation of these two pathways is presented in Fig. 1. As shown, binding of norepinephrine to the $\alpha_{2D/A}$-AR activates the pathway. This results in an elevation in cellular calcium, which is driven by entry of calcium through voltage-gated calcium channels and/or release from intracellular stores. These events are common to both pathways. Once the calcium levels are elevated, in the GCAP1-linked pathway, present in the majority of the pinealocyte neurons, binding of calcium to GCAP1 causes inhibition of ROS-GC1 activity and, consequently, depletion of cyclic GMP. This results in the closure of cyclic GMP-gated channels and hyperpolarization. On the other hand, in the S100β-linked pathway, calcium-bound S100β stimulates the cyclase, causing enhanced cyclic GMP synthesis, opening of cyclic GMP-gated channels and, consequently, depolarization. This model assumes that, like photoreceptors, all sensory neurons have a cyclic GMP-gated channel. In the case of pinealocytes, the existence of cyclic GMP-gated channels is well documented [40, 41]. This model clearly explains how the same ROS-GC, through the use of different

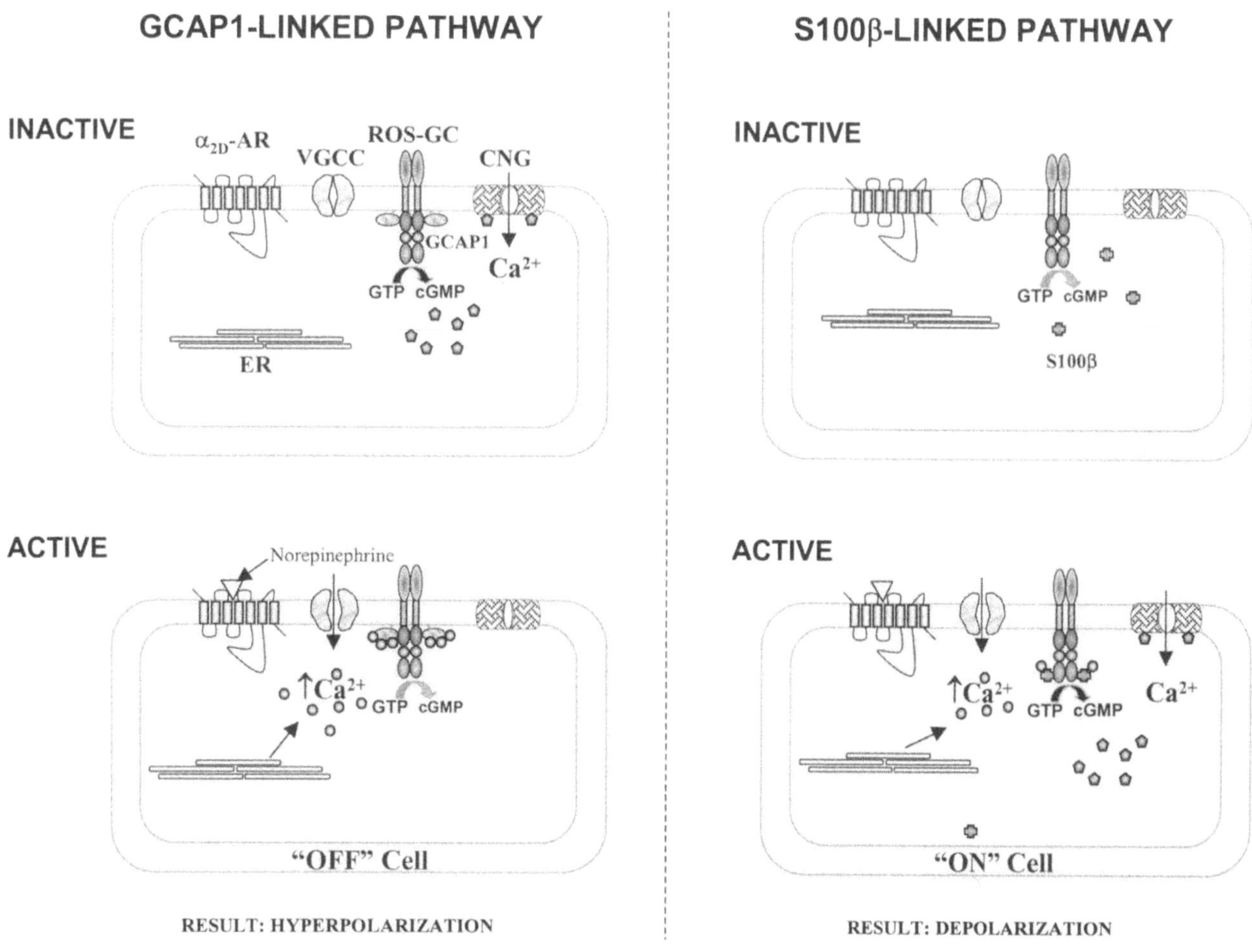

Fig. 1. Two opposite Ca^{2+}-modulated ROS-GC pathways linked to α_{2D}-AR. The model depicts the linkage of α_{2D}-AR to two opposite modes of Ca^{2+}-modulated ROS-GC signaling mediated by GCAP1 and S100β in separate panels. The signal, in both cases, is the binding of norepinephrine to the α_{2D}-AR, which causes an elevation in cellular Ca^{2+}, presumably through voltage-sensitive calcium channel and/or release from intracellular stores (typically the endoplasmic reticulum). In the GCAP1-linked pathway, this results in the binding of Ca^{2+} to the ROS-GC1-bound GCAP1 and results in inhibition of guanylate cyclase activity. The consequent decline in cyclic GMP levels results in the closure of cyclic GMP-gated channels and hyperpolarization. In the S100β-linked pathway, the elevated Ca^{2+} levels lead to the binding of Ca^{2+}-bound S100β to the cyclase. This causes activation of the cyclase, increase in cyclic GMP concentration and, consequently, opening of the cyclic GMP-gated channels causing depolarization. α_{2D}-AR – α_{2D}-adrenergic receptor; VGCC – voltage-gated calcium channel; ER – endoplasmic reticulum; ROS-GC – rod outer segment guanylate cyclase; CNG – cyclic GMP (nucleotide)-gated channel.

Ca^{2+}-sensitive modulators, can produce two entirely opposite physiological responses, hyper- and de-polarization, in response to a single signal. In response to norepinephrine, there will be hyper-polarization in a GCAP1-linked ROS-GC system; and depolarization, in a S100β linked system. It is also reasonable to propose that any signal other than norepinephrine, which is capable of altering intracellular calcium levels, would have a similar effect on the ROS-GC transduction system depicted in Fig. 1.

Physiological implications

S100β linked system

One attractive possibility for the S100β-driven ROS-GC1 transduction system is that it might play a role in the light-regulated melatonin secretion [37]. Synthesis and secretion of melatonin in the pineal gland is light regulated. Light inhibits the synthesis of melatonin, and melatonin secretion is regulated by epinephrine (reviewed in [42]). At night, norepinephrine levels in the pineal gland significantly rise, and the sympathetic nerves in the pineal gland are capable of producing norepinephrine [43]. In the current model of light-regulated melatonin secretion [37], the generated norepinephrine via $\alpha_{2D/A}$-AR causes an elevation in intracellular Ca^{2+} levels, Ca^{2+} binds to S100β, which then activates the cyclase. The cyclic GMP thus produced mediates melatonin synthesis and secretion.

GCAP1 linked system

To deduce the physiology of the GCAP1-linked ROS-GC1 transduction system in the pinealocyte neurons, its analogy is drawn from the retinal phototransduction system. Current operational principals of phototransduction involve regulation of dark and light states of photoreceptors by ROS-GC1 via its Ca^{2+}-sensing modulator GCAP1. In the dark state, when Ca^{2+} levels are around 500 nM, GCAP1 bound to ROS-GC1 keeps the cyclase in its basal state. In the light state, Ca^{2+} levels decrease to 50 nM due to closure of cyclic GMP-gated channels resulting from amplified hydrolysis of cyclic GMP. Under these conditions, GCAP1 bound to ROS-GC1 stimulates the cyclase activity, resulting in elevation of cyclic GMP leading to the recovery state. Thus, in photoreceptors, Ca^{2+} via GCAP1 oscillates ROS-GC1 activity between basal (inhibited) and stimulated states, depending on the presence of a light stimulus (reviewed in [1]). In the mammalian pineal gland, where there is no direct light stimulus, it is possible that this role is taken over by photosensitive oscillations of epinephrine/norepinephrine [37]. Elevation of Ca^{2+} from norepinephrine signaling would induce a dark-like state in pinealocytes and the removal of this signal, a light-like state. Thus, in this hypothetical scheme [37], cycling between dark and light in the pineal gland is mediated by photosensitive oscillations of epinephrine/norepinephrine levels, which provides a plausible sensing mechanism of light by the pineal gland termed photoentrainment [44]. It is noteworthy that the postganglionic nerve fibers that innervate the pineal gland and release norepinephrine in circadian fashion arise from the superior cervical ganglion, which receives input from the ganglion cells in the retina through the suprachiasmatic nucleus, and intermediolateral column (reviewed in [42]). It is conceivable that the GCAP1-modulated ROS-GC1 machinery might be the one that is propelled by photo-entrainment.

General pineal function

In addition to distinct physiological roles for the two types of pinealocytes, these observations have other implications in terms of general function of pineal gland. These studies, for the first time, clearly show that two separate and distinct subtypes of pinealocyte neurons are identifiable based on their response to norepinephrine stimulus: one subtype of pinealocytes (GCAP1-linked) would exhibit hyperpolarization, while the other (S100β-linked) would exhibit depolarization. An identical situation is observed in retinal bipolar and ganglion neurons: the 'ON' type respond with depolarization to light signal, while the 'OFF' type, with hyperpolarization. Based on this analogy, the GCAP1-containing pinealocytes would be 'OFF' pinealocytes, while the S100β-containing pinealocytes would be 'ON' pinealocytes (Fig. 1). Why are the pinealocyte neurons designed to exhibit opposing responses to a single signal? One plausible answer is that this 'check and balance' provides a fine, graded response, rather than a crude 'all or none' response. Another possibility is an accentuation of the signal to noise ratio upon integration of the two responses. In the case of the visual system, for example, integration of the 'ON' and 'OFF' responses would provide a better 'contrast' to heighten the perception of the object. It is axiomatic that such integration would benefit each neurosensory process. The unique bimodal calcium switch in the ROS-GC1 signaling system makes it an ideal component for such dual processing of input signals. Is this signal transduction system a common feature of all sensory neurons?

Calcium-modulated ROS-GC1 signaling exists in the olfactory bulb

In an attempt to scrutinize the presence of the ROS-GC1 signaling system in the olfactory system, the olfactory epithelium – the site of odorant transduction – and the olfactory bulb, which receives input from the epithelial sensory neurons, were tested for the presence of a calcium-sensitive mGC. The results of the functional, biochemical, immunocytochemical and molecular analyses were reported recently [45] and demonstrated that (1) An mGC with the biochemical characteristics of ROS-GC1 is expressed in the olfactory bulb, but not the olfactory epithelium; (2) the GCAP1-ROS-GC1 signaling system, but not the GCAP2-ROS-GC2 system, is expressed in the olfactory bulb; (3) the expression of GCAP1 and ROS-GC1 mRNA as well as the encoded protein is colocalized in the mitral cells of the olfactory bulb. Expression of ROS-GC1, however, is not restricted to the mitral cells. Thus, ROS-GC1 and its Ca^{2+}-sensitive modulator, GCAP1, are synthesized and expressed in the mitral neurons wherein, they are the sole the components of Ca^{2+}-dependent mGC system.

Implications

What could be the physiological role of the ROS-GC1: GCAP1 signaling system in mitral cells? The mitral cell soma is the site of integration of a variety of input signals such as the excitatory signals from the olfactory sensory neurons (reviewed in [46]), dendrodendritic inhibitory signals from neighboring cells [47, 48] and autoexcitatory signals [49]. As a result of integration of these inputs in mitral cells, the initial excitatory signal from the olfactory sensory neurons may be strengthened, dampened or stopped. This process leads to a rapidly changing calcium levels in the mitral cell soma. The ROS-GC1:GCAP1 signal transduction system in the mitral cell body, is ideally placed to sense these rapid changes in Ca^{2+} levels which will dampen or activate the cyclase. It is known that mitral cells express both voltage-dependent as well as cyclic nucleotide-gated Ca^{2+} channels [50–53]. Thus, proportional levels of cyclic GMP synthesized by the cyclase would regulate the channel activity, which, in turn, would determine the polarization state of the soma.

In the specific case of autoexcitation, for example, release of glutamate from the cell body is triggered by calcium influx. In a low calcium environment, GCAP1-bound ROS-GC1 will be activated resulting in enhanced synthesis of cyclic GMP. This will facilitate regenerative calcium spikes by increasing the entry of calcium through cyclic GMP-gated channels. The resulting calcium influx will cause two events: (1) release of glutamate from the cell body; and (2) inhibition of the cyclase catalytic activity, leading to decrease in cyclic GMP. Thus, the ROS-GC1:GCAP1 signaling system provides an excellent autoregulatory loop to sense and modulate Ca^{2+} spikes. It is conceivable that the voltage-gated Ca^{2+} channels are involved in this process in addition to the cyclic GMP-gated channels. The model is, necessarily, speculative since this is the first report of a calcium-sensitive mGC in the mitral cells. However, it is also a testable one because the features of the ROS-GC1:GCAP1 signaling system are well established.

Significantly, these observations provide the next tier of evidence that the ROS-GC1 signal transduction system is widespread in neurosensory cells and lays to rest the belief held hitherto that it is unique to photoreceptors. While the mitral cell signaling system corresponds to the GCAP1-linked pathway in Fig. 1, it also expands the scope of the model by highlighting the fact that the ROS-GC1:GCAP1 system can serve as the focus of integration of a wide variety of signals that involve fluctuations in cellular calcium. Thus, the calcium-modulated ROS-GC1 transduction system can be recruited to accomplish different physiological goals in different systems: for example, phototransduction in photoreceptors, possibly photoentrainment and melatonin secretion in pinealocytes and signal integration in mitral cells.

ROS-GC and neuronal signaling – fortuity or design?

The family of mGCs is a diverse one: they are involved in a wide variety of cellular signaling pathways. There are also many members whose function is not yet known. In order to gain insight into the precise structural and evolutionary position of ROS-GC1 in this family tree, the protein sequences of known membrane guanylate cyclases were retrieved from the GenBank based on homology to ROS-GC1 and aligned. The phylogenetic tree based on the alignment of full-length protein is presented in Fig. 2. The sequences included represent mGCs from a wide range of species – from the nematode *C. elegans* to human. It is evident that there is extensive diversification of mGCs during evolution, as indicated by the number of branches. It is also clear that most invertebrate mGCs segregate into a group towards the lower half of the tree (Fig. 2; CELEMGCs, APUNGC, SPERGC, and HPULGC). However, that function was the primary driving force rather than speciation is indicated by the clustering of the natriuretic peptide receptor mGCs from vertebrates (ANFRGCs and CNPRGCs) with the mGCs from echinoderms (HPULGC, APUNGC, SPERGC), fly (DMELMGC) and nematode (CELEMGC). The segregation of the natriuretic peptide cluster from the ROS-GC cluster occurred quite early (indicated by an arrow). The STARGCs, which are known to be receptors for stable enterotoxins (STa), have diverged early from the ROS-GC cluster. Interestingly, GCG, the rat mGC designated as an orphan receptor, occupies an independent position. In the top half of tree reside the members of ROS-GC family, which are modulated by calcium. Two distinct clusters are formed by ROS-GC1s and ROS-GC2s. It is also clear that the segregation into ROS-GC1 and ROS-GC2 occurred earlier than segregation of an individual ROS-GC into species-specific forms. Thus, the phylogenetic tree supports the idea that the ROS-GC family

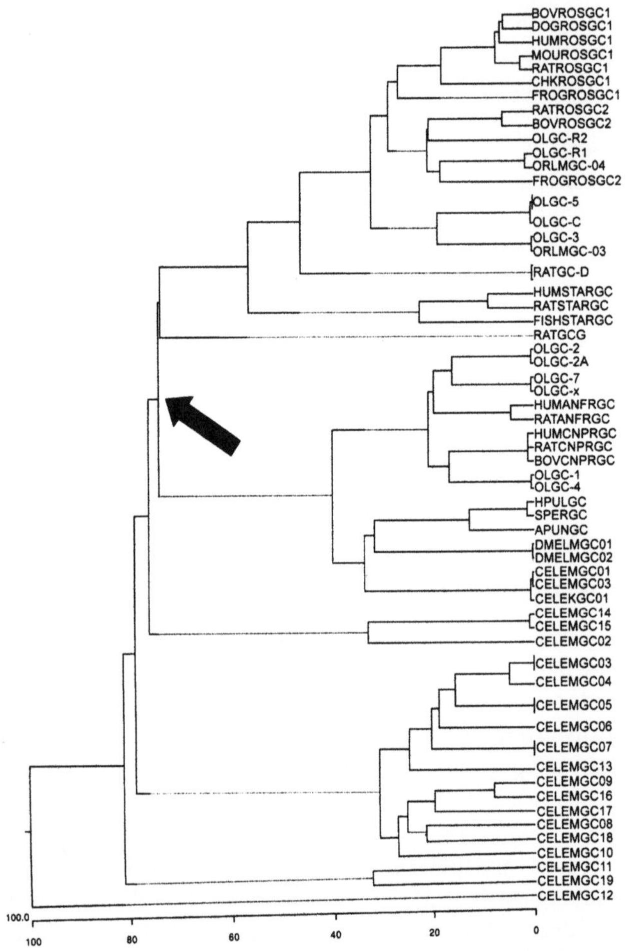

Fig. 2. Phylogenetic tree of membrane guanylate cyclases. A phylogenetic tree of known membrane guanylate cyclases was constructed using Clustal (Megalign program of DNASTAR). The results were confirmed by using GCG software. The alignment was performed with a gap length of 10 and a gap penalty of 10. Membrane guanylate cyclases from the invertebrates *C. elegans* (CELEMGC), *Drosophila melanogaster* (DMEMGC), *Strongylocentratus purpuratus* (SPERGC), *Hemicentrotus pulcherrimus* (HPULGC), and *Arbacia puctulata* (APUNGC) were aligned with those from vertebrates: fish (OLGC), frog (FROG), chicken (CHK), mouse (MOU), rat (RAT), bovine (BOV), and human (HUM). Wherever known, the cyclase has been named. Otherwise, numbers have been provided. The sequences were obtained from GenBank through a FASTA search, using a sequence from the ROS-GC1 catalytic domain.

in general, and ROS-GC1 in particular, was derived from an ancestral form early in evolution and that ROS-GCs are designed for a basic common function, which has undergone diversification over time. This conclusion is further supported by the fact that a similar tree is obtained irrespective of which region (extracellular region alone, catalytic domain alone, entire protein) of mGCs were compared (Fig. 2 is based on alignment of the full-length proteins; individual domain-based alignments have not been shown).

That calcium-sensitive mGC signaling is an ancient phenomenon is also substantiated from a functional perspective: (1) ROS-GC homologs in Oryzias latipes are expressed in retina, pineal organ and olfactory system [54, 55]; (2) calcium-modulated mGCs have been shown in *Paramecium* and *Tetrahymena* [56]. Interestingly, in *Paramecium*, the calcium-modulated mGC plays a role in regulating the circadian behavioral rhythms of the organism [57]. It is natural to see that the ancient linkage between calcium signaling and mGC transduction is manifested in the one type of cell where calcium signaling is critical – neurons.

Conclusion

To date, there are two systems outside of photoreceptors where ROS-GC signaling has been demonstrated unequivocally: in the pinealocyte neurons and the mitral cells of the olfactory bulb. Fragmentary evidence for the presence of ROS-GC1 transcripts in regions of the brain and auditory system also exist [58, 59]. As a result, the idea that the ROS-GC1 transduction system is restricted to photoreceptors is no longer valid. It is fitting that such an exquisitely designed bimodal calcium switch is used in multiple neurosensory systems to achieve distinct physiological goals. Its linkage to calcium signaling and modulation by neurotransmitters confers upon it versatility and an ability to respond to a wide variety of signals. A clear demonstration of the role of ROS-GC1 in processes other than vision remains to be accomplished, but has been indicated. It is noteworthy that, to date, no report describing the expression of the other calcium-modulated retinal ROS-GC, ROS-GC2, outside of photoreceptors exists. However, absence of evidence is not evidence of absence. A set of ROS-GCs, each with a distinct complement of calcium-dependent modulators and modulatory profiles, provides a rich permutation of options to channel a given signal. Thus, while individual molecules may vary in different neurons and may achieve different goals therein, a common theme of calcium-dependent modulation of an mGC is predicted to exist in the neurosensory cells.

Acknowledgements

The author (V.V.) wishes to acknowledge the cherished guidance and support of Drs. Rameshwar K. Sharma and Teresa Duda in this endeavor. He also wishes to thank all his teachers at various stages in his career for their dedication, help and encouragement, without which this would not have been possible.

References

1. Pugh EN Jr, Duda T, Sharma RK, Sitaramayya A: Photoreceptor guanylate cyclases: A review. Biosci Rep 17: 429–473, 1997
2. Duda T, Venkataraman V, Goraczniak R, Lange C, Koch K-W, Sharma RK: Functional consequences of a rod outer segment membrane guanylate cyclase (ROS-GC1) gene mutation linked with Leber's congenital amaurosis. Biochemistry 38: 509–515, 1999
3. Duda T, Krishnan A, Venkataraman V, Lange C, Koch KW, Sharma RK: Mutations in the rod outer segment membrane guanylate cyclase in a cone-rod dystrophy cause defects in calcium signaling. Biochemistry 38: 13912–13919, 1999
4. Duda T, Venkataraman V, Jankowska A, Lange C, Koch KW, Sharma RK: Impairment of the rod outer segment membrane guanylate cyclase dimerization in a cone-rod dystrophy results in defective calcium signaling. Biochemistry 39: 12522–12533, 2000
5. Gregory-Evans K, Kelsell RE, Gregory-Evans CY, Downes SM, Fitzke FW, Holder GE, Simunovic M, Mollon JD, Taylor R, Hunt DM, Bird AC, Moore AT: Autosomal dominant cone-rod retinal dystrophy (CORD6) from heterozygous mutation of GUCY2D, which encodes retinal guanylate cyclase. Ophthalmology 107: 55–61, 2000
6. Kelsell RE, Gregory-Evans K, Payne AM, Perrault I, Kaplan J, Yang RB, Garbers DL, Bird AC, Moore AT, Hunt DM: Mutations in the retinal guanylate cyclase (RETGC-1) gene in dominant cone-rod dystrophy. Hum Mol Genet 7: 1179–1184, 1998
7. Perrault I, Rozet JM, Calvas P, Gerber S, Camuzat A, Dollfus H, Chatelin S, Souied E, Ghazi I, Leowski C, Bonnemaison M, Le Paslier D, Frezal J, Dufier JL, Pittler S, Munnich A, Kaplan J: Retinal-specific guanylate cyclase gene mutations in Leber's congenital amaurosis. Nature Genet 14: 461–464, 1996
8. Tucker CL, Woodcock SC, Kelsell RE, Ramamurthy V, Hunt DM, Hurley JH: Biochemical analysis of a dimerization domain mutation in Ret-GC1 associated with dominant cone-rod dystrophy. Proc Natl Acad Sci USA 96: 9039–9044, 1999
9. Dizhoor AM, Lowe DG, Olshevskaya EV, Laura RP, Hurley JB: The human photoreceptor membrane guanylyl cyclase, RetGC, is present in outer segments and is regulated by calcium and a soluble activator. Neuron 12: 1–20, 1994
10. Frins S, Bonigk W, Muller F, Kellner R, Koch K-W: Functional characterization of a guanylyl cyclase-activating protein from vertebrate rods. Cloning, heterologous expression, and localization. J Biol Chem 271: 8022–8027, 1996
11. Duda T, Goraczniak R, Surgucheva I, Rudnicka-Nawrot M, Gorczyca WA, Palzcewski K, Sitaramayya A, Baehr W, Sharma RK: Calcium modulation of bovine photoreceptor guanylate cyclase. Biochemistry 35: 8478–8482, 1996
12. Krishnan A, Goraczniak RM, Duda T, Sharma RK: Third calcium-modulated rod outer segment membrane guanylate cyclase transduction mechanism. Mol Cell Biochem 178: 251–259, 1998
13. Gorczyca WA, Gray-Keller MP, Detwiler PB, Palczewski K: Purification and physiological evaluation of a guanylate cyclase activating protein from retinal rods. Proc Natl Acad Sci USA 91: 4014–4018, 1994
14. Palczewski K, Subbaraya I, Gorczyca WA, Helekar BS, Ruiz CC, Ohguro H, Huang J, Zhao X, Crabb JW, Johnson RS, Walsh KA, Gray-Keller MP, Detweiler P, Baehr W: Molecular cloning and characterization of retinal photoreceptor guanylate cyclase-activating protein. Neuron 13: 395–404, 1994
15. Pozdnyakov N, Yoshida A, Cooper NG, Margulis A, Duda T, Sharma RK, Sitaramayya A: A novel calcium-dependent activator of retinal rod outer segment membrane guanylate cyclase. Biochemistry 34: 14279–14283, 1995
16. Margulis A, Pozdnyakov N, Sitaramayya A: Activation of bovine photoreceptor guanylate cyclase by S100 proteins. Biochem Biophys Res Commun 218: 243–247, 1996
17. Duda T, Goraczniak RM, Sharma RK: Molecular characterization of S100A1-S100B protein in retina and its activation mechanism of bovine photoreceptor guanylate cyclase. Biochemistry 35: 6263–6266, 1996
18. Kumar VD, Vijay-Kumar S, Krishnan A, Duda T, Sharma RK: A second calcium regulator of rod outer segment membrane guanylate cyclase, ROS-GC1: Neurocalcin. Biochemistry 38: 12614–12620, 1999
19. Sitaramayya A, Pozdnyakov N, Margulis A, Yoshida A: Calcium-dependent activation of membrane guanylate cyclase by S100 proteins. Meth Enzymol 315: 730–742, 2000
20. Lange C, Duda T, Beyermann M, Sharma RK, Koch K-W: Regions in vertebrate photoreceptor guanylyl cyclase ROS-GC1 involved in Ca($^{2+}$)-dependent regulation by guanylyl cyclase-activating protein GCAP-1. FEBS Lett 460: 27–31, 1999
21. Sokal I, Haeseleer F, Arendt A, Adman ET, Hargrave PA, Palczewski, K: Identification of a guanylyl cyclase-activating protein-binding site within the catalytic domain of retinal guanylyl cyclase 1. Biochemistry 38: 1387–1393, 1999
22. Perchellet J-P, Sharma RK: Ectopic-adrenergic accumulation of guanosine 3′:5′-monophosphate in isolated adrenocortical carcinoma cells. Endocrinology 106: 1589–1593, 1980
23. Shanker G, Sharma RK: Characterization of ectopic-adrenergic binding receptors of adrenocortical carcinoma cells. Endocrinology 106: 1594–1598, 1980
24. Nambi P, Aiyar NV, Sharma RK: Solubilization of epinephrine-specific α_2-adrenergic receptors from adrenocortical carcinoma. FEBS Lett 140: 98–102, 1982
25. Jaiswal RK, Sharma RK: Purification and characterization of α_2-adrenergic receptors from the rat adrenocortical carcinoma. Biochem Biophys Res Commun 130: 58–64, 1985
26. Chalberg SC, Duda T, Rhine JA, Sharma RK: Molecular cloning, sequencing, and expression of an α_2-adrenergic receptor complementary DNA from rat brain. Mol Cell Biochem 97: 161–172, 1990
27. Lanier SM, Downing S, Duzic E, Homcy CJ: Isolation of rat genomic clones encoding subtypes of the α_2-adrenergic receptor. J Biol Chem 266: 10470–10478, 1991
28. O'Rourke MF, Iversen LJ, Lomasney JW, Bylund DB: Species orthologs of the alpha-2A adrenergic receptor: The pharmacological properties of the bovine and rat receptors differ from the human and porcine receptors. J Pharmacol Exp Ther 271: 735–740, 1994
29. Wypijewski K, Duda T, Sharma RK: Structural, genetic and pharmacological identity of the rat α_2-adrenergic receptor subtype cA2-47 and its molecular characterization in rat adrenal, adrenocortical carcinoma and bovine retina. Mol Cell Biochem 144: 181–190, 1995
30. Regan JW, Nakata H, DeMarinis RM, Caron M, Lefkowitz RJ: Purification and characterization of the human platelets α_2-adrenergic receptor. J Biol Chem 261: 3894–3900, 1986
31. Kobilka BK, Matsui H, Kobilka TS, Yang-Feng TL, Francke U, Caron MG, Lefkowitz RJ, Regan JW: Cloning, sequencing and expression of the gene coding for the human platelet α_2-adrenergic receptor. Science 238: 650–656, 1987
32. Bylund DB, Eikenberg DC, Hieble JP, Langer SZ, Lefkowitz RJ, Minneman KP, Molinoff PB, Ruffolo RR Jr, Trendelenburg U: International union of pharmacology nomenclature of adrenoceptors. Pharmacol Rev 46: 121–136, 1994
33. Jaiswal N, Sharma RK: Dual regulation of adenylate cyclase and guanylate cyclase: α_2-Adrenergic signal transduction in adrenocortical carcinoma cells. Arch Biochem Biophys 249: 616–619, 1986
34. Goraczniak RM, Duda T, Sitaramayya A, Sharma RK: Structural and functional characterization of the rod outer segment membrane guanylate cyclase. Biochem J 302: 455–461, 1994

35. Yang, R-B, Foster DC, Garbers DL, Fulle HJ: Two membrane forms of guanylate cyclase found in the eye. Proc Natl Acad Sci USA 92: 602–606, 1995
36. Venkataraman V, Duda T, Sharma RK: The $\alpha_{2D/A}$-adrenergic receptor linked membrane guanylate cyclase: A new signal transduction system in the pineal gland. FEBS Lett 427: 69–73, 1998
37. Venkataraman V, Nagele R, Duda T, Sharma RK: Rod outer segment membrane guanylate cyclase type 1-linked stimulatory and inhibitory calcium signaling systems in the pineal gland: Biochemical, molecular, and immunohistochemical evidence. Biochemistry 39: 6042–6052, 2000
38. Goraczniak RM, Duda T, Sharma RK: Structural and functional characterization of a second subfamily member of the calcium-modulated bovine rod outer segment membrane guanylate cyclase, ROSGC2. Biochem Biophys Res Commun 234: 666–670, 1997
39. Otto-Bruc A, Fariss RN, Haeseleer F, Huang J, Buczylko J, Surgucheva I, Baehr W, Milam AH, Palczewski K: Localization of guanylate cyclase-activating protein 2 in mammalian retinas. Proc Natl Acad Sci USA 94: 4727–4732, 1997
40. Sautter A, Biel M, Hofmann F: Molecular cloning of cyclic nucleotide-gated cation channel subunits from rat pineal gland. Brain Res Mol Brain Res 48: 171–175, 1997
41. Bonigk W, Muller F, Middendorff R, Weyand I, Kaupp UB: Two alternatively spliced forms of the cGMP-gated channel alpha-subunit from cone photoreceptor are expressed in the chick pineal organ. J Neurosci 16: 7458–7468, 1996
42. Korf HW, Schomerus C, Stehle JH: The pineal organ, its hormone melatonin, and the photoneuroendocrine system. Adv Anat Embryol Cell Biol 46: 1–100, 1998
43. Racke K, Krupa H, Schroder H, Vollrath L: *In vitro* synthesis of dopamine and noradrenaline in the isolated rat pineal gland: Day-night variations and effects of electrical stimulation. J Neurochem 53: 354–361, 1989
44. Takahashi JS, DeCoursey PJ, Bauman L, Menaker M: Spectral sensitivity of a novel photoreceptive system mediating entrainment of mammalian circadian rhythms. Nature 308: 186–188, 1984
45. Duda T, Venkataraman V, Krishnan A, Nagele RG, Sharma RK. Negatively calcium-modulated membrane guanylate cyclase signaling system in the rat olfactory bulb. Biochemistry 40: 4654–4662, 2001
46. Mori K, Nagao H, Yoshihara Y: The olfactory bulb: Coding and processing of odor molecule information. Science 286: 711–715, 1999
47. Jahr CE, Nicoll RA: Dendrodendritic inhibition: Demonstration with intracellular recording. Science 207: 1473–1475, 1980
48. Yokoi M, Mori K, Nakanishi S: Refinement of odor molecule tuning by dendrodendritic synaptic inhibition in the olfactory bulb. Proc Natl Acad Sci USA 92: 3371–3375, 1995
49. Friedman D, Strowbridge BW: Functional role of NMDA autoreceptors in olfactory mitral cells. J Neurophysiol 84: 39–50, 2000
50. Tanaka O, Sakagami H, Kondo H: Localization of mRNAs of voltage-dependent Ca(2+)-channels: Four subtypes of alpha 1- and beta-subunits in developing and mature rat brain. Brain Res Mol Brain Res 30: 1–16, 1995
51. Ludwig A, Flockerzi V, Hofmann F: Regional expression and cellular localization of the alpha1 and beta subunit of high voltage-activated calcium channels in rat brain. J Neurosci 17: 1339–1349, 1997
52. Park SH, Suh YS, Kim H, Rhyu IJ, Kim HL: Chromosomal localization and neural distribution of voltage dependent calcium channel beta 3 subunit gene. Mol Cells 7: 200–203, 1997
53. Hobom M, Dai S, Marais E, Lacinova L, Hofmann F, Klugbauer N: Neuronal distribution and functional characterization of the calcium channel alpha2delta-2 subunit. Eur J Neurosci 12: 1217–1226, 2000
54. Kusakabe T, Suzuki N: Photoreceptors and olfactory cells express the same retinal guanylyl cyclase isoform in medaka: Visualization by promoter transgenics. FEBS Lett 483: 143–148, 2000
55. Hisatomi O, Honkawa H, Imanishi Y, Satoh T, Tokunaga F: Three kinds of guanylate cyclase expressed in medaka photoreceptor cells in both retina and pineal organ. Biochem Biophys Res Commun 255: 216–220, 1999
56. Schultz JE, Klumpp S: Calcium/calmodulin-regulated guanylate cyclases in the ciliary membranes from Paramecium and Tetrahymena. Adv Cycl Nucl Prot Phos Res17: 275–283, 1984
57. Hasegawa K, Kikuchi H, Ishizaki S, Tamura A, Tsukahara Y, Nakaoka Y, Iwai E, Sato T: Simple fluctuation of Ca^{2+} elicits the complex circadian dynamics of cyclic AMP and cyclic GMP in Paramecium. J Cell Sci 112: 201–207, 1999
58. Veske A, Nilsson SE, Gal A: Organization of the canine gene encoding the E isoform of retinal guanylate cyclase (cGC-E) and exclusion of its involvement in the inherited retinal dystrophy of the Swedish Briard and Briard-beagle dogs. Biochim Biophys Acta 1372: 69–77, 1998
59. Seebacher T, Beitz E, Kumagami H, Wild K, Ruppersberg JP, Schultz JE: Expression of membrane-bound and cytosolic guanylyl cyclases in the rat inner ear. Heart Res 127: 95–102, 1999

Molecular and Cellular Biochemistry **230**: 125–128, 2002.

A novel role of RGS9: Inhibition of retinal guanylyl cyclase

Vladimir A. Bondarenko,[1,2] Hao Yu,[1,2] Russell K. Yamazaki[3] and Akio Yamazaki[1,2,3]

[1]*Kresge Eye Institute and Departments of* [2]*Ophthalmology and* [3]*Pharmacology, Wayne State University, School of Medicine, Detroit, MI, USA*

Abstract

Cyclic GMP plays a key role in retinal phototransduction and its photoreceptor concentration is precisely controlled by the cooperative action of cGMP phosphodiesterase (PDE) and retinal guanylyl cyclase (retGC). However, studies of the relationship between these two systems have focused only on a Ca^{2+}-mediated, indirect connection. This article summarizes our studies strongly suggesting that RGS9-1 is directly involved in the cooperative action of PDE and retGC, and that this ingenious mechanism plays an important role in tuning of cGMP concentration in photoreceptors. (Mol Cell Biochem **230**: 125–128, 2002)

Key words: RGS9, retinal guanylyl cyclase, phototransduction

Introduction

In retinal phototransduction, cGMP plays the key mediator and its concentration in photoreceptors is precisely controlled by the cooperative action of cGMP phosphodiesterase (PDE) and retinal guanylyl cyclase (retGC). However, studies of the relationship between these two systems have focused only on a Ca^{2+}-mediated, indirect connection. We believe that in addition to the Ca^{2+}-mediated connection, direct interactions mediated by proteins involved in these two systems are also comprised in the cooperative action of these two systems. In this article, we propose that RGS9 is a candidate protein for a direct mediator between these two systems.

Regulation of PDE and a role of RGS9 in the recovery phase of phototransduction

Since Bitensky and Miller suggested the involvement of cyclic nucleotides in retinal phototransduction [1], many investigators have contributed to establish the role of cGMP in phototransduction [1–4]. In rod outer segments (ROS), bleached rhodopsin stimulates GTP/GDP exchange on Tα (the α subunit of transducin) followed by dissociation of GTP/Tα from Tβγ (the βγ subunits of transducin) [5]. GTP/Tα interacts with inactive PDE. Inactive PDE is composed of Pαβ (the catalytic subunits) and two Pγs (the regulatory subunits) and is represented by Pαβγγ. The GTP/Tα-Pαβγγ interaction interrupts Pγ regulatory effects on Pαβ. One of the Pγ effects on Pαβ is the inhibition of cGMP hydrolysis activity of Pαβ [6]. Thus, GTP/Tα releases the Pγ inhibitory effect from Pαβ and PDE is activated [7]. The resulting decrease in cytoplasmic cGMP concentration leads to reduction in the activity of cGMP-gated channels and the hyperpolarization of plasma membranes [1–4].

Restoration of the dark membrane potential requires recovery of the dark level of cytoplasmic cGMP concentration in ROS. The recovery of the dark level of cGMP concentration requires two processes: interruption of the PDE activation process and replenishment of the cytoplasmic cGMP concentration. Several steps in the PDE activation cascade are deactivated for the turnoff of PDE activation process. In particular, mechanisms related to the GTP/GDP exchange by illuminated rhodopsin and the reassociation of Pαβ with Pγ have

Address for offprints: A. Yamazaki, Kresge Eye Institute, Wayne State University, School of Medicine, 4717 St. Antoine Street, Detroit, MI 48201, USA (E-mail: ayamazak@med.wayne.edu)

been studied extensively. In the reassociation of Pαβ with Pγ, GTP-hydrolysis by Tα has been considered to be the main mechanism [5]. After activation of PDE, GTP bound to Tα is hydrolyzed by its intrinsic GTPase activity. The resulting GDP/Tα loses the affinity for Pγ and releases Pγ. The released Pγ reassociates with Pαβ to inhibit cGMP hydrolytic activity of Pαβ [7]. However, the intrinsic GTPase activity of Tα is much slower than the rate of termination of phototransduction [8]. Subsequently, a protein was isolated to stimulate the GTPase activity of Tα from photoreceptor outer segments [9]. The protein is a member of a new gene family called 'regulators of G protein signaling' (RGS) and is termed RGS9.

Functions of RGS9 domains in the stimulation of GTP hydrolysis by Tα

RGS proteins (Fig. 1) are characterized by a homologous 120-amino acid region, referred to as the RGS domain, that is responsible for binding the α subunit of G protein and stimulating its GTPase activity [10, 11]. However, outside of this RGS domain, the more than 30 members of the family are structurally diverse. Depending on these outside domains, RGS proteins can be subclassified into small and large proteins. The large proteins contain additional sequences that may mediate subcellular targeting and assemble signaling complexes [10, 11]. For example, RGS9, a member of large RGS proteins, contains the G protein γ-like (GGL) domain, which is responsible for the specific interaction with the neurospecific G protein β subunit, Gβ5, [12, 13] and the relatively longer C-terminus [9]. Recent studies have suggested that RGS9 forms a complex with Gβ5L through the GGL domain, and that the complex acts as an accelerator of the Tα GTPase in concert with Pγ [14, 15]. The C-terminus is also involved in the

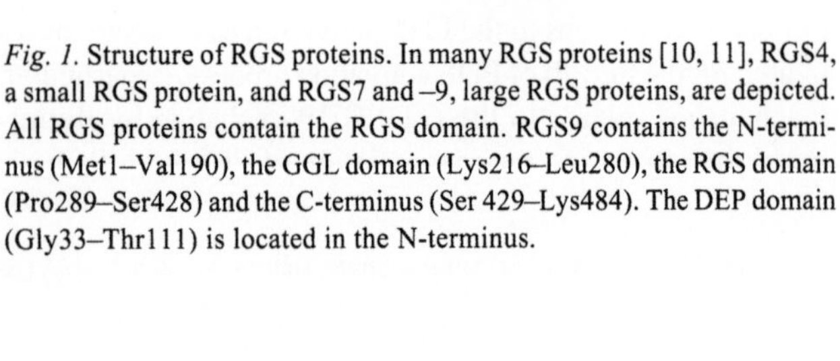

Fig. 1. Structure of RGS proteins. In many RGS proteins [10, 11], RGS4, a small RGS protein, and RGS7 and –9, large RGS proteins, are depicted. All RGS proteins contain the RGS domain. RGS9 contains the N-terminus (Met1–Val190), the GGL domain (Lys216–Leu280), the RGS domain (Pro289–Ser428) and the C-terminus (Ser 429–Lys484). The DEP domain (Gly33–Thr111) is located in the N-terminus.

acceleration of GTPase by the complex [15]. RGS9 also contains a large N-terminus and a Disheveled/Egl-10/pleckstrin (DEP) domain is located in the N-terminus [11]. The N-terminus appears to be involved in the acceleration of the Tα GTPase [15]. However, the function is not critical [15], suggesting that the N-terminus may have an unknown function(s).

Activation of retGC in the recovery phase of phototransduction

RetGC increases the cytoplasmic cGMP concentration in photoreceptors during the recovery phase of phototransduction. After activation of PDE, the low cGMP concentration reduces cGMP-gated channel activity [1–4]. The reduction of cGMP-gated channel activity blocks Na^{2+} and Ca^{2+} influx, and allows a Na^{2+}/Ca^{2+}, K^{+} exchanger to decrease cytoplasmic Ca^{2+} concentration. When Ca^{2+} concentration is low, retGC is stimulated [16, 17]. This Ca^{2+}-sensitive stimulation of retGC is mediated by at least two calmodulin-like Ca^{2+}-binding proteins termed by GCAP1 and GCAP2 [18, 19]. Thus, PDE and retGC systems are indirectly connected by the Ca^{2+}-mediated mechanism. Other interactions of these two systems, including direct interactions mediated by proteins involved in these two systems, have not been thoroughly investigated.

Identification of RGS9 as a protein interacting with retGC and its effect on retGC activity

We have been searching for direct protein–protein interactions between PDE and retGC systems because we believe that precise regulation of the cGMP concentration in photoreceptors should be based on a firm foundation. First, we focused on isolating proteins interacting with retGC in a Triton X-100-insoluble fraction, because retGC appears to be located in a Triton X-100-insoluble fraction [20, 21] and a protein interacting retGC, if present, should be in the Triton X-100-insoluble fraction. After resolution of proteins in the Triton X-100-insoluble fraction by two-dimensional gel electrophoresis, overlay of retGC on these proteins showed that several proteins in the fraction bound to retGC [22]. The major protein (~55 kDa) was isolated in the alkaline region. The amino acid sequences of peptides derived from the 55 kDa protein indicated that the protein was RGS9 [22].

In addition, the following observations also support the idea that the 55 kDa protein and RGS9 are identical [22]. (a) The molecular mass of the 55 kDa protein is similar to that of RGS9 (~57 kDa). (b) The pI value of the 55 kDa protein is estimated to be more than 8.0 (under our experimental conditions, it is difficult to measure a pI value of more than 8.0). Calculation from the amino acid sequence indicates that the pI value of RGS9 is 9.54. (c) Both the 55 kDa protein [22] and RGS9 [23] are solubilized by detergents in a similar way. (d) The localization of the 55 kDa protein has been shown in rod and cone photoreceptors in mammalian retinas by both light and electron microscopes [24]. RGS9 is also reported to be located in ROS, although higher expression was observed in cone [23]. (f) Later, we showed that retGC was precipitated with a 55 kDa protein-specific antibody in a ROS homogenate [25]. Together, these results strongly indicate that the 55 kDa protein is identical with RGS9, and that RGS9 interacts with retGC.

In a previous study [22], it was found that RGS9 (the 55 kDa protein) inhibited slightly, but constantly, retGC purified from bovine ROS. Recently we found that recombinant RGS9 also slightly inhibited GCAP-activated retGC activity in ROS membranes [25]. Together, these observations suggest that with or without GCAPs, RGS9 interacts with retGC and inhibits retGC activity although the inhibition is not great, and that the retGC inhibition by RGS9 is not due to the blocking of active conformational change of retGC by GCAPs, such as retGC dimerization [26]. The inhibition may be resulted from simple blocking by RGS9 of access of GTP to the catalytic cavity.

Roles of RGS9 domains in the inhibition of retGC

Recently, we expressed RGS9 and its fragments (Fig. 1) in *E. coli* and solubilized them using a new method [25]. Using these recombinant proteins, we found that GCAP-activated retGC was clearly inhibited by RGS9, RGS9N (a RGS9 fragment containing only its N-terminus) and RGS9NR (a RGS9 fragment with the C-terminus deleted). The order of inhibition is RGS9N > RGS9 > RGS9NR. The RGS9N inhibitory activity was much higher than that of RGS9 and RGS9NR. In contrast, a fragment containing only the RGS domain did not show any inhibitory activity for retGC. Overlay of retGC also showed that RGS9N directly interacted with retGC; however, RGS9 domains other than the N-terminus, including the C-terminus, did not bind to retGC [25]. These observations strongly suggest that the N-terminus is responsible for the retGC inhibition, and that addition of the GGL domain and/or the RGS domain to the N-terminus weakens the inhibitory effect of the N-terminus on retGC activity. This implies that the GGL domain and/or the RGS domain function as an internal suppressor domain against the N-terminus.

Conclusion – future trend

Our studies clearly suggested that RGS9 interacts with retGC [22, 25], and that the N-terminus of RGS9 functions as an inhibitor of retGC [25]. Moreover, the GGL domain and/or the RGS domain function as an internal suppressor against the inhibitory activity of the N-terminus. In other words, the inhibitory activity of the N-terminus is regulated by the GGL domain and/or the RGS domain. Since Gβ5L and GTP/Tα bind to the GGL domain and the RGS domain [11–15], respectively, it is speculated that binding of these proteins to these domains affects the N-terminal inhibition on retGC activity. We expect that binding of these proteins to RGS9 releases the inhibitory effect of the GGL domain and/or the RGS domain on the N-terminus, and that the released N-terminus inhibits retGC activity in ROS membranes.

Recent studies have indicated that many mechanisms are involved in the recovery of cytoplasmic cGMP concentration to the dark level in photoreceptors. In addition to mechanisms investigated in the prevailing model of phototransduction, several new mechanisms have been revealed. In the direct interaction between PDE and retGC systems, it was reported that Tβγ inhibited retGC activity [27]. These authors have claimed that under saturating light conditions, retGC might be regulated by Tβγ and cations, providing a possible mechanism of desensitization and light adaptation. Our study [25] suggests the possibility that retGC may be regulated by GTP/Tα through binding to the RGS domain and releasing the inhibitory effect of the RGS domain on the N-terminus. Experiments are in progress; however, the relationship between regulations of retGC by GTP/Tα and Tβγ is unknown now. Moreover, we have recently proposed a totally new mechanism for the turnoff of the GTP/Tα-activated PDE [28–31]. In this mechanism, Pγ complexed with GTP/Tα is phosphorylated by cyclin-dependent protein kinase 5 (Cdk5) and the phosphorylated Pγ loses the ability to interact with GTP/Tα but gains increased inhibitory activity as compared with non-phosphorylated Pγ. Thus, this mechanism does not required GTP hydrolysis for the turnoff of GTP/Tα-activated PDE. Moreover, in this mechanism, it is possible that one GTP/Tα interacts with more than two effectors. RGS9 is also expected to be involved in this mechanism because Pγ complexed with GTP/Tα, the substrate for Cdk5, interacts with RGS9. Furthermore, we have suggested that release of cGMP from non-catalytic sites on Pαβ may be involved in the recovery of the dark level of cytoplasmic cGMP concentration [32]. Together, recent studies strongly suggest that the mechanism for the recovery of cytoplasmic cGMP concentration to the dark level is much more complicated than we expected. The complicated mechanism may be required for the precise regulation of cGMP concentration in photoreceptors, especially because the regulation of cGMP concentration is also involved in desensitization and light adaptation of photoreceptors.

Acknowledgements

This work was supported by a grant from the National Institute of Health (EY 09631) and an unrestricted grant from Research to Prevent Blindness.

References

1. Miller WH: Dark Mimic. Invest Ophthalmol Visual Sci 31: 1664–1673, 1990
2. Yau K-W, Baylor DA: Cyclic GMP-activated conductance of retinal photoreceptor cells. Ann Rev Neurosci 12: 289–327, 1989
3. Pugh EN Jr, Lam TD: Amplification and kinetics of the activation steps in phototransduction. Biochim Biophys Acta 1141: 111–149, 1993
4. Yarfitz S, Hurley JB: Transduction mechanisms of vertebrate and invertebrate photoreceptors. J Biol Chem 269: 14329–14332, 1994
5. Fung BK-K, Hurley JB, Stryer L: Flow of information in the light-triggered cyclic nucleotide cascade of vision. Proc Natl Acad Sci USA 78: 152–156, 1981
6. Hurley JB, Stryer L: Purification and characterization of the gamma regulatory subunit of the cyclic GMP phosphodiesterase from retinal rod outer segments. J Biol Chem 257: 11094–11099, 1982
7. Yamazaki A, Stein PJ, Chernoff N, Bitensky MW: Activation mechanism of rod outer segment cyclic GMP phosphodiesterase – release of inhibitor by the GTP/GTP-binding protein. J Biol Chem 258: 8188–8194, 1983
8. Ting TD, Ho Y-K: Molecular mechanism of GTP hydrolysis by bovine transducin: Pre-steady-state kinetic analysis. Biochemistry 30: 8996–9007, 1991
9. He W, Cowan CW, Wensel TG: RGS9, a GTPase accelerator for phototransduction. Neuron 20: 95–102, 1998
10. Koelle MR: A new family of G-protein regulators – the RGS proteins. Curr Opin Cell Biol 9: 143–147, 1997
11. De Vries L, Farquhar MG: RGS proteins: More than just GAPs for heterotrimrric G proteins. Trends Cell Biol 9: 138–144, 1999
12. Makino ER, Handy JW, Li T, Arshavsky VY: The GTPase activating factor for transducin in rod photoreceptors is the complex between RGS9 and type 5 G protein β subunit. Proc Natl Acad Sci USA 96: 1947–1952, 1999
13. Levay K, Cabrera JL, Satpaev DK, Slepak VZ: Gβ5 prevents the RGS7-Gαo interaction through binding to a distinct Gγ-like domain found in RGS7 and other RGS proteins. Proc Natl Acad Sci USA 96: 2503–2507, 1999
14. Skiba NP, Hopp JA, Arshavsky VY: The effector enzyme regulates the duration of G protein signaling in vertebrate photoreceptors by increasing the affinity between transducin and RGS protein. J Biol Chem 275: 32716–32720, 2000
15. He W, Lu L, Zang X, El-Hodiri HM, Chen C-K, Slep KC, Simon MI, Jamrich M, Wensel TG: Modules in the photoreceptor RGS9-1·$G_{\beta 5L}$ GTPase-accelerating protein complex control effector coupling, GTPase acceleration, protein folding, and stability. J Biol Chem 275: 37093–37100, 2000
16. Lolley RN, Racz E: Calcium modulation of cyclic GMP synthesis in rat visual cells. Vision Res 22: 1481–1486, 1982
17. Koch K-W, Stryer L: Highly cooperative feedback control of retinal rod guanylate cyclase by calcium ions. Nature 334: 64–66, 1988
18. Palczewski K, Subbaraya I, Gorczyca WA, Helekar BS, Ruiz CC, Ohguro H, Huang J, Zhao X, Crabb JW, Johnson RS, Walsh KA, Gray-Keller MP, Detwiler PB, Baehr W: Molecular cloning and characterization of retinal photoreceptor guanylyl cyclase-activating protein. Neuron 13: 395–404, 1994
19. Dizhoor AM, Lowe DG, Olshevskaya EV, Laura RP, Hurley JB: The human photoreceptor membrane guanylyl cyclase, RetGC, is present in outer segments and is regulated by calcium and a soluble activator. Neuron 12: 1345–1352, 1994
20. Hakki S, Sitaramayya A: Guanylate cyclase from bovine rod outer segments: Solubilization, partial purification, and regulation by inorganic pyrophosphate. Biochemistry 29: 1088–1094, 1990
21. Hayashi F, Hutson LD, Kishigami A, Nagao S, Yamazaki A: Preparation and characterization of guanylate cyclase from vertebrate rod photoreceptors. Methods Neurosci 15: 237–247, 1993
22. Seno K, Kishigami A, Ihara S, Maeda T, Bondarenko VA, Nishizawa Y, Usukura J, Yamazaki A, Hayashi F: A possible role of RGS9 in phototransduction – a bridge between the cGMP phosphodiesterase system and the guanylyl cyclase system. J Biol Chem 273: 22169–22172, 1998
23. Cowan CW, Fariss RN, Sokal I, Palczewski K, Wensel TG: High expression levels in cones of RGS9, the predominant GTPase accelerating protein of rods. Proc Natl Acad Sci USA 95: 5351–5356, 1998
24. Nishizawa Y, Seno K, Ihara S, Kachi S, Watanabe N, Hayashi F, Yamazaki A, Usukura J: Immunocytochemical characterization of GCBP in mammalian retinas. Invest Ophthalmol Visual Sci 39: S1051, 1998
25. Yu H, Bondarenko VA, Yamazaki A: The N-terminus of RGS9 is involved in its inhibition of retinal guanylate cyclase. Biochem Biophys Res Commun 286: 12–19, 2001
26. Yu H, Olshevskaya E, Duda T, Seno K, Hayashi F, Sharma RK, Dizhoor AM, Yamazaki A: Activation of retinal guanylyl cyclase-1 by Ca^{2+}-binding proteins involves its dimerization. J Biol Chem 274: 15547–15555, 1999
27. Wolbring G, Baehr W, Palczewski K, Schnetkamp PPM: Light inhibition of bovine retinal rod guanylyl cyclase mediated by βγ-transducin. Biochemistry 38: 2611–2616, 1999
28. Tsuboi S, Matsumoto H, Jackson KW, Tsujimoto K, Williams T, Yamazaki A: Phosphorylation of an inhibitory subunit of cGMP phosphodiesterase in *Rana catesbiana* rod photoreceptors – I. Characterization of the phosphorylation. J Biol Chem 269: 15016–15023, 1994
29. Tsuboi S, Matsumoto H, Yamazaki A: Phosphorylation of an inhibitory subunit of cGMP phosphodiesterase in *Rana catesbiana* rod photoreceptors – II. A possible mechanism for the turnoff of cGMP phosphodiesterase without GTP hydrolysis. J Biol Chem 269: 15024–15029, 1994
30. Matsuura I, Bondarenko VA, Maeda T, Kachi S, Yamazaki M, Usukura J, Hayashi F, Yamazaki A: Phosphorylation by cyclin-dependent protein kinase 5 of the regulatory subunit of retinal cGMP phosphodiesterase – I. Identification of the kinase and its role in the turnoff of phosphodiesterase *in vitro*. J Biol Chem 275: 32950–32957, 2000
31. Hayashi F, Matsuura I, Kachi S, Maeda T, Yamamoto M, Fujii Y, Liu H, Yamazaki M, Usukura J, Yamazaki A: Phosphorylation by cyclin-dependent protein kinase 5 of the regulatory subunit of retinal cGMP phosphodiesterase – II. Its role in the turnoff of phosphodiesterase *in vivo*. J Biol Chem 275: 32958–32965, 2000
32. Yamazaki A, Bondarenko VA, Dua S, Yamazaki M, Usukura J, Hayashi F: Possible stimulation of retinal rod recovery to dark state by cGMP release from a cGMP phosphodiesterase noncatalytic site. J Biol Chem 271: 32495–32498, 1996

Molecular and Cellular Biochemistry **230**: 129–138, 2002.

Retinal diseases linked with photoreceptor guanylate cyclase

Teresa Duda[1] and Karl-Wilhelm Koch[2]

[1]*The Unit of Regulatory and Molecular Biology, Department of Cell Biology and Ophthalmology, SOM and NJMS, University of Medicine and Dentistry of New Jersey, Stratford, New Jersey, USA;* [2]*Institut für Biologische Informationsverarbeitung 1, Forschungszentrum Jülich, Jülich, Germany*

Abstract

Inherited retinal dystrophies are the main causes of progressive visual impairment often leading to blindness. They represent a clinically and genetically heterogenous group of disorders. Continuously increasing body of evidence links retinal dystrophies to mutations in numerous genes. These genes code for retinal proteins of various function (phototransduction, visual cycle, transcription factors, structural and metabolic functions). Mutations in the gene coding for photoreceptor specific guanylate cyclase type 1, ROS-GC1, were found to be the cause for the type 1 Leber's congenital amaurosis (LCA1) and cone-rod dystrophy type 6 (CORD6). The LCA1-linked mutations are distributed over almost the entire ROS-GC1 coding sequence but the CORD6-linked mutations are restricted to three positions, E786, R787 and T788, located within the putative ROS-GC1 dimerization domain. A linkage between the biochemical effect of the mutation and its phenotypic manifestation was provided for only one LCA1 mutation, F514S. This was followed by biochemical analyses of the consequences of the CORD6-causing mutations. Here, an overview on the existing results and a discussion of the possible physiological implications are presented. (Mol Cell Biochem **230**: 129–138, 2002)

Key words: rod outer segment membrane guanylate cyclase, phototransduction, inherited retinal diseases, LCA1, CORD6

Introduction

Inherited retinal dystrophies are a heterogeneous group of eye diseases that affect approximately one in 2000 individuals worldwide and are a major cause of inherited blindness. Different disease forms have been classified according to clinical and molecular genetic criteria. For example, retinitis pigmentosa (RP) is the synonymous name for a rather wide group of various retinal disorders that lead to the progressive loss of photoreceptor cells and the accompanying appearance of pigment granules in the retina [1–3]. The most common causes of RP are mutations in the photopigment rhodopsin, but other mutated genes also correlate with RP [4–14]. These genes code for retinal proteins of various function. For example, there are proteins of the visual transduction cascade [4–14] and of the visual cycle [15–18], cell transcription factors [19, 20] and proteins of structural function [21–24]. In addition, there are other inherited retinal diseases that are related to RP according to clinical symptoms and diagnostic criteria. These include Stargadt disease [25], Leber congenital amaurosis (LCA) [26], cone-rod dystrophies (CORD) [27] and Refsum disease [28].

Most mutations correlated with RP and RP-allied diseases are found in genes encoding proteins, that serve key functions in vertebrate phototransduction. Among these proteins are guanylate cyclase [29–34] and phosphodiesterase [7–10], the enzymes that synthesize and hydrolyze cyclic GMP, the transmitter of visual excitation. The downstream target of cyclic GMP, the cyclic nucleotide-gated cation channel (CNG-channel) [11, 35, 36] is also susceptible to the disease-causing mutations; in rod cells they result in a form of autosomal recessive RP, and in cones, in total color-blindness.

Mutations in the gene coding for photoreceptor specific guanylate cyclase type1, ROS-GC1 or synonymously retGC1, GC-E, were found in patients suffering from LCA1 [29, 30] and a particular form of cone-rod dystrophy, CORD6 [31–

Address for offprints: T. Duda, The Unit of Regulatory and Molecular Biology, Department of Cell Biology and Ophthalmology, NJMS, University of Medicine and Dentistry of New Jersey, Stratford, New Jersey 08084, USA (E-mail: dudatm@umdnj.edu)

34]. So far, no mutation was found in the second photoreceptor specific guanylate cyclase ROS-GC2 (retGC2, GC-F). Photoreceptor specific guanylate cyclases are regulated by small Ca^{2+}-binding proteins named GCAPs (guanylate cyclase-activating proteins) [37–43]. These proteins mediate the increase of cyclase activity when the cytoplasmic [Ca^{2+}] in a photoreceptor cell drops below 100 nM after illumination [39–45]. Recently two mutations were found in GCAP1 (P50L and Y99C) [46, 47] that cause a form of autosomal dominant cone dystrophy. Retinopathies related to mutations in GCAP1 are discussed in a review by Dizhoor *et al.* in this issue. This review is focused on the biochemical aspects of the ROS-GC1 mutations related to two retinal disorders, LCA1 and CORD6; it gives an overview on recent results and discusses the possible physiological implications.

Chromosomal localization and clinical features of LCA1 and CORD6

LCA is a heterogeneous retinal disease [48]. It accounts for ~5% of all inherited retinal diseases and is the most frequent cause of inherited blindness in children. Heterogeneity of LCA is manifested in clinical symptoms and in molecular genetic features. Four genes have been associated with LCA. One codes for RPE65, a protein expressed in the retinal pigment epithelium with a probable function in the vitamin-A cycle [15, 48]. The RPE65 locus was mapped to chromosome 1p31. A second gene codes for CRX, a cone-rod homeobox gene that is located on chromosome 19q13.3 and encodes a transcription factor [19]. Two genes associated with LCA map to two loci in close proximity on chromosome 17p13.1 [29, 30, 49]. One gene codes for a protein named AIPL1 (arylhydrocarbon interacting protein-like 1) [49], the other codes for the photoreceptor specific membrane bound guanylate cyclase ROS-GC1 (retGC1, GC-E) [29, 30]. Initially, through an analysis of polymorphic markers the LCA gene in 15 unrelated families was located on the short arm of chromosome 17. Homozygosity mapping revealed genetic heterogeneity of LCA and careful analysis allowed to locate the gene of 5 families of Magherabian origin (LCA1 gene) on chromosome 17p13.1 between markers D17S938 and D17S1353 [50]. Using YAC clones on chromosome 17p13.1 the ROS-GC1 gene was identified as a candidate gene for LCA1 [50]. Although the organization of the human ROS-GC1 gene was not known, based on available mouse and bovine gene sequences [51–53] a strategy for sequencing the exons only was developed. The search for point mutations in genomic DNA from LCA1 patients resulted in an identification of the first four ROS-GC1 mutations responsible for the disease (A52S; delC460; delC693, F514S) [29]. The identified mutations were absent in healthy individuals. Thus, ROS-GC1 mutations were linked to LCA1 [29].

Patients with LCA suffer from a severe loss of visual function. Although the LCA1-affected children are born blind, the dystrophy is diagnosed only a few months later when it is apparent that the newborn does not follow light or moving objects. Roving eye movements and eye poking are additional symptoms of the disease. It is generally accepted that LCA1 is a consequence of impaired development of photoreceptors or their very early degeneration [48]. The initial clinical examination of an infant gives two contrasting results. The back of the eye (fundus) is normal but there is no recordable ERG. The last indicates early onset of impaired functioning of rod and cone photoreceptors. Visible degeneration of photoreceptors appears later and the changes are progressive. Finally, after several years, decreasing number of retinal vessels, atrophy of retinal pigment epithelium and choriocapilaries are found [48]. Patients may also develop keratoconus [48]. Mental retardation or neurological abnormalities can also be developed in association with LCA1 although it is not clear whether such a case would represent the same or another genetic form of the disease [54].

Cone-rod dystrophies (CORDs) are caused by mutations in a variety of genes which loci have been mapped to chromosomes 6 [46, 55–57], 17 [29], 18 [58] and 19 [59]. A specific form, CORD6, has been linked to mutations in the ROS-GC1 gene [31–34]. CORD6 differs from LCA1 by its delayed onset. Typical phenotypic characteristic of CORD6 is first deterioration of cone photoreceptors and then of rod photoreceptors [31]. At the early stage of the disease there is a decrease of visual acuity, loss of color recognition and photophobia. Electrophysiological examination shows significant loss of photopic function without changes in scotopic function. These symptoms are the result of degeneration of cone photoreceptors and appear usually within the first decade of life. During the second and third decade of life rod photoreceptor cells start to degenerate also. This causes that in addition to severely deteriorated color vision (achromatopsia) patients develop dramatic decrease in visual acuity, night blindness and loss of peripheral vision. At the beginning of the fourth decade of life all the earlier observed symptoms are greatly deepened and the ERG becomes unrecordable [31, 33].

Regulation of ROS-GC1 activity by GCAP1

In order to better understand the biochemical consequences of LCA1 and CORD6 mutations, we will start with a brief description of the regulatory aspects of the wild type ROS-

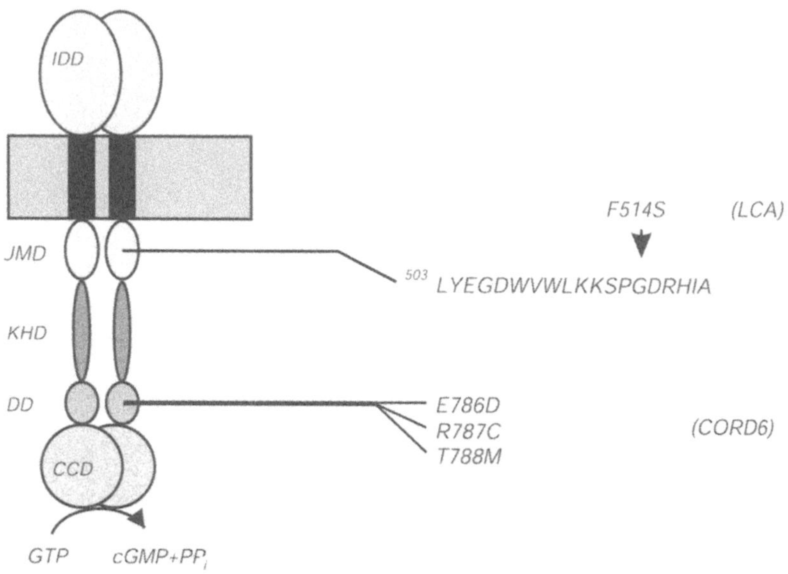

Fig. 1. Schematic representation of ROS-GC1 and localization of mutations linked to retinal diseases. ROS-GC1 consists of an intradiskal domain (IDD), that corresponds to the extracellular domain (ECD) in peptide hormone receptor guanylate cyclases, a transmembrane domain (block part spanning the membrane), a juxtamembrane domain (JMD), a kinase homology domain (KHD), a dimerization domain (DD) and a cytoplasmic catalytic domain (CCD). Only those point mutations are shown that are discussed in more detail in the text, F514S connected with LCA1 and E786D, R787C and T788M connected with CORD6. The catalytically active form of ROS-GC1 is a dimer.

GC1. The enzyme is activated at low free [Ca^{2+}] by all three GCAP forms described so far (GCAP1, GCAP2 and GCAP-3) [37–43]. Increasing [Ca^{2+}] concentrations inhibit the GCAP-stimulated cyclase activity, and the half-maximal inhibition occurs between 100 and 400 nM [37–43]. The target regions of GCAP1 in ROS-GC1 have been identified ([60, 61] and Fig. 1). Initially the GCAP1 binding site was broadly located within the ROS-GC1 region aa 441–730 [44]. Through peptide competition experiments the GCAP1 targeted regions were narrowed down to two short amino acid stretches, M445–L456 and L503–I522 located in the juxtamembrane domain (JMD) [60]. The site M445–L456 was indicated to function as a transducer unit and the site L503–I522 as a binding site [60]. These results were confirmed through deletion-mutagenesis/expression experiments. In another study, a main site of GCAP1 interaction with ROS-GC1 was located within the cyclase catalytic domain (CCD) based on peptide competition experiments [61]. The GCAP2 target site was broadly located within the cyclase C-terminal region aa 731–1054 [45].

In addition to transducing the low [Ca^{2+}] signal mediated by GCAPs, ROS-GC1 is also able to transduce the high [Ca^{2+}] signal [62–69]. In this case the signal is conveyed to ROS-GC1 by a member of the S100 family of Ca^{2+}-binding proteins, the S100β protein [62–68] or neurocalcin [69]. In micromolar Ca^{2+} concentrations ROS-GC1 activity is stimulated by S100β or neurocalcin in a dose-dependent fashion [62–67, 69]. This mode of ROS-GC1 regulation was postulated to be operative outside the photoreceptor outer segments [62, 63, 69].

The cyclase catalytic domain (CCD) catalyzes the synthesis of cyclic GMP from GTP. The functional unit of ROS-GC1 is a dimer and the dimerization most probably occurs through the aa segment 766–810 (DD) [70–72]. In the dimer, the constituent monomers of the catalytic domain are in an antiparallel orientation [73]. A ROS-GC1 molecule that is bound to and regulated by GCAP1 would therefore fold into a twisted conformation as it is shown in Fig. 2.

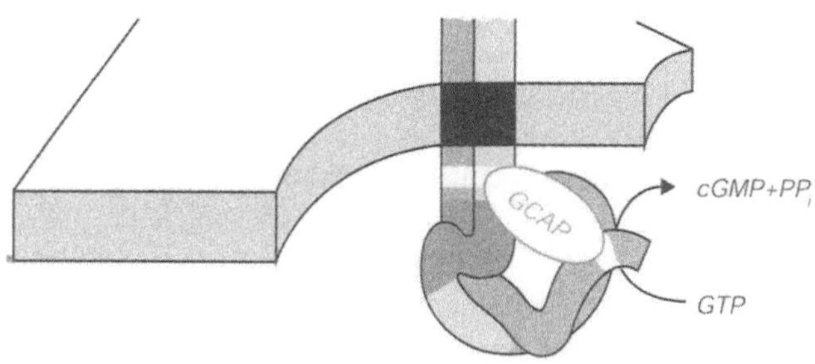

Fig. 2. Model of a ROS-GC1 dimer that is regulated by GCAP1. Three regulatory regions have been identified in ROS-GC1, two are located in the JMD (white and light grey) and one is located in the CCD. ROS-GC1 probably forms a twisted conformation in order to bind GCAP-1 by a multipoint attachment.

Biochemical consequences of LCA1 and CORD6 mutations

LCA1

So far 22 mutations in the ROS-GC1 gene have been found in patients affected with LCA [30]. While reasonable predictions on the biochemical consequences of some mutations have been made, the biochemical consequences of only one mutation has been experimentally tested [74]. At position 565 in the human sequence occurs a substitution of phenylalanine to serine (F→S) [29]. This corresponds to the bovine ROS-GC1 position F514S of the mature protein. Heterologous expression of the F514S mutant resulted in the same expression level as for the wild type, but the measured basal (i.e. non-stimulated) cyclase activity of the mutant was less than 10% of that of the wild type [74]. The F514S mutation is located within the cytoplasmic part (JMD) in a region that has been identified as a main target site for GCAP1 ([60]; Figs 1 and 3A). This automatically suggested that the crucial property of ROS-GC1, activation by GCAP1 in low Ca^{2+} might be affected by this mutation. And indeed, when the ROS-GC1 carrying the F514S substitution was expressed and tested for its response to GCAP1 no stimulation was observed even at saturating concentrations of GCAP1 [74]. Thus, the mutation causes total loss of the regulation by GCAP1. A conclusive control experiment of the study was a test with another modulator of ROS-GC1 activity, the Ca^{2+}-binding protein S100β (also named CD-GCAP). Opposite to regulation by GCAP1, ROS-GC1 is activated by S100β at micromolar free [Ca^{2+}]. The target site for S100β action is located in the ROS-GC1 cytoplasmic domain at a region distant from the regulatory site of GCAP1 [64, 67]. The profile of the F514S mutant response to S100β at high [Ca^{2+}] is almost identical to that of the wild type ROS-GC1 [74]. Taken together the results obtained with GCAP1 and S100β clearly show that the F514S substitution selectively destroys the phototransduction linked ROS-GC1 stimulation by GCAP1.

These results provided not only an insight into the biochemical consequences of the LCA1-linked ROS-GC1 mutation, they have also important structural implications. They strongly confirm the localization of the target site of GCAP-1 in ROS-GC1 [60]. They, however, do not support the model in which the major GCAP1 binding site is localized exclusively within the cyclase catalytic domain [61]. If the catalytic domain housed the major site of ROS-GC1 interaction with GCAP1, a mutation distal to that site would not abolish the GCAP1 stimulatory effect.

CORD6

The first linkage of CORD6 with the ROS-GC1 gene was through the identification of two missense mutations leading

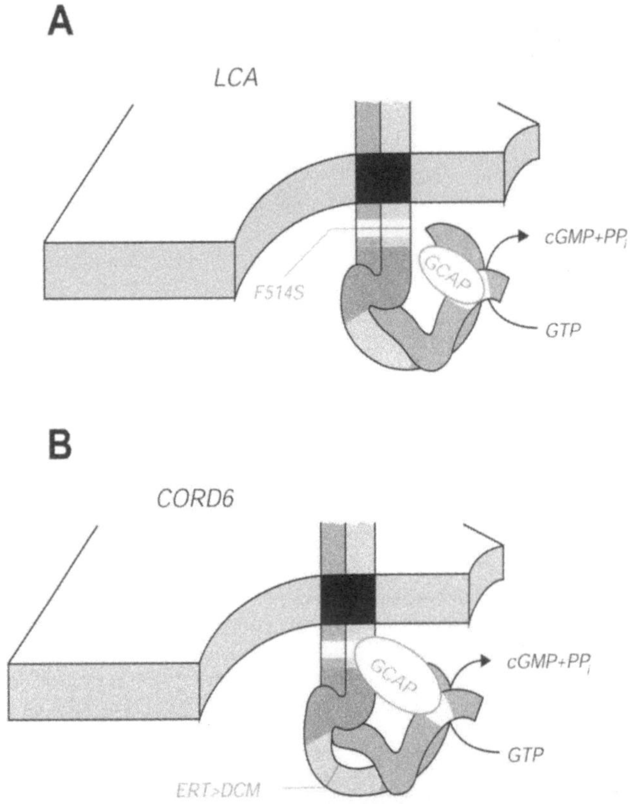

Fig. 3. Consequences of LCA1 (A) and CORD6 (B) mutations. The point mutation F514S leads to a complete loss of GCAP1 sensitivity. Since this mutation is located in the binding site for GCAP1 in the JMD, GCAP1 probably has lost contact to this site. The triple mutation ERT > DCM disturbs the dimerization of ROS-GC1. Because the regulatory sites of GCAP1 are intact, this mutant responds to GCAP1 (although in a modified manner), but has a reduced basal activity. For a more detailed description see the text.

to single amino acid substitutions – E786D and R787C – that were found independently in different families [31].

The consequences of these mutations were tested experimentally by analyses of the activities of the mutated cyclases expressed in heterologous systems of COS-7 or human embryonic kidney (HEK) 293 cells [75, 76]. Dramatic effects were observed with the R788C mutant. Basal GC activities decreased 2.3-fold with manganese as cofactor (assayed in HEK 293 cells; [76]) and more than 4-fold with magnesium as cofactor (assayed in COS cells; [75]). Most interestingly, the mutant became hyperactive to modulation by saturating concentrations of GCAP1 (25-fold instead of 5-fold stimulation) [75]. Activation by GCAP2 was lower, however, differences were observed in two different studies that use either COS-7 cells (only 10–20% lower activity) and HEK 293 cells (14-fold lower activity) [76]. No change in the Ca^{2+}-sensitivity was observed in the COS-7 cell study, but in the HEK 293 cell study the half-maximal activation of the mutant was shifted to higher free [Ca^{2+}] when assayed with GCAP1 [75,

76]. It is not clear, what accounts for the observed differences in these two studies, but the experimental conditions were not identical, e.g. different heterologous expression systems and different guanylate cyclase genes (bovine and human).

Activation by S100β at micromolar Ca^{2+} concentrations was lowered in the R787C mutant to about 50% of the control (wild type) [75]

The single amino acid substitution E786D had no effect on the basal cyclase activity [75]. Concerning the activation by GCAPs there was no difference in the mutant's response to GCAP1 when compared with the wild type ROS-GC1. The response to GCAP2, however, changed and was up to 70% higher than that of the wild-type. Based on these results it was predicted that E786D is likely to be a polymorphism rather than a cause for CORD6 [75]. This prediction was subsequently confirmed in the study which reported that the E786D families also carry the R787S mutation [33]. The biochemical consequences of the R787S mutation are of the same nature as those of R787C but quantitatively they are slightly more profound; thus, it has been suggested that R787S physiologically may have more severe effects [77].

In addition to the single amino acid substitutions (R787C and R787H) and the double substitution (E786D, R787S) [33, 78], a triple amino acid substitution – E786D, R787C, T788M [79] – was also detected in patients suffering from the disease.

The biochemical characteristics of the triple mutant ERT → DCM were in some aspects similar to those of R787C mutant [80]. When compared to the wild type, the mutant's basal activity decreased over 4-fold and its responses to saturating concentrations of GCAP1 and GCAP2 at low free [Ca^{2+}] increased. The cyclase affinity for GCAP2 was also slightly higher. Activation by S100β was lowered to about 70% of the control (wild type) [80]. Direct binding studies with the use of overlapping peptides covering the ROS-GC1 region encompassing the CORD6-related amino acids (744–807) showed that this region is not involved in binding of GCAP1, GCAP2 or S100β [75]. Alignment of the ROS-GC1 sequence with the sequences of other membrane guanylate cyclases indicated that the region aa 744–807 may be involved in the dimerization process. Two types of experiments were employed to test whether the ERT→DCM substitution affects the dimerization; cross-linking of heterologously expressed full-length ROS-GC1 or the ERT→DCM mutant, and gel filtration chromatography of the purified fragments corresponding to aa 733–1054 of ROS-GC1 and of the ERT → DCM mutant [80]. The obtained experiments clearly showed that the triple mutation profoundly affects dimerization of the cyclase. Thus, distortion of the dimer formation (Fig. 3B) will be a reason for the lower basal GC activity, since a correct dimer interface is necessary for efficient catalysis.

Proposed physiological consequences of mutations in ROS-GC1

LCA1

Although the mutations in the ROS-GC1 gene occur in rare eye disorders, determination of their distinct biochemical consequences will deepen our understanding as to how single point mutations in one gene lead to impaired vision or retinal cell degeneration. The LCA1-linked ROS-GC1 mutation leads to over 10-fold reduction in cyclic GMP synthesis and to a complete suppression of the cyclase modulation by GCAP1. This would have several drastic effects. In the dark state (Fig. 4) in a healthy photoreceptor outer segment the cyclic GMP concentration of 3–4 μM keeps about 5% of the CNG-channels open [reviewed in: 81, 82, 83]. A permanent reduction of the cyclic GMP concentration by more than 90% would keep only a very small fraction of CNG-channels open and thus would reduce the dark current that enters the photoreceptor cell during the dark adapted state. Since about 10% of the dark current is carried by Ca^{2+}-ions, less Ca^{2+}-ions will flow into the cell. Due to the continuous operation of a Na^+/K^+,Ca^{2+}-exchanger the cytoplasmic [Ca^{2+}] in an affected cell in the dark would be persistently lower than the experimentally observed value of 400–600 nM (vertebrate rod cells) [84–87]. The cell would sense a state of light adaptation or even light saturation without being illuminated. The decrease in cytoplasmic [Ca^{2+}] in turn triggers several negative feedback loops that would readjust the photoreceptor dark current (Fig. 4). However a key step in the Ca^{2+} feedback is the regulation of ROS-GC1 by GCAP1 which is completely absent in the case of the F514S mutant. Therefore, pathways of the Ca^{2+}-feedback that replenish the cyclic GMP pool are ineffective and will leave the dark current at a persistently low level. The situation turns worse under illumination. Due to the low level of cyclic GMP complete closure of CNG-channels will occur at lower light intensities and the rod cell will saturate after absorption of a few photons. Since a crucial part of the Ca^{2+}-feedback is absent the recovery to the dark state would be delayed. A summary of effects connected with the LCA1 mutation F514S is presented in Table 1.

Secondary effects could arise from a permanently different Ca^{2+}-homeostasis in affected cells. It is not known whether these changes in [Ca^{2+}] trigger photoreceptor degeneration and how this would occur.

CORD6

The conservative substitution E786D does not affect the basal ROS-GC1 activity [75] and therefore would not change the dark state of a photoreceptor cell. Since only the regulation by GCAP2 is slightly modified (~70% increase) the restora-

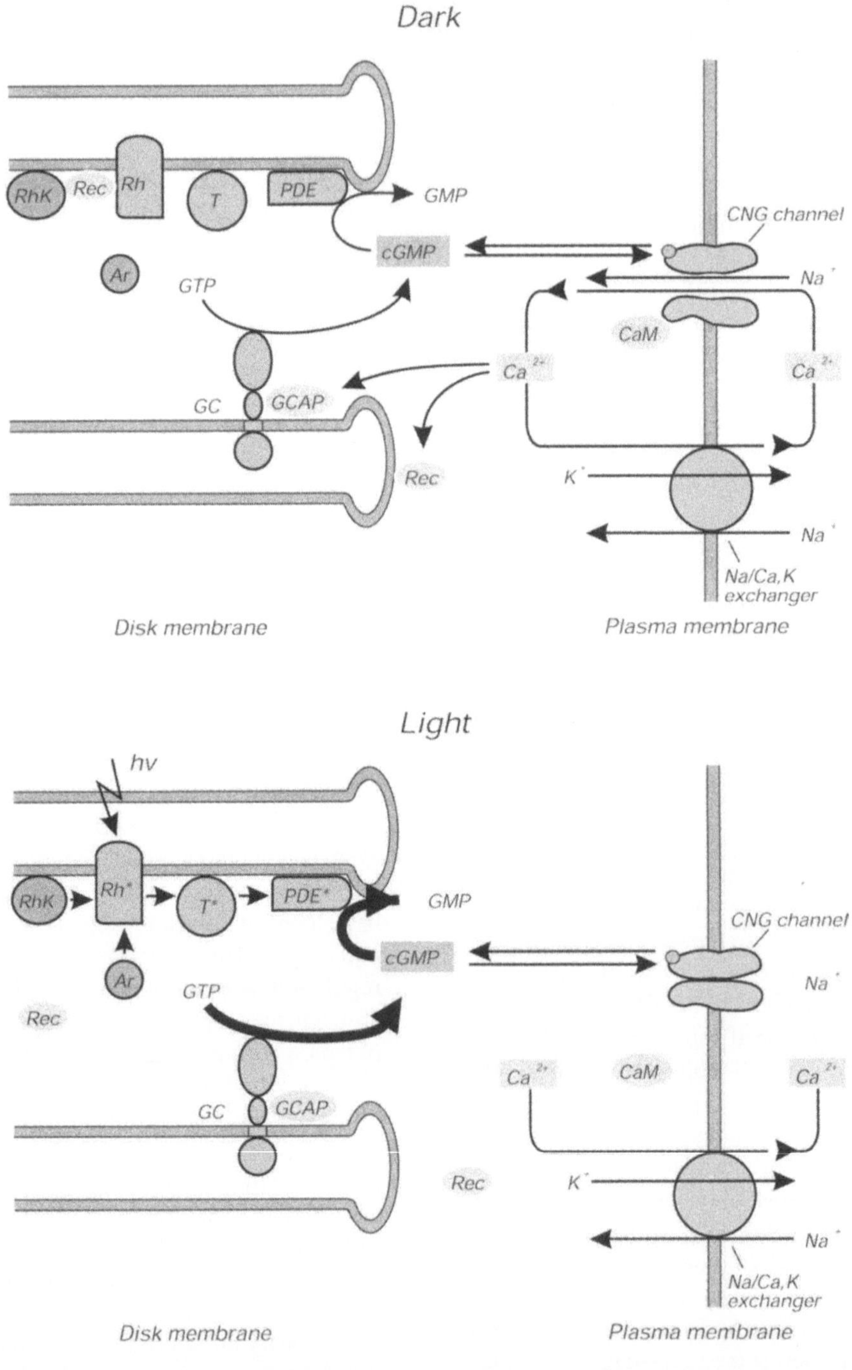

Fig. 4. Main steps in phototransduction and in Ca^{2+}-mediated feedback reactions. In the dark state (upper part) the signalling cascade consisting of rhodopsin (Rh), the G-protein transducin (T) and the phosphodiesterase (PDE) is not activated. Due to high concentration of cyclic GMP in the cytoplasm the CNG-channels are open and Na^+- and Ca^{2+}-ions flow into the cell. Ca^{2+} is continuously extruded via the $Na^+/Ca^{2+},K^+$-exchanger. The balance of Ca^{2+}-influx and efflux maintains the free $[Ca^{2+}]$ at 500–600 nM in the dark. Guanylate cyclase activity is basal (i.e. at the non-stimulated level). The overall turnover of cyclic GMP is low. Illumination (lower part) leads to activation of the visual transduction cascade and to efficient hydrolysis of cyclic GMP (activated components are marked with an asterisk). The subsequent closure of the CNG-channels stops the influx of Ca^{2+}, but Ca^{2+} is still extruded via the exchanger. The resulting decrease of cytoplasmic $[Ca^{2+}]$ is sensed by several Ca^{2+}-binding proteins, calmodulin (CaM), recoverin (Rec) and GCAPs. CaM modulates the cyclic GMP sensitivity of the CNG-channel, recoverin regulates rhodopsin kinase activity and guanylate cyclase is regulated by GCAPs. After illumination the rate of synthesis and hydrolysis of cyclic GMP is increased (bold arrows). At low $[Ca^{2+}]$ CaM dissociates from the CNG-channel and the channel becomes more sensitive to cyclic GMP. Inhibition of rhodopsin kinase is relieved and guanylate cyclase is activated. The Ca^{2+}-dependent feedback loops restore the dark state and regulate the light sensitivity of the cell. For a more detailed description please refer to [81–83, 88].

Table 1. Summary of effects of LCA1 linked mutation in ROS-GC1

Effects	LCA1 *F514S*
Biochemical consequences	decrease of basal cyclase activity loss of modulation by GCAP-1 no loss of modulation by S100β
Physiological consequences	restoration of dark state is slower impairment of recovery and light adaptation impairment of Ca^{2+}-homeostasis

tion of the circulating dark current will be minimally faster. Since the Ca^{2+}-feedback is almost not affected, light adaptation in a mutant cell is almost unchanged. In contrast the R787C mutation has several impacts on ROS-GC1 activity, its physiological implications are particularly interesting if we compare it with the LCA1 mutation F514S. Reduction of basal cyclase activity is 5-fold, which is significantly lower than in the wildtype, but is still much higher than in the F514S mutant [74, 75]. The cytoplasmic concentration of cyclic GMP in the dark will be largely reduced and in consequence the circulating dark current. This will also keep the influx of Ca^{2+} into the cell at a lower rate and will significantly reduce the cytoplasmic [Ca^{2+}]. In turn the Ca^{2+}-mediated feedback reactions involving GCAPs, recoverin and calmodulin (Fig. 4) will operate and adjust the photoreceptor to a light adapted state without prior illumination. The response-intensity curve will be shifted to lower light intensities.

The main difference of the R787C mutant and the wildtype lies in the GCAP-mediated feedback reaction. Reduction of dark current and thus reduction of cytoplasmic [Ca^{2+}] will trigger activation by GCAPs. The hyperresponsive activation by GCAP1 [75] can compensate the reduced rate of basal activity, but it can not reach the same high level of cyclic GMP synthesis as in the wildtype. Two principally different mechanisms of guanylate cyclase activity contribute to light adaptation in photoreceptor cells [88]. The *steady* increase in cyclic GMP synthesis rate extends the operating range, it increases the flash sensitivity and thus contributes to the restoration of the circulating dark current. A *transient* increase in cyclic GMP synthesis rate triggered by a *transient* decrease of [Ca^{2+}] during a flash response has no effect on the operating range, would decrease the flash sensitivity and thus contributes to the acceleration of recovery [88]. The defect in GCAP1 signaling leads to a lower steady synthesis rate. We predict that this will shorten the operating range, diminish the increase in sensitivity and thus will impart the restoration of the dark current. However the transient increase in cyclic GMP synthesis rate will be increased, which would further decrease the flash sensitivity and accelerate recovery.

Similar predictions can be made for the triple mutant ERT →DCM, for example a shift in the response-light intensity curve. The GCAP-mediated feedback reactions will work less powerfully (as in the R787C mutant), so every distortion of cyclic GMP synthesis is more severe and leads to chronically insufficient cyclic GMP concentration. Photoreceptors are driven in a state of permanent photoexcitation and hyperpolarization. A summarizing comparison of effects linked with CORD6 mutations is listed in Table 2.

Acknowledgements

This study was supported by the facilities provided by the Unit of Regulatory and Molecular Biology supported by USPHS awards EY 10828 and HL 58151; by the affiliated support of the Research to Prevent Blindness Inc., New York; the Lions Eye Research Foundation, New Jersey, and by the facilities provided by UMDNJ-SOM.

Table 2. Summary of effects of mutations in ROS-GC1 linked to CORD6

Effects	CORD6		
	E786D	*R787C*	*ERT > DCM*
Biochemical consequences	no change of basal cyclase activity no effect on modulation by GCAP-1	decrease of basal cyclase activity response to GCAP-1 is 5-fold higher	decrease of basal cyclase activity response to GCAP-1 is 70% higher
	maximal stimulation by GCAP-2 is doubled	no change of modulation by GCAP-2	response to GCAP-2 is 70% higher
	no effect on modulation by S100β	decrease of response to S100β	large decrease of response to S100β
		no change of EC_{50} for [Ca^{2+}]	no change of EC_{50} for [Ca^{2+}]
Physiological consequences	no change of dark state	reduction in dark current	reduction in dark current
	recovery state and light adaptation are almost not affected	trigger of Ca^{2+}-feedback is changed shift in response-intensity curve to lower light intensities acceleration of recovery phase	trigger of Ca^{2+}-feedback is changed shift in response-intensity curve to lower light intensities acceleration of recovery phase

References

1. Berson EL: Retinitis pigmentosa. Invest Ophthalmol Vis Sci 34: 1659–1676, 1993
2. Dryja TP, Berson EL: Retinitis pigmentosa and allied diseases. Invest Ophthal Vis Sci 36: 1197–1200, 1995
3. Fulton AB, Breton ME: Clinical physiology of heritable photoreceptor diseases. Arch Ophthalmol 111: 1479–1480, 1993
4. Dryja TP, McGee TL, Reichel E, Han LA, Cowley GS, Yandell DW, Sandberg MA, Berson EL: A point mutation of the rhodopsin gene in one form of retinitis pigmentosa. Nature 343: 364–366, 1990
5. Shastry BS: Retinitis pigmentosa and related disorders: Phenotypes of rhodopsin and peripherin/RDS mutations. Am J Med Genet 52: 467–474, 1994
6. Macke JP, Davenport CM, Jacobson SG, Hennessey JC, Gonzalez-Fernandez F, Conway BP, Heckenlively J, Palmer R, Maumenee IH, Sieving P: Identification of novel rhodopsin mutations responsible for retinitis pigmentosa: Implications for the structure and function of rhodopsin. Am J Hum Genet 53: 80–89,1993
7. Huang SH, Pittler SJ, Huang X, Oliveira L, Berson EL, Dryja TP: Autosomal recessive retinitis pigmentosa caused by mutations in the alpha subunit of rod cGMP phosphodiesterase. Nat Genet 11: 468–471,1995
8. McLaughlin ME, Sandberg MA, Berson EL, Dryja TP: Recessive mutations in the gene encoding the beta-subunit of rod phosphodiesterase in patients with retinitis pigmentosa. Nat Genet 4: 130–134, 1993
9. Dancinger M, Blaney J, Gao UQ, Zhao DY, Heckenlively JR, Jacobson SG, Farber DB: Mutations in the PDE6B gene in autosomal recessive retinitis pigmentosa. Genomics 30: 1–7, 1995
10. McLaughlin ME, Ehrhart TL, Berson EL, Dryja TP: Mutation spectrum of the gene encoding beta subunit of rod phosphodiesterase among patients with autosomal recessive retinitis pigmentosa. Proc Natl Acad Sci USA 92: 3249–3253, 1995
11. Dryja TP, Finn JT, Peng YW, McGee TL, Berson EL, Yau KW: Mutations in the gene encoding alpha subunit of the rod cGMP-gated channel in autosomal recessive retinitis pigmentosa. Proc Natl Acad Sci USA 92: 10177–10181, 1995
12. Nakamachi Y, Nakamura M, Fujii S, Yamamoto M, Okubo K: Oguchi disease with sectoral retinitis pigmentosa harboring adenine deletion at position 1147 in the arrestin gene. Am J Ophthalmol 125: 249–251, 1998
13. Nakazawa M, Wada Y, Tamai M: Arrestin gene mutations in autosomal recessive retinitis pigmentosa. Arch Ophthalmol 116: 498–501, 1998
14. Yoshii M, Murakami A, Akeo K, Nakamura A, Shimoyama M, Ikeda Y, Kikuchi Y, Okisaka S, Yanashima K, Oguchi Y: Visual function and gene analysis in a family with Oguchi's disease. Ophthalmic Res 30: 394–401, 1998
15. Morimura H, Fishman GA, Grover SA, Fulton AB, Berson EL, Dryja TP: Mutations in the RPE65 gene in patients with autosomal recessive retinitis pigmentosa or Leber congenital amaurosis. Proc Natl Acad Sci USA 95: 3088–3093, 1998
16. Maw MA, Kennedy B, Knight A, Bridges R, Roth KE, Mani EJ, Mukkadan JK, Nancarrow D, Crabb JW, Denton MJ: Mutation in the gene encoding cellular retinaldehyde-binding protein in autosomal recessive retinitis pigmentosa. Nat Genet 17: 198–200, 1997
17. Martinez-Mir A, Paloma E, Allikmets R, Ayuso C, del Rio T, Dean M, Vilageliu L, Gonzales-Duarte R, Bacells S: Retinitis pigmentosa caused by a homozygogous mutation in the Stargardt disease gene ABCR. Nat Genet 18: 11–12, 1998
18. Cremers FP, van der Pol DJ, van Driel M, den Hollander AI, van Haren FJ, Knoers NV, Tijmes N, Bergen AA, Rohrschneider K, Blankenagel A, Pinckers AJ, Deutman AF, Hoyng CB: Autosomal recessive retinitis pigmentosa and cone-rod dystrophy caused by splice site mutations in the Stargardt's disease gene ABCR. Hum Mol Genet 7: 355–356, 1998
19. Sohocki MM, Sullivan LS, Mintz-Hittner HA, Birch D, Heckenlively JR, Freund CL, McInnes RR, Daiger SP: A range of clinical phenotypes associated with mutations in CRX, a photoreceptor transcription-factor gene. Am J Hum Genet 63: 1307–1315, 1998
20. Bessant DA, Payne AM, Mitton KP, Wang QL, Swain PK, Plant C, Bird AC, Zack DJ, Swaroop A, Bhattacharya SS: A mutation in NRL is associated with autosomal dominant retinitis pigmentosa. Nat Genet 21: 355–356, 1999
21. Kajiwara K, Berson EL, Dryja TP: Digenic retinitis pigmentosa due to mutations at the unlinked peripherin/RDS and ROM1 loci. Science 264: 1604–1608, 1994
22. Dryja TP, Hahn LB, Kajiwara K, Berson EL: Dominant and digenic mutations in the peripherin/RDS and ROM1 genes in retinitis pigmentosa. Invest Ophthalmol Vis Sci 38: 1972–1982, 1997
23. Bascom RA, Liu L, Heckenlively JR, Stone EM, McInnes RR: Mutation analysis of the ROM1 gene in retinitis pigmentosa. Hum Mol Genet 4: 1895–1902, 1995
24. Sakuma H, Inana G, Murakami A, Yajima T, Weleber T, Murphey WH, Gass JD, Hayakawa M, Fujiki K: A heterozygous putative null mutation in ROM1 without a mutation in peripherin/RDS in a family with retinitis pigmentosa. Genomics 27: 384–386, 1995
25. Shroyer NF, Lewis RA, Allikmets R, Singh N, Dean M, Leppert M, Lupski JR: The rod photoreceptor ATP-binding cassette transporter gene, ABCR, and retinal disease: From monogenic to multifactorial. Vision Res 39: 2537–2544, 1999
26. Grieshaber MC, Niemeyer G: Leber congenital amaurosis: Diagnosis, follow-up and differential diagnosis. Klin Monatsbl Augenheilkd 212: 309–310, 1998
27. Weber BH: Recent advances in the molecular genetics of hereditary retinal dystrophies with primary involvement of the macula. Acta Anat 162: 65–74, 1998
28. Wierzbicki AS, Mitchell J, Lambert-Hammill M, Hancock M, Greenwood J, Sidey MC, de Belleroche J, Gibberd FB: Identification of genetic heterogeneity in Refsum's disease. Eur J Hum Genet 8: 649–651, 2000
29. Perrault I, Rozet JM, Calvas P, Gerber S, Camuzat A, Dollfus H, Chatelin S, Souied E, Ghazi I, Leowski C, Bonnemaison M, Le Paslier D, Frezal J, Dufier JL, Pittler S, Munnich A, Kaplan J: Retinal-specific guanylate cyclase gene mutations in Leber's congenital amaurosis. Nat Genet 14: 461–464, 1996
30. Perrault I, Rozet JM, Gerber S, Ghazi I, Duroc D, Souied E, Leowski C, Bonnemaison M, Dufier JL, Munnich A, Kaplan J: Spectrum of retGC1 mutations in Leber's congenital amaurosis. Eur J Hum Genet 8: 578–582, 2000
31. Kelsell RE, Gregory-Evans K, Payne AM, Perrault I, Kaplan J, Yang RB, Garbers DL, Bird AC, Moore AT, Hunt DM: Mutations in the retinal guanylate cyclase (RETGC-1) gene in dominant cone-rod dystrophy. Hum Mol Genet 7: 1179–1184, 1998
32. Perrault I, Rozet JM, Gerber S, Kelsell RE, Souied E, Cabot A, Hunt DM, Munnich A, Kaplan J: A retGC-1 mutation in autosomal dominant cone-rod dystrophy. Am J Hum Genet 63: 651–654, 1998
33. Gregory-Evans K, Kelsell RE, Gregory-Evans CY, Downes SM, Fitzke FW, Holder GE, Simunovic M, Mollon JD, Taylor R, Hunt DM, Bird AC, Moore AT: Autosomal dominant cone-rod retinal dystrophy (CORD6) from heterozygous mutation of GUCY2D, which encodes retinal guanylate cyclase. Ophthalmology 107: 55–61, 2000
34. Weigell-Weber M, Fokusten S, Torok B, Niemeyer G, Schnizel A, Hergersberg M: Codons 837 and 838 in the retinal guanylate cyclase gene on chromosome 17p: Hot spots for mutations in autosomal dominant cone-rod dystrophy. Arch Ophthalmol 118: 300, 2000

35. Kohl S, Marx T, Giddings I, Jagle H, Jacobson SG, Apfelstedt-Sylla E, Zrenner E, Sharpe LT, Wissinger B: Total colour blindness is caused by mutations in the gene encoding the alpha-subunit of the cone photoreceptor cGMP-gated cation channel. Nat Genet 19: 257–259, 1998
36. Kohl S, Baumann B, Broghammer M, Jagle H, Sieving P, Kellner U, Spegal R, Anastasi M, Zrenner E, Sharpe LT, Wissinger B: Mutations in the CNGB3 gene encoding the beta-subunit of the cone photoreceptor cGMP-gated channel are responsible for achromatopsia (ACHM3) linked to chromosome 8q21. Hum Mol Genet 9: 2107–2116, 2000
37. Gorczyca WA, Gray-Keller MP, Detwiler PB, Palczewski K: Purification and physiological evaluation of a guanylate cyclase activating protein from retinal rods. Proc Natl Acad Sci USA 91: 4014–4018, 1994
38. Gorczyca WA, Polans AS, Surgucheva IG, Subbaraya I, Baehr W, Palczewski K: Guanylate cyclase activating protein. A calcium-sensitive regulator of phototransduction. J Biol Chem 270: 22029–22036, 1995
39. Palczewski K, Subbaraya I, Gorczyca WA, Helekar BS, Ruiz CC, Ohguro H, Huang J, Zhao X, Crabb JW, Johnson RS: Molecular cloning and characterization of retinal photoreceptor guanylyl cyclase-activating protein. Neuron 13: 395–404, 1994
40. Frins S, Bonigk W, Muller F, Kellner R, Koch K-W: Functional characterization of a guanylate cyclase-activating protein from vertebrate rods. Cloning, heterologous expression, and localization. J Biol Chem 271: 8022–8027, 1996
41. Dizhoor AM, Lowe DG, Olshevskaya EV, Laura RP, Hurley JB: The human photoreceptor membrane guanylate cyclase, RetGC, is present in outer segments and is regulated by calcium and a soluble activator. Neuron 12: 1345–1352, 1994
42. Dizhoor AM, Olshevskaya EV, Henzel WJ, Wong SC, Stults JT, Ankoudinova I, Hurley JB: Cloning, sequencing, and expression of a 24-kDa Ca^{2+}-binding protein activating photoreceptor guanylyl cyclase. J Biol Chem 270: 25200–25206, 1995
43. Haeseleer F, Sokal I, Li N, Pettenati M, Rao N, Bronson D, Wechter R, Baehr W, Palczewski K: Molecular characterization of a third member of the guanylyl cyclase-activating protein subfamily. J Biol Chem 274: 6526–6535, 1999
44. Duda T, Goraczniak R, Rudnicka-Nawrot M, Surguheva IG, Gorczyca WA, Sitaramayya A, Palczewski K, Baehr W, Sharma RK: Calcium modulation of bovine photoreceptor guanylate cyclase. Biochemistry 35: 8478–8482, 1996
45. Krishnan A, Goraczniak R, Duda T, Sharma RK: Third calcium-modulated rod outer segment membrane guanylate cyclase transduction mechanism. Mol Cell Biochem 178: 251–259, 1998
46. Payne AM, Downes SM, Bessant DA, Taylor R, Holder GE: A mutation in guanylate cyclase activator 1A (GUCA1A) in an autosomal dominant cone dystrophy pedigree mapping to a new locus on chromosome 6p21.1. Hum Mol Genet 7: 273–277, 1998
47. Newbold RJ, Deery EC, Walker CE, Wilkie SE, Srinivasan N, Hunt DM, Bhattacharya SS, Warren MJ: The destabilization of human GCAP1 by proline to leucine mutation might cause cone-rod dystrophy. Hum Mol Genet 10: 47–54, 2001
48. Perrault I, Rozet JM, Gerber S, Ghazi I, Leowski C, Ducroq D, Souied E, Difier JL, Munnich A, Kaplan J: Leber congenital amaurosis. Minireview. Mol Genet Metabol 68: 200–208, 1999
49. Sohocki MM, Bowne SJ, Sullivan LS, Blackshaw S, Cepko CL, Payne AM, Bhattacharya SS, Khaliq S, Qasim Mehdi S, Birch DG, Harrison WR, Elder FF, Heckenlively JR, Daiger SP: Mutations in a new photoreceptor-pineal gene on 17p cause Leber congenital amaurosis. Nat Genet 24: 79–83, 2000
50. Camuzat A, Dollfus H, Rozet JM, Gerber S, Bonneau D, Bonnemaison M, Briard ML, Dufier JL, Ghazi I, Leowski C, Weissenbach J, Frezal J, Munnich A, Kaplan J: A gene for leber congenital amaurosis maps to chromosome 17p. Hum Mol Genet 8: 1447–1452, 1995
51. Yang RB, Fulle HJ, Garbers DL: Chromosomal localization and genomic organization of genes encoding guanylyl cyclase receptors expressed in olfactory sensory neurons and retina. Genomics 31: 367–372, 1995
52. Johnston JP, Farhangfar F, Aparicio JG, Nam SH, Applebury ML: The bovine guanylate cyclase GC-E gene and 5' flanking region. Gene 193: 219–227, 1997
53. Duda T, Venkataraman V, Krishnan A, Sharma RK: Rod outer segment membrane guanylate cyclase type 1 (ROS-GC1) gene: Structure, organization and regulation by phorbol ester, a protein kinase C activator. Mol Cell Biochem 189: 63–70, 1998
54. Dollfus H, Rozet JM, Delrieu O, Vignal A, Ghazi I, Dufier JL, Mattei MG, Weissenbach J, Frezal J, Kaplan J, Munnich A: Clinical and genetic heterogeneity of Leber congenital amaurosis. In: J.G. Hollyfield *et al.* (eds), Retinal Degeneration. Plenum Press, New York, pp 143–152, 1993
55. Nakazawa M, Naoi N, Wada Y, Nakazaki S, Maruiwa F, Sawada A, Tamai M: Autosomal dominant cone-rod dystrophy associated with a Val200Glu mutation of the peripherin/rds gene. Retina 16: 405–410, 1996
56. Nakazawa M, Kikawa E, Chida Y, Wada Y, Shiono T, Tamai M: Autosomal dominant cone-rod dystrophy associated with mutations in codon 244 (Asn244His) and codon 184 (Tyr184Ser) of the peripherin/rds gene. Arch Ophthalmol 114: 72–78, 1996
57. Jacobson SG, Cideciyan AV, Kemp CM, Sheffield VC, Stone EM: Photoreceptor function in heterozygotes with insertion or deletion mutations in the RDS gene. Invest Ophthalmol Vis Sci 37: 1662–1674, 1996
58 Warburg M, Sjo O, Tranebjaerg L, Fledelius HC: Deletion mapping of a retinal cone-rod dystrophy: Assignment to 18q211. Am J Med Genet 39: 288–293, 1991
59. Freund CL, Gregory-Evans CY, Furukawa T, Papaioannou M, Looser J, Ploder L, Bellingham J, Ng D, Herbrick JA, Duncan A, Scherer SW, Tsui LC, Loutradis-Anagnostou A, Jacobson SG, Cepko CL, Bhattacharya SS, McInnes RR: Cone-rod dystrophy due to mutations in a novel photoreceptor-specific homeobox gene (CRX) essential for maintenance of the photoreceptor. Cell 91: 543–553, 1997
60. Lange C, Duda T, Beyerman M, Sharma RK, Koch K-W: Regions in vertebrate photoreceptor guanylyl cyclase ROS-GC1 involved in Ca(2+)-dependent regulation by guanylyl cyclase-activating protein GCAP-1. FEBS Lett 460: 27–31, 1999
61. Sokal I, Haeseleer F, Arendt A, Adman ET, Hargrave PA, Palczewski K: Identification of a guanylyl cyclase-activating protein-binding site within the catalytic domain of retinal guanylyl cyclase 1. Biochemistry 38: 1387–1393, 1999
62. Pozdnyakov N, Yoshida A, Cooper NFG, Margulis A, Duda T, Sharma RK, Sitaramayya A: A novel calcium-dependent activator of retinal rod outer segment guanylate cyclase. Biochemistry 34: 14279–14283, 1995
63. Cooper NGF, Liu L, Yoshida A, Pozdnyakov N, Margulis A, Sitaramayya A: The bovine rod outer segment guanylate cyclase, ROS-GC, is present in both outer segment and synaptic layers of the retina. J Mol Neurosci 6: 211–222, 1995
64. Duda T, Goraczniak R, Sharma RK: Molecular characterization of S100A1–S100B protein in retina and its activation mechanism of bovine photoreceptor guanylate cyclase. Biochemistry 35: 6263–6266, 1996
65. Margulis A, Pozdnyakov N, Sitaramayya A: Activation of bovine photoreceptor guanylate cyclase by S100 proteins. Biochem Biophys Res Commun 194: 855–861, 1996
66. Pozdnyakov N, Goraczniak R, Margulis A, Duda T, Sharma RK, Yoshida A, Sitaramayya A: Biochemistry 36: 14159–14166, 1997
67. Duda T, Goraczniak R, Pozdnyakov N, Sitaramayya A, Sharma RK: Differential activation of rod outer segment membrane guanylate cyclases, ROS-GC1 and ROS-GC2, by CD-GCAP and identification of the signaling domain. Biochem Biophys Res Commun 242: 118–122, 1998

68. Rambotti MG, Giambanco I, Spreca A, Donato R: S100B and S100A1 proteins in bovine retina: Their calcium-dependent stimulation of a membrane-bound guanylate cyclase activity as investigated by ultracytochemistry. Neuroscience 92:1089–1101, 1999
69. Kumar V, Senadhi V-K, Krishnan A, Duda T, Sharma RK: A second regulator of rod outer segment membrane guanylate cyclase: Neurocalcin. Biochemistry 38: 12614–12620, 1999
70. Hao Y, Olshevskaya EV, Duda T, Seno K, Hayashi F, Sharma RK, Dizhoor AM, Yamazaki A: J Biol Chem 274: 15547–15555, 1999
71. Yang RB, Garbers DL: Two eye guanylyl cyclases are expressed in the same photoreceptor cells and form homomers in preference to heteromers. J Biol Chem 272: 13738–13742, 1997
72. Olshevskaya EV, Ermilov AN, Dizhoor AM: Dimerization of guanylyl cyclase-activating protein and a mechanism of photoreceptor guanylyl cyclase activation. J Biol Chem 274: 25583–25587, 1999
73. Liu Y, Ruoho AE, Rao VD, Hurley JH: Catalytic mechanism of the adenylyl and guanylyl cyclases: Modeling and mutational analysis. Proc Natl Acad Sci USA 94: 13414–13419, 1997
74. Duda T, Venkataraman V, Goraczniak R, Lange C, Koch K-W, Sharma RK: Functional consequences of a rod outer segment membrane guanylate cyclase (ROS-GC1) gene mutation linked with Leber's congenital amaurosis. Biochemistry 38: 509–515, 1999
75. Duda T, Krishnan A, Venkataraman V, Lange C, Koch K-W, Sharma RK: Mutations in the rod outer segment membrane guanylate cyclase in a cone-rod dystrophy cause defects in calcium signaling. Biochemistry 38: 13912–13919, 1999
76. Tucker CL, Woodcock SC, Kelsell RE, Ramamurthy V, Hunt DM, Hurley JB: Biochemical analysis of a dimerization domain mutation in RetGC-1 associated with dominant cone-rod dystrophy. Proc Natl Acad Sci USA 96: 9039–9044, 1999
77. Wilkie S, Newbold RJ, Deery E, Walker CE, Stinton I, Ramamurthy V, Hurley JB, Bhattacharya SS, Warren MJ, Hunt DM: Functional characterization of missense mutations at codon 838 in retinal guanylate cyclase correlates with disease severity in patients with autosomal dominant cone-rod dystrophy. Hum Mol Genet 9: 3065–3073, 2000
78. Gregory-Evans K, Kelsell RE, Gregory-Evans CY, Downes SM, Fitzke FW, Holder GE, Simunovic M, Mollen JD, Taylor R, Hunt DM, Bird AC, Moore AT: Abnormal cone synapses in human cone-rod dystrophy. Ophthalmology 107: 55–61, 2000
79. Perrault I, Rozet J-M, Gerber S, Kelsell RE, Souied E, Cabot A, Hunt DM, Munnich A, Kaplan J: A ret-GC1 mutation in autosomal dominant cone-rod dystrophy. Am J Hum Genet 63: 651–654, 1998
80. Duda T, Venkataraman V, Jankowska A, Lange C, Koch K-W, Sharma RK: Impairment of the rod outer segment membrane guanylate cyclase dimerization in a cone-rod dystrophy results in defective calcium signaling. Biochemistry 39: 12522–12533, 2000
81. Yau KW: Phototransduction mechanism in retinal rods and cones. The Friedenwald Lecture. Vis Sci 35: 9–32, 1994
82. Pugh EN Jr, Lamb TD: Amplifications and kinetics of the activation steps in phototransduction. Biochem Biophys Acta 1141: 111–149, 1993
83. Pugh EN Jr, Duda T, Sitaramayya A, Sharma RK: Photoreceptor guanylate cyclases: A review. Biosci Rep 17: 429–473, 1997
84. Ratto GM, Payne R, Owen WG, Tsien RY: The concentration of cytosolic free calcium in vertebrate rod outer segments measured with fura-2. J Neurosci 8: 3240–3246, 1988
85. Korenbrot JI, Miller DL: Cytoplasmic free calcium concentration in dark-adapted retinal rod outer segments. Vision Res 29: 939–958, 1989
86. Lagnado AL, Cervetto L, McNaughton PA: Calcium homeostasis in the outer segments of retinal rods from the tiger salamander. J Physiol 455: 111–142, 1992
87. Gray-Keller MP, Detwiler PB: Ca^{2+} dependence of dark- and light-adapted flash responses in rod photoreceptors. Neuron 17: 323–331, 1996
88. Pugh EN Jr, Nikonov S, Lamb TD: Molecular mechanisms of vertebrate photoreceptor light adaptation. Curr Opin Neurobiol 9: 410–418, 1999

Molecular and Cellular Biochemistry **230**: 139–147, 2002.

Factors that affect regulation of cGMP synthesis in vertebrate photoreceptors and their genetic link to human retinal degeneration

Elena V. Olshevskaya, Alexandre N. Ermilov and Alexander M. Dizhoor

Departments of Ophthalmology/Kresge Eye Institute, Pharmacology and Anatomy and Cell Biology, Wayne State University School of Medicine, Detroit, MI, USA

Abstract

Cyclic GMP is essential for the ability of rods and cones to respond to the light stimuli. Light triggers hydrolysis of cGMP and stops the influx of sodium and calcium through the cGMP-gated ion channels. The consequence of this event is 2-fold: first, the decrease in the inward sodium current plays the major role in an abrupt hyperpolarization of the cellular membrane; secondly, the decrease in the Ca^{2+} influx diminishes the free intracellular Ca^{2+} concentration. While the former constitutes the essence of the phototransduction pathway in rods and cones, the latter gives rise to a potent feedback mechanism that accelerates photoreceptor recovery and adaptation to background light. One of the most important events by which Ca^{2+} feedback controls recovery and light adaptation is synthesis of cGMP by guanylyl cyclase. Two isozymes of membrane photoreceptor guanylyl cyclase (retGC) have been identified in rods and cones that are regulated by Ca^{2+}-binding proteins, GCAPs. At low intracellular concentrations of Ca^{2+} typical for light-adapted rods and cones GCAPs activate RetGC, but concentrations above 500 nM typical for dark-adapted photoreceptors turn them into inhibitors of retGC. A variety of mutations found in GCAP and retGC genes have been linked to several forms of human congenital retinal diseases, such as dominant cone degeneration, cone-rod dystrophy and Leber congenital amaurosis. (Mol Cell Biochem **230**: 139–147, 2002)

Key words: photoreceptors, cGMP, retinal degeneration, guanylyl cyclase, GCAP, calcium-binding proteins

Cyclic GMP synthesis regulation in recovery and light adaptation

The key role of cGMP in phototransduction is defined by its ability to regulate circulating current carried primarily by Na^+ and in much lesser extent by Ca^{2+} through the channels located in the plasma membrane of photoreceptor cells. The estimated 2–4 μM free cGMP concentrations in dark adapted rods maintain several percent of cGMP gated channels remain in the open state, and the changes in the rate of hydrolysis and synthesis of cGMP in the light versus dark mediate hyperlpolarization of the cellular membrane after illumination or its partial depolarization during recovery and light adaptation. Photoisomerized rhodopsin triggers the hydrolysis of cGMP by activation of a G protein, transducin, that stimulates cGMP phosphodiesterase and thus causes cGMP-gated channels to close (see refs [1–3] for review). Rods and cones rapidly recover to their resting potential after excitation induced by a non-saturating flash of light. Another process, called light adaptation, partially restores the light sensitivity of the cells exposed to a constant background illumination. Compared to a dark-adapted retina, light-adapted photoreceptors decrease the amplitude and accelerate the kinetics of their electrical responses to a standard test flash, and the degree of such change depends on the intensity of the

Address for offprints: A. Dizhoor, WSU, Kresge Eye Institute, 4717 St. Antoine Boulevard, Detroit, MI 48201, USA (E-mail: adizhoor@med.wayne.edu)

background light (see ref. [4] for review). Reopening of the cGMP-gated channels during recovery and light adaptation occurs as a result of multiple mechanisms that include re-synthesis of cGMP by guanylyl cyclase, and the major factor that controls the activity of guanylyl cyclase in photoreceptors is a dramatic fluctuation in the level of free Ca^{2+} that occurs inside the photoreceptor between light and dark conditions.

Closure of cGMP gated channels in the light caused by activation of cGMP hydrolysis decreases the intracellular Ca^{2+} concentrations in rods and cones (Fig. 1). In the dark, Ca^{2+} extruded from the photoreceptor outer segment by a Na^+/K^+, Ca^{2+}-exchanger re-enters the outer segment through the open cGMP-gated channels, so the average free intracellular Ca^{2+} level lies between 500–600 nM [5]. In the light, when cGMP is hydrolyzed and the channels are closed, the exchanger continues to extrude Ca^{2+} ions from the photoreceptor outer segment, therefore, the intracellular Ca^{2+} concentrations in illuminated rods and cones can decrease to below 50 nM [5, 6]. Such a decrease in the intracellular Ca^{2+} concentrations contributes to various reactions of recovery and adaptation processes, from stimulation of cGMP binding to the cGMP-gated channels to inactivation of photoexcited rhodopsin [7–9], but most importantly, it accelerates cGMP resynthesis by guanylyl cyclase (for review see refs [1, 10–12]).

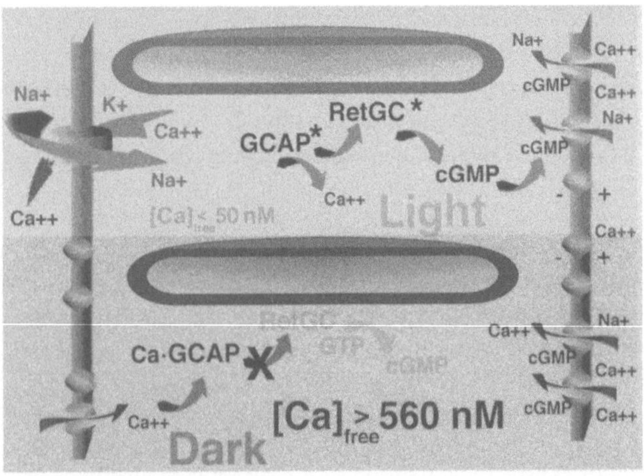

Fig. 1. Regulation of cGMP synthesis in photoreceptors. In response to light-activated hydrolyis of cGMP, cGMP gated cation channels close and the intercellular Ca^{2+}concentrations drop to ~ 50 nM, because Na/K,Ca-exchanger constantly extrudes Ca^{2+} from the cyctoplasm. The lowered Ca^{2+} concentrations activates Ca^{2}-sensor proteins, GCAPs, that stimulate membrane guanylyl cyclase, retGC, in photoreceptor disk membrane. The increased rate of cGMP synthesis facilitates re-opening of the cGMP-gated channels and thus contributes to restoration of photoreceptor sensitivity during light adaptation. In dark adapted photoreceptors (bottom part) stimulation of cGMP synthesis stops because hydrolysis of cGMP by light-dependent phosphodiesterase is suppressed, and several percent of the cation channels are opened so that the influx of Ca^{2+} increases its intracellular concentration to the point (500–600 nM) when GCAP becomes loaded with Ca^{2+} and can no longer activate retGC.

Earlier reports from Lolley and Racz [13] implicated Ca^{2+} as a potential regulator of the cGMP synthesis in photoreceptors, but the most important contribution to our present understanding of the cyclase regulation was made later by Koch and Stryer [13] who defined the principal role of calcium as a regulator of photoreceptor guanylyl cyclase. They have made the following two major observations: (1) guanylyl cyclase activity in preparations of photoreceptor membranes becomes both low and insensitive to Ca^{2+} when cytoplasmic proteins are extracted by low-salt buffer solution; (2) reconstitution of the washed membranes with the cytoplasm increases cyclase activity in the absence of Ca^{2+}, while increase in Ca^{2+} concentrations above 100 nM inhibits the activation in a cooperative manner. These findings initiated a search for the identity of both the cyclase and the unknown activator protein and thus led to present understanding of the molecular mechanisms for the cyclase regulation.

Photoreceptor guanylyl cylcases (ret GC) and their Ca^{2+}-sensor proteins (GCAPs)

In the initial attempts to purify photoreceptor guanylyl cyclases reported by Fleishman [14] who determined that photoreceptor guanylyl cyclase was associated with particulate fraction. Successful purification procedures developed in early 90's [15, 16] allowed to purify the enzyme from outer segment membranes. Yet, the identity of the cyclase remain obscure until its isozymes were cloned and expressed in heterologous systems.

The cDNA for the first retina-specific cyclase, retGC-1 (GC-E), was cloned in laboratory of D. Low in 1992 and named retGC [17]. The original nucleotide sequence of retGC was later revised and corrected [18] after comparing it to its homolog, ROS-GC1, purified from bovine retina [19]. The identity of retGC as a Ca^{2+}-regulated cyclase was finally established, when its Ca^{2+}-sensor protein, p24 (GCAP-2), was purified from bovine retina and reconstituted with the recombinant retGC expressed in a mammalian cell culture [20]. These experiments demonstrated that recombinant retGC, that showed only low basal activity and was insensitive to Ca^{2+}, was efficiently activated by highly purified GCAP-2 in a Ca^{2+}-sensitive manner.

It became obvious that the situation with retGC was more complex, when its second homolog, retGC-2 (GC-F), was cloned from human, rat and bovine retinal cDNA libraries [18, 21, 22]. The antibodies against retGC-1 produce a strong signal in cones, but they give only a weak, if any, signal in rods [20, 23] (Fig. 2). Disruption of a mouse retGC-1 gene results in degeneration of cones but has no significant effect on viability or electrophysiological behavior of rods [24].

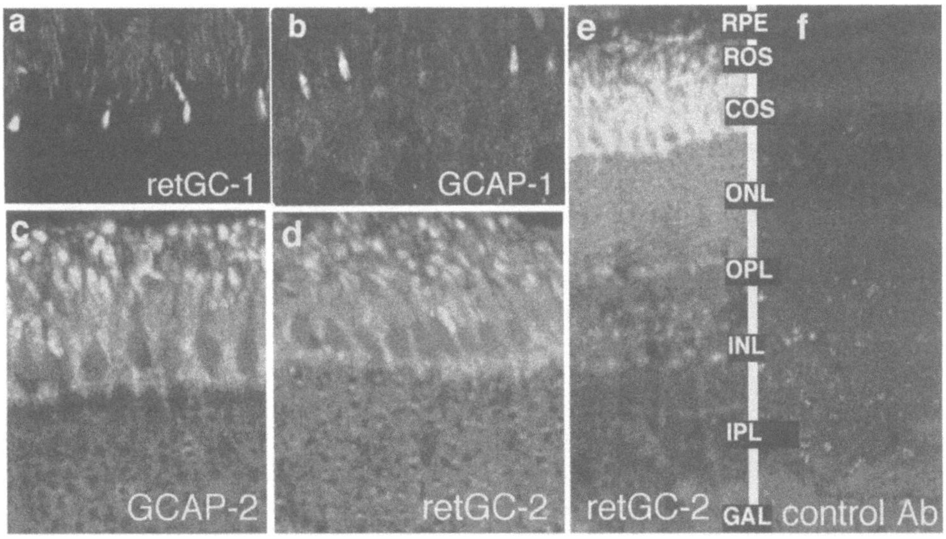

Fig. 2. Localization of isoforms for retGC and GCAPs in bovine retina. Immunostaining for both retGC-1 (a) and GCAP-1 (b) is strong in cones, while both GCAP-2 (c) and retGC-2 (d, e) are strongly expressed in rods (Olshevskaya *et al.*, unpublished data). Preimmune antiserum (f) from the rabbit used for immunization against retGC-2 produced background fluorescence in the inner nuclear and ganglion layers, but not in photoreceptors. Cryosections of bovine retina fixed in 4% paraformaldehyde solution were labeled with polyclonal antibodies directed against synthetic protein fragments. The signal from FITC-conjugated secondary antibody was detected using confocal microscopy. Obj.: × 100,000 (a–d); × 40,000(e–f). RPE – pigment epithelium; ROS – rod outer segments; COS – cone outer segments; ONL – outer nuclear layer; OPL – outer plexiform layer; INL – inner nuclear layer; IPL – inner plexiform layer; GAL – ganglion layer.

This observation is also consistent with the retGC-1 being predominantly a cone-specific isozyme based on the immunostaining, while retGC-2 appears more abundant in rods.

RetGC-1 and retGC-2 are closely related to each other and other membrane guanylyl cyclases: natriuretic peptide receptor cyclases (GC-A, GC-B); guanylin- and enterotoxin-activated intestinal cyclase (GC-C), peptide receptor cyclases in sea urchin sperm cells, olfactory guanylyl cyclase (GC-D) and others (for review see refs [25–27]).

RetGC-1 and -2 have three major homology regions common to transmembrane guanylyl cyclases. In receptor guanylyl cyclase, the extracellular domain (ECD) is responsible for binding the extracellular activator ligand (i.e. cyclic peptides). The homology between retGC-1, retGC-2 and other members of this family is rather weak within this domain compared to other regions of the molecule [18, 21].

Similarly to other membrane guanylyl cyclases, the intracellular portion of retGC contains a catalytic domain and a region, KHD (*k*inase *h*omology *d*omain), homologous to protein kinases (reviewed in refs [26, 27]. The actual consensus sequence implicated in protein kinase activity has not been found in KHD of membrane guanylyl cyclases, and it remains unclear whether or not KHD in retGC has any kinase activity [26].

There is also a short region, 'dimerization domain', located between the KHD and the catalytic domain in membrane GCs that is essential for the ability of the cyclase to form a homodimer [26, 28]. Although there is no direct structural evidence that it actually creates a direct contact between two retGC molecules, the peptide fragment that contains putative dimerization domain in retGC reportedly can form a dimer in solution [29]. The dimerization domain plays an important functional role, because several mutations in this short region of the molecule have been linked to a con-rod dystrophy [30].

Unlike other members of the membrane guanylyl cyclase family that in order to be active require an extracellular peptide ligand bound to their ECD, both retGC isozymes are regulated by Ca^{2+}-binding proteins, GCAPs (guanylyl cyclase activator proteins), that interact with the cytoplasmic portion of the cyclase molecule (for review see refs [11, 12, 31]).

Two homologous 24-kDa proteins GCAP-1 and GCAP-2 that impart Ca^{2+}-sensitivity to RetGC were isolated in 1994 from a bovine retina [20, 32]. Amino acid sequences of GCAP-1 and GCAP-2 are over 40% identical. They include three distinct EF-hand Ca^{2+} -binding domains, and their N-terminal Gly residues are fatty acylated [11, 12]. A cDNA for the third homologue, GCAP-3, has been found exclusively in a human retinal cDNA library [33].

GCAP-1 and GCAP-2 are photoreceptor-specific proteins. Anti-GCAP-1 antibodies detect expression of GCAP-1 primarily in cones, while GCAP-2 has been detected primarily in outer and inner segments of rods and, at a lower level, in cones (Fig. 2, and refs [34–36]). The signal for GCAP-1 and GCAP-2 within the intracellular compartments of rods and cones may partially depend on the strength of a particular antibody and on how different methods of fixation affect their epitopes exposure in various specimens. However, as an additional evidence that GCAP-1 is specific for cones comes from the fact that a mutation, Y99C, in GCAP-1 in humans results in cone degeneration, but has little or no effect on the viability or electrophysiological behavior of rods [37]. The role and localization of GCAP-3 in photoreceptors is not immediately apparent, since the disruption of both GCAP-1 and GCAP-2 genes in transgenic mice seems to be sufficient to decelerate the recovery in rods in a manner consistent with the lack of the Ca^{2+}-feedback regulation of the cGMP synthesis [38].

GCAPs are members of the recoverin family, a separate group within the EF-hand superfamily that includes proteins remotely homologous to calmodulin. However, unlike the ubiquitous calmodulin, recoverin-like proteins are N-fatty acylated and are predominantly localized to neurons (see ref. [39] for review).

The recoverin-like proteins have four EF-hand motifs (named EF-1 through EF-4), but unlike calmodulin, only three (or sometimes just two) of them form true Ca^{2+}-binding domains. The N-terminal EF-hand-related motif, EF-1, deviates from the consensus sequence and lacks several key side chain residues required for coordinating Ca^{2+} ions [11, 12]. The solution structure of the Ca^{2+}-bound GCAP-2 recently established by Ames *et al.* [40] is also drastically different from calmodulin and closely resembles recoverin and neurocalcin. The GCAP-2 molecule consists of two globular halves connected by a flexible 'hinge' region between the exiting helix of the EF-2 and the entering helix of the EF-3. Based on functional study in vitro, both GCAP-1 and GCAP-2 have three functional EF-hands, and inactivation of these EF-hands by point mutations renders GCAPs in a constitutively active form that efficiently stimulates RetGC both at low and high Ca^{2+} concentrations [41, 42]. GCAP-2 binds three Ca^{2+} ions with a Hill coefficient of ~ 2 and an apparent dissociation constant of 300 nM [40]. These values closely resemble the cooperativity and the EC_{50} values for the Ca^{2+} effects on retGC regulation by GCAP-2 [20, 41]. Binding of Ca^{2+}, makes GCAPs undergo an 'activator-to-inhibitor' transition, and fully Ca^{2+} loaded GCAPs inhibit basal activity of RetGC in photoreceptor membranes [41]. However, at the free Ca^{2+} concentrations typical for the dark-adapted photoreceptors, the ratio between the Ca^{2+}-free (activator) and the Ca^{2+} bound (inhibitor) forms of GCAP-2 is such that the activity of RetGC is fairly close to its basal activity measured in the absence of GCAPs [41]. The presence of the N-fatty acyl group significantly increases activity of GCAP-1 [42], but makes only minor contribution to the activity or Ca^{2+}-sensitivity of GCAP-2 [43].

The relative contribution of each EF-hand to the overall Ca^{2+}-sensitivity of the cyclase regulation is slightly different between GCAP-1 and GCAP-2. Inactivation of EF-3 or EF-4 by point mutations in GCAP-1 produces a strong decrease in Ca^{2+}-sensitivity of the cyclase activation, while inactivation of the EF-2 has only a small effect. Contrary to this, inactivation of the third Ca^{2+} binding loop, EF-3, in GCAP-2 has relatively small effect on Ca^{2+}-sensitivity of RetGC compared to EF-2 and EF-4. Inactivation of any pair of EF-hands is sufficient to shift the Ca^{2+}-sensitivity of RetGC regulation outside the physiological submicromolar range, and inactivation of all three EF-hands completely eliminates Ca^{2+}-sensitivity of GCAPs and turns them into constitutive activators of RetGC [12, 41].

GCAP-2 can activate both recombinant RetGC-1 and RetGC-2 *in vitro* [18, 20, 44]. The matter of selectivity between the two RetGC isozymes in case of GCAP-1 remains unclear. Some authors observed only activation of recombinant RetGC-1 by GCAP-1 expressed from baculoviral vectors in cultured insect cells [33, 45]. However, fully myristoylated GCAP-1 produced in *E. coli* can also activate recombinant RetGC-2 *in vitro* [46]. Such a variance may possibly reflect variations in a degree of the fatty acylation of GCAP-1 produced in different expression systems. Yet, based on immunohistochemical analysis of localization of RetGC and GCAPs, it is likely that *in vivo* retGC-1 serves primarily as a target for GCAP-1 in cones while GCAP-2 may play a major role in activation of retGC-2 in rods (Fig. 2). This agrees with the fact that retGC-1 gene disruption in mice causes cone degeneration, but does not significantly affect rods [24].

Putative mechanism of RetGC regulation by GCAP

The catalytic domain of RetGC and other membrane GCs is closely related to the adenylyl cyclase II, in which the active center of the enzyme is formed by its dimerized catalytic domains [47, 48], and to peptide receptor cyclases that also known to form homodimers [26]. Recent observations also strongly indicate that retGC complex likely consists of a dimerized cyclase and dimerized GCAP-2: (1) Both RetGC-1 and RetGC-2 have a putative 'dimerization domain' typical for peptide receptor guanylyl cyclases (for review see also refs [25–27]). Co-expression of recombinant RetGC-1 and RetGC-2 in transfected cells can result in a complex that contains both isozymes, but they are likely to form only homodimers *in vivo* [49]. (2) RetGC in outer segment membranes isolated from the retina can be chemically cross-linked into a covalent dimer in a Ca^{2+}-sensitive manner, and (3) The efficiency of the covalent cross-linking of retGC into a dimer increases in the presence of the Ca^{2+}-free recombinant GCAP-1 and GCAP-2, while Ca^{2+}-loaded GCAPs inhibit it [50].

The functional role of GCAP is likely to stimulate closer interactions between the catalytic domains within the RetGC homodimer. GCAP-2 can also form functional homodimer.

The structure of GCAP-2 dimer is affected by Ca^{2+} binding, and its Ca^{2+}-loaded form completely dissociates into monomers during gel-chromatography [51]. Moreover, the ability of GCAP-2 and its chimerical mutants to activate RetGC at low free Ca^{2+} concentrations correlates with their ability to dimerize in the absence of Ca^{2+}. Thus, not only GCAP:RetGC, but GCAP:GCAP interactions, as well, are apparently important for the cyclase regulation [51].

Not only the Ca^{2+}-free GCAPs activate RetGC, but the Ca^{2+}-loaded GCAPs also inhibit the intrinsic basal activity of the cyclase in photoreceptor membranes *in vitro*. Also, it was directly demonstrated that GCAP-2 binds to the cyclase in a Ca^{2+}-independent manner and protects the intracellular fragment of retGC from proteolysis both in the presence and in the absence of Ca^{2+} [52]. This clearly indicates that GCAP-2 molecules remain bound to retGC both when in their Ca^{2+}-free and Ca^{2+}-loaded forms.

There is a putative 'dimer-adapter' model [11, 51] that can fit with these observations. According to this model, GCAP-2 monomers bound to each molecule of the cyclase form a GCAP:GCAP dimer that modulates interactions between the catalytic domains of the cyclase subunits. The Ca^{2+}-loaded GCAP molecules attached to the individual retGC subunits may interfere with dimerization of the catalytic domains, and therefore inhibit the function of the active center of the enzyme. The dissociation of Ca^{2+} in the light tightens the GCAP dimer conformation and thus increases the affinity between the two GCAP-2 molecules. Forces that 'pull' together two GCAP molecules bound to the cyclase subunits can bring the retGC subunits (either directly or by causing a secondary conformational change in retGC) into a proper contact between their catalytic domains and thus activate the catalytic center in a retGC homodimer.

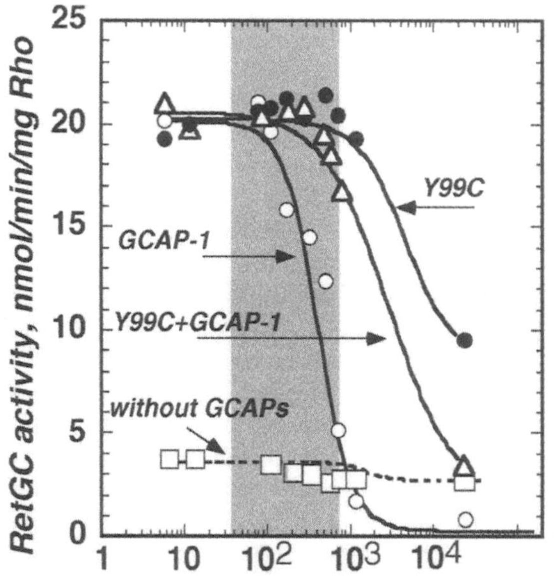

Fig. 3. The Y99C GCAP-1 causes the abnormal Ca^{2+} sensitivity of cyclase regulation. The shaded area marks the range for the free Ca^{2+} concentrations in photoreceptors [5, 6]. When reconstituted with photoreceptor membranes, normal GCAP-1 regulates retGC activity in a Ca^{2+} sensitive manner within this range. The Y99C GCAP-1 has much lower sensitivity to Ca^{2+} and continues to activate the cyclase outside this range even in the presence of the equal amount of normal GCAP-1. Composed using the data by Dizhoor *et al.* [56].

The places were the contacts between retGCs and GCAPs occur, have not yet been clearly determined. One of the putative sites for binding of GCAP-1 and GCAP-2 may lie within the cyclase catalytic domain [53]. However, there can be alternative sites in retGC that interact with GCAPs. Based on the study of the chimerical retGC-1 and retGC-2, their affinity for GCAPs may be determined by the KHD rather than the cyclase catalytic domain [54]. Also, in ROS-GC1, a bovine homologue of retGC-1, two different fragments, M445-L456 and L503-I522, have been implicated as possible sites for the interaction with GCAP [55].

Mutations in GCAP and RetGC associated with retinal diseases

A case of congenital dominant cone degeneration has been linked to a mutation, Y99C, within the entering helix of the third EF-hand structure in GCAP-1 [**37**]. This mutation causes gradually progressing degeneration of cones and results in a complete loss of color perception and daylight vision, yet rods are unaffected and their responses to dim light remain virtually normal. Although the Y99C GCAP-1 activates retGC *in vitro* to the same extent as the wild type GCAP-1, in the presence of the Y99C mutant the cyclase remains active even at the free Ca^{2+} concentrations above 1 μM (Fig. 3) [56, 57]. Unlike wild type GCAP-1, the Y99C mutant promotes dimerization of the cyclase not only in the absence but also in the presence of micromolar Ca^{2+} [50]. The Y99C GCAP-1 efficiently competes with the Ca^{2+}-loaded wild type GCAP-1 and GCAP-2 *in vitro*, and that may account for the dominant nature of this mutation in humans [56] (Fig. 3).

High intracellular Ca^{2+} concentrations may alter the permeability of rod mitochondria and promote apoptosis in photoreceptors [58]. Since activation of RetGC *in vitro* by the Y99C GCAP-1 is not fully suppressed even at ~ 10 μM free Ca^{2+}, it is very tempting to suggest that in live photoreceptors it also remains active when the free Ca^{2+} concentrations reach their normal 'dark' level (Fig. 4). In normal photoreceptors, the 'dark' free concentrations of the circulating Ca^{2+} are at ~500–600 nM level because the rate of cGMP synthesis is lowered by C-GCAP. This maintains the free cGMP concentrations near 2–4 μM level [59], so that just near 3% of the total cGMP-gated channels are open to allow the influx of

Na^+ and Ca^{2+}. The increased rate of cGMP synthesis by the Y99C GCAP-1 may lead to opening of the excessive number of cGMP-gated channels in cones and therefore cause much higher than normal influx of Na^+/Ca^{2+}. It may take the intracellular Ca^{2+} concentrations to reach > 10 μM before the Y99C GCAP-1 would finally stop activating the cyclase, and such concentrations of Ca^{2+} may turn to be sufficient for trigger the apoptosis in cones. Similar scenario can take place in another case of human autosomal dominant cone dystrophy recently linked to a mutation, E155G, in GCAP-1 that inactivates Ca^{2+}-binding loop of the EF-hand 4 [65]. It may also be relevant to a case of mutations in R838 of RetGC-1 linked to human cone/rod dystrophy [52], as will be discussed further in this section.

It should be noted that at present there is no direct experimental evidence that the Y99C GCAP-1 increases the free Ca^{2+} level in cones *in vivo*. Transgenic models need to be developed in order to verify the actual biochemical links between the mutant GCAP-1 and the photoreceptor degeneration.

Severe cases of Leber congenital amaurosis (LCA1 - photoreceptor degeneration in early childhood) have been linked to frame shift and missense mutations in retGC-1 [60]. The

Fig. 4. Putative scenario of abnormal regulation of RetGC by the Y99C GCAP-1 mutant. In the dark, when $[Ca^{2+}]_{free}$ is above 500 nM, synthesis of cGMP is inhibited by normal Ca/GCAP-1 to the level that is sufficient to maintain only a few percent of the channels in the open state. Contrary to that, its mutant form, Y99C GCAP-1, associated with congenital cone degeneration has lower affinity for Ca^{2+} and therefore continues to activate the cyclase in the dark. Consequently, the excessive number of the channels keep opening, but cGMP synthesis by retGC does not stop until the intracellular $[Ca^{2+}]_{free}$ raises to a much higher than normal level (> 10 μM) that becomes sufficient to inhibit activation of the cyclase by the mutant GCAP.

frame shift results in truncated inactive retGC-1. The missense mutation, F514S, also decreases activity of RetGC-1 *in vitro* [61].

Defects in the retGC-1 gene were also found in patients with dominant cone-rod dystrophy (CORD6), a disease that causes degeneration of cones followed by the death of rods. The genetic defects were initially described as two single missense mutations, E837D and R838C, and a triple mutation, E837D; R838C; T839M, in the putative dimerization domain of retGC-1 [30, 62, 63]. These substitutions suppress the maximal level of stimulation produced in vitro by reconstitution with GCAP-2 [52, 64]. The mutations in R838 can also alter the kinetics of retGC-1 regulation by GCAP-1 [29, 30, 64]. However, the results of this study with regard to GCAP-1 reported by two different groups are at significant variance with each other.

According to one series of reports [30, 52], substitutions of T838 found in several cases of cone-rod dystrophy decreased Ca^{2+} sensitivity of the cyclase activation by GCAP-1 *in vitro* so that the mutant cyclase remains stimulated by GCAP-1 at higher than micromolar Ca^{2+} concentrations, which in case of wild type retGC-1 should have inhibited its activation. The authors suggest [30, 52] that the disease-linked mutations at the position 838 of the dimerization domain may stimulate tighter association between the cyclase subunits, and thus influence the affinity for Ca^{2+} of GCAP-1 bound to the cyclase. According to Wilkie *et al.*[30], computational models of the three-dimensional structure of retGC-1 dimerization domain suggest that naturally occurring mutations, R838C, R838H and R838S, all result in stronger coil-coil interactions between the dimerizations domains in retGC-1 homodimers. Still, the reason why would such changes cause the loss of GCAP-1 affinity for Ca^{2+}, remains unclear.

Different results have been reportedly observed when corresponding double – or triple mutations found in human retGC-1 gene were introduced into its bovine homolog, ROS-GC1 [29, 64]. In this case the mutations significantly (by approximately 70%) damage the catalytic activity of the cyclase, but they did not affect the Ca^{2+}-sensitivity of its regulation by GCAP-1 or GCAP-2. According to the data presented by this group, the triple mutation interfered with dimerization of the recombinant retGC-1, so that chemical cross-linking of the cyclase dimer in the presence of GCAP was drastically suppressed. Moreover, in contrast to the conclusions made by Wilkie *et al.* [30], a recombinant fragment of the mutated dimerization domain reportedly failed to form a stable dimer, unlike the corresponding fragment of dimerization domain derived from the wild type ROS-GC1 [29].

The reason for this controversy is not immediately apparent. It could be one of the possible explanations that the structure of bovine and human retGC-1 is different enough to account for the opposite behavior of the two cyclase homologs even if they carry very similar mutations. However, the amino acid sequences in the regions of the dimerization domain in both retGC-1 and ROS-GC1, where the mutations were introduced, are virtually identical. Hence, if there is any structural difference between the two cyclases that may account for their different kinetic properties, it must be in some other part(s) of the cyclase molecule.

Why would various mutations in retGC-1 result in photoreceptor degeneration? The first scenario is based on loss of function by retGC-1. Although it is not quite clear why should the absence of cGMP be lethal for the photoreceptor cells, in the transgenic mice that lack retGC-1 [24] cones quickly degenerate. This genetic model directly confirms that mutations causing truncation of retGC-1 can account for the cell death in severe cases of Leber congenital amaurosis [60].

If the mutations in dimerization domain of retGC-1 result in a dramatic loss of the cyclase basal activity *in vivo*, similar scenario may takes place in the photoreceptor degeneration associated with congenital rod-cone dystrophy, as well [29]. A different hypothesis was offered by Tucker and colleagues [52] in order to explain the biological effect of the mutations, in particular of T838, in retGC-1 dimerization domain. They suggest that lowered sensitivity of retGC/GCAP-1 complex to inhibition by Ca^{2+} observed in their experiments may, on the contrary, cause abnormally high cGMP concentrations in the cells. The consequence of the events in this case would be very similar to the scenario with the Y99C GCAP-1 that was discussed earlier in this section and Fig. 4. This hypothesis would also explain the dominant character of the CORD6 cases linked to the mutations in R838.

It is not quite clear to what extent biochemical properties of recombinant retGC-1 may reflect the actual properties of the native enzyme. Some skepticism may be raised because in all cited reports recombinant retGC and its homologs typically have much lower basal activity than retGC in photoreceptor membranes. Also, its apparent affinity to GCAPs is approximately 50-fold lower compared to the native enzyme. This may reflect, for example, lower efficiency of folding, assembly or/and incorporation into membrane of recombinant cyclase, and that could potentially contribute to the variance between the reported results. It appears that to make such study more reliable it would require transgenic models that carry mutant cyclases.

There is another important issue that remains to be understood. The GC-E (retGC-1) gene knock-out that leads to cone degeneration in transgenic mice has no prominent physiological effects on rods [24], perhaps, because retGC-1 is primarily expressed in cones (Fig. 2). And yet, in humans the null mutations in retGC-1 have been linked to both cone and rod degeneration in several cases of LCA1 [60]. Apparently, either the death of rods in human retina has little to do with the mutation in retGC-1 *per se* (i.e. it may be a human retina tissue response to the presence of dying cones), or there is a significant difference between various mammals with regard to the relative importance of the two retGC isozymes for the physiology of rods and cones.

Acknowledgements

The financial support for A.M.D. has been provided in part by the National Institutes of Health (EY11522 from NEI) and Research to Prevent Blindness, Inc.

References

1. Pugh EN Jr, Lamb TD: Amplification and kinetics of the activation steps in phototransduction. Biochim Biophys Acta 1141: 111–149, 1993
2. Baylor D: How photons start vision. Proc Natl Acad Sci USA 93: 560–565, 1996
3. Molday RS: Photoreceptor membrane proteins, phototransduction, and retinal degenerative diseases. The Friedenwald Lecture. Invest Ophthalmol Vis Sci 39: 2491–2513, 1998
4. Pugh EN Jr, Nikonov S, Lamb TD: Molecular mechanisms of vertebrate photoreceptor light adaptation. Curr Opin Neurobiol 9: 410–418, 1999
5. Gray-Keller MP, Detwiler PB: The calcium feedback signal in the phototransduction cascade of vertebrate rods. Neuron 13: 849–861, 1994
6. Sampath AP, Matthews HR, Cornwall MC, Bandarchi J, Fain GL: Light-dependent changes in outer segment free-Ca^{2+} concentration in salamander cone photoreceptors. J Gen Physiol 113: 267–277, 1999
7. Hsu YT, Molday RS: Modulation of the cGMP-gated channel of rod photoreceptor cells by calmodulin. Nature 361: 76–79, 1993
8. Hackos DH, Korenbrot JI: Calcium modulation of ligand affinity in the cyclic GMP-gated ion channels of cone photoreceptors. J Gen Physiol 110: 515–528, 1997
9. Kawamura S: Rhodopsin phosphorylation as a mechanism of cyclic GMP phosphodiesterase regulation by S-modulin. Nature 362: 855–857, 1993
10. Pugh EN Jr, Duda T, Sitaramayya A, Sharma RK: Photoreceptor guanylate cyclases: A review. Biosci Rep 17: 429–473, 1997
11. Dizhoor AM: Regulation of cGMP synthesis in photoreceptors: Role in signal transduction and congenital diseases of the retina. Cell Sig 12: 711–719, 2000
12. Palczewski K, Polans AS, Baehr W, Ames JB: Ca(2+)-binding proteins in the retina: Structure, function, and the etiology of human visual diseases. Bioessays 22: 337–350, 2000
13. Koch KW, Stryer L: Highly cooperative feedback control of retinal rod guanylate cyclase by calcium ions. Nature 334: 64–66, 1988
14. Fleischman D, Denisevich M, Raveed D, Pannbacker RG: Association of guanylate cyclase with the axoneme of retinal rods. Biochim Biophys Acta 630: 176–186, 1980
15. Hayashi F, Yamazaki A: Polymorphism in purified guanylate cyclase from vertebrate rod photoreceptors. Proc Natl Acad Sci USA 88: 4746–4750, 1991
16. Koch KW: Purification and identification of photoreceptor guanylate cyclase. J Biol Chem 266: 8634–8637, 1991
17. Shyjan AW, de Sauvage FJ, Gillett NA, Goeddel DV, Lowe DG: Molecular cloning of a retina-specific membrane guanylyl cyclase. Neuron 9: 727–737, 1992
18. Lowe DG, Dizhoor AM, Liu K, Gu Q, Spencer M, Laura R, Lu L, Hurley JB: Cloning and expression of a second photoreceptor-specific membrane retina guanylyl cyclase (RetGC), RetGC-2. Proc Natl Acad Sci USA 92: 5535–5539, 1995

19. Margulis A, Goraczniak RM, Duda T, Sharma RK, Sitaramayya A: Structural and biochemical identity of retinal rod outer segment membrane guanylate cyclase. Biochem Biophys Res Commun 194: 855–861, 1993
20. Dizhoor AM, Lowe DG, Olshevskaya EV, Laura RP, Hurley JB: The human photoreceptor membrane guanylyl cyclase, RetGC, is present in outer segments and is regulated by calcium and a soluble activator. Neuron 12: 1345–1352, 1994
21. Yang RB, Foster DC, Garbers DL, Fulle HJ: Two membrane forms of guanylyl cyclase found in the eye. Proc Natl Acad Sci USA 92: 602–606, 1995
22. Goraczniak R, Duda T, Sharma RK: Structural and functional characterization of a second subfamily member of the calcium-modulated bovine rod outer segment membrane guanylate cyclase, ROS-GC2. Biochem Biophys Res Commun 234: 666–670, 1997
23. Liu X, Seno K, Nishizawa Y, Hayashi F, Yamazaki A, Matsumoto H, Wakabayashi T, Usukura J: Ultrastructural localization of retinal guanylate cyclase in human and monkey retinas. Exp Eye Res 59: 761–768, 1994
24. Yang RB, Robinson SW, Xiong WH, Yau KW, Birch DG, Garbers DL: Disruption of a retinal guanylyl cyclase gene leads to cone-specific dystrophy and paradoxical rod behavior. J Neurosci 19: 5889–5897, 1999
25. Garbers DL, Lowe DG: Guanylyl cyclase receptors. J Biol Chem 269: 30741–30744, 1994
26. Garbers DL: The guanylyl cyclase receptors. Methods 19: 477–484, 1999
27. Lucas KA, Pitari GM, Kazerounian S, Ruiz-Stewart I, Park J, Schulz S, Chepenik KP, Waldman SA: Guanylyl cyclases and signaling by cyclic GMP. Pharmacol Rev 52: 375–414, 2000
28. Foster DC, Wedel BJ, Robinson SW, Garbers DL: Mechanisms of regulation and functions of guanylyl cyclases. Rev Physiol Biochem Pharmacol 135: 1–39, 1999
29. Duda T, Venkataraman V, Jankowska A, Lange C, Koch KW, Sharma RK: Impairment of the rod outer segment membrane guanylate cyclase dimerization in a cone-rod dystrophy results in defective calcium signaling. Biochemistry 39: 12522–12533, 2000
30. Wilkie SE, Newbold RJ, Deery E, Walker CE, Stinton I, Ramamurthy V, Hurley JB, Bhattacharya SS, Warren MJ, Hunt DM: Functional characterization of missense mutations at codon 838 in retinal guanylate cyclase correlates with disease severity in patients with autosomal dominant cone-rod dystrophy. Hum Mol Genet 9: 3065–3973, 2000
31. Dizhoor AM, Hurley JB: Regulation of photoreceptor membrane guanylyl cyclases by guanylyl cyclase activator proteins. Methods 19: 521–531, 1999
32. Gorczyca WA, Gray-Keller MP, Detwiler PB, Palczewski K: Purification and physiological evaluation of a guanylate cyclase activating protein from retinal rods. Proc Natl Acad Sci USA 91: 4014–4018, 1994
33. Haeseleer F, Sokal I, Li N, Pettenati M, Rao N, Bronson D, Wechter R, Baehr W, Palczewski K: Molecular characterization of a third member of the guanylyl cyclase- activating protein subfamily. J Biol Chem 274: 6526–6535, 1999
34. Dizhoor AM, Olshevskaya EV, Henzel WJ, Wong SC, Stults JT, Ankoudinova I, Hurley JB: Cloning, sequencing, and expression of a 24-kDa Ca(2+)-binding protein activating photoreceptor guanylyl cyclase. J Biol Chem 270: 25200–25206, 1995
35. Howes K, Bronson JD, Dang YL, Li N, Zhang K, Ruiz C, Helekar B, Lee M, Subbaraya I, Kolb H, Chen J, Baehr W: Gene array and expression of mouse retina guanylate cyclase activating proteins 1 and 2. Invest Ophthalmol Vis Sci 39: 867–875, 1998
36. Kachi S, Nishizawa Y, Olshevskaya E, Yamazaki A, Miyake Y, Wakabayashi T, Dizhoor A, Usukura J: Detailed localization of photoreceptor guanylate cyclase activating protein-1 and -2 in mammalian retinas using light and electron microscopy. Exp Eye Res 68: 465–473, 1999
37. Payne AM, Downes SM, Bessant DA, Taylor R, Holder GE, Warren MJ, Bird AC, Bhattacharya SS: A mutation in guanylate cyclase activator 1A (GUCA1A) in an autosomal dominant cone dystrophy pedigree mapping to a new locus on chromosome 6p21.1. Hum Mol Genet 7: 273–277, 1998
38. Mendez A, Burns ME, Sokal I, Baehr J, Palczewski K, Baylor D: Invest Ophthalmol Vis Sci 40: 2056, 1999
39. Nef P: In: M.R. Celio, T. Pauls, B. Schwaller (eds). Guidebook to the Calcium-binding Proteins. Oxford University Press, New York, 1996, pp 94–120
40. Ames JB, Dizhoor AM, Ikura M, Palczewski K, Stryer L: Three-dimensional structure of guanylyl cyclase activating protein-2, a calcium-sensitive modulator of photoreceptor guanylyl cyclases. J Biol Chem 274: 19329–19337, 1999
41. Dizhoor AM, Hurley JB: Inactivation of EF-hands makes GCAP-2 (p24) a constitutive activator of photoreceptor guanylyl cyclase by preventing a Ca^{2+}-induced 'activator- to-inhibitor' transition. J Biol Chem 271: 19346–19350, 1996
42. Otto-Bruc A, Buczylko J, Surgucheva I, Subbaraya I, Rudnicka-Nawrot M, Crabb JW, Arendt A, Hargrave PA, Baehr W, Palczewski K: Functional reconstitution of photoreceptor guanylate cyclase with native and mutant forms of guanylate cyclase-activating protein 1. Biochemistry 36: 4295–4302, 1997
43. Olshevskaya EV, Hughes RE, Hurley JB, Dizhoor AM: Calcium binding, but not a calcium-myristoyl switch, controls the ability of guanylyl cyclase-activating protein GCAP-2 to regulate photoreceptor guanylyl cyclase. J Biol Chem 272: 14327–14333, 1997
44. Olshevskaya EV, Boikov S, Ermilov A, Krylov D, Hurley JB, Dizhoor AM: Mapping functional domains of the guanylate cyclase regulator protein, GCAP-2. J Biol Chem 274: 10823–10832, 1999
45. Goraczniak RM, Duda T, Sharma RK: Calcium modulated signaling site in type 2 rod outer segment membrane guanylate cyclase (ROS-GC2). Biochem Biophys Res Commun 245: 447–453, 1998
46. Krylov DM, Niemi GA, Dizhoor AM, Hurley JB: Mapping sites in guanylyl cyclase activating protein-1 required for regulation of photoreceptor membrane guanylyl cyclases. J Biol Chem 274: 10833–10839, 1999
47. Beuve A: Conversion of a guanylyl cyclase to an adenylyl cyclase. Methods 19: 545–550, 1999
48. Tucker CL, Hurley JH, Miller TR, Hurley JB: Two amino acid substitutions convert a guanylyl cyclase, RetGC-1, into an adenylyl cyclase. Proc Natl Acad Sci USA 95: 5993–5997, 1998
49. Yang RB, Garbers DL: Two eye guanylyl cyclases are expressed in the same photoreceptor cells and form homomers in preference to heteromers. J Biol Chem 272: 13738–13742, 1997
50. Yu H, Olshevskaya E, Duda T, Seno K, Hayashi F, Sharma RK, Dizhoor AM, Yamazaki A: Activation of retinal guanylyl cyclase-1 by Ca^{2+}-binding proteins involves its dimerization. J Biol Chem 274: 15547–15555, 1999
51. Olshevskaya EV, Ermilov AN, Dizhoor AM: Dimerization of guanylyl cyclase-activating protein and a mechanism of photoreceptor guanylyl cyclase activation. J Biol Chem 274: 25583–25587, 1999
52. Tucker CL, Woodcock SC, Kelsell RE, Ramamurthy V, Hunt DM, Hurley JB: Biochemical analysis of a dimerization domain mutation in RetGC-1 associated with dominant cone-rod dystrophy. Proc Natl Acad Sci USA 96: 9039–9044, 1999
53. Sokal I, Haeseleer F, Arendt A, Adman ET, Hargrave PA, Palczewski K: Identification of a guanylyl cyclase-activating protein-binding site within the catalytic domain of retinal guanylyl cyclase 1. Biochemistry 38: 1387–1393, 1999
54. Laura RP, Hurley JB: The kinase homology domain of retinal guanylyl cyclases 1 and 2 specifies the affinity and cooperativity of interaction

with guanylyl cyclase activating protein-2. Biochemistry 37: 11264–11271, 1998
55. Lange C, Duda T, Beyermann M, Sharma RK, Koch KW: Regions in vertebrate photoreceptor guanylyl cyclase ROS-GC1 involved in Ca(2+)-dependent regulation by guanylyl cyclase-activating protein GCAP-1. FEBS Lett 460: 27–31, 1999
56. Dizhoor AM, Boikov SG, Olshevskaya EV: Constitutive activation of photoreceptor guanylate cyclase by Y99C mutant of GCAP-1. Possible role in causing human autosomal dominant cone degeneration. J Biol Chem 273: 17311–17314, 1998
57. Sokal I, Li N, Surgucheva I, Warren MJ, Payne AM, Bhattacharya SS, Baehr W, Palczewski K: GCAP1 (Y99C) mutant is constitutively active in autosomal dominant cone dystrophy. Mol Cell 2: 129–133, 1998
58. He L, Poblenz AT, Medrano CJ, Fox DA: Lead and calcium produce rod photoreceptor cell apoptosis by opening the mitochondrial permeability transition pore. J Biol Chem 275: 12175–12184, 2000
59. Pugh EN Jr, Lamb TD: Cyclic GMP and calcium: The internal messengers of excitation and adaptation in vertebrate photoreceptors. Vision Res 30: 1923–1948, 1990
60. Perrault I, Rozet JM, Calvas P, Gerber S, Camuzat A, Dollfus H, Chatelin S, Souied E, Ghazi I, Leowski C, Bonnemaison M, Le Paslier D, Frezal J, Dufier JL, Pittler S, Munnich A, Kaplan J: Retinal-specific guanylate cyclase gene mutations in Leber's congenital amaurosis. Nat Genet 14: 461–464, 1996
61. Duda T, Venkataraman V, Goraczniak R, Lange C, Koch KW, Sharma RK: Functional consequences of a rod outer segment membrane guanylate cyclase (ROS-GC1) gene mutation linked with Leber's congenital amaurosis. Biochemistry 38: 509–515, 1999
62. Kelsell RE, Gregory-Evans K, Payne AM, Perrault I, Kaplan J, Yang RB, Garbers DL, Bird AC, Moore AT, Hunt DM: Mutations in the retinal guanylate cyclase (RETGC-1) gene in dominant cone-rod dystrophy. Hum Mol Genet 7: 1179–1184, 1998
63. Perrault I, Rozet JM, Gerber S, Kelsell RE, Souied E, Cabot A, Hunt DM, Munnich A, Kaplan J: A retGC-1 mutation in autosomal dominant cone-rod dystrophy. Am J Hum Genet 63: 651–654, 1998
64. Duda T, Krishnan A, Venkataraman V, Lange C, Koch KW, Sharma RK: Mutations in the rod outer segment membrane guanylate cyclase in a cone- rod dystrophy cause defects in calcium signaling. Biochemistry 38: 13912–13919, 1999
65. Wilkie SE, Li Y, Deery EC, Newbold RJ, Garibaldi D, Bateman JB, Zhang H, Lin W, Zack DJ, Bhattacharya SS, Warren MJ, Hunt DM, Zhang K: Identification and functional consequences of a new mutation (E155G) in the gene for GCAP1 that causes autosomal dominant cone dystrophy. Am J Hum Genet 69: 471–480, 2001

Molecular and Cellular Biochemistry **230**: 149–158, 2002.

Guanylyl cyclases in unicellular organisms

Jürgen U. Linder and Joachim E. Schultz
Abteilung Pharmazeutische Biochemie, Pharmazeutisches Institut, Universität Tübingen, Morgenstelle, Tübingen, Germany

Abstract

Guanylyl cyclases in eukaryotic unicells were biochemically investigated in the ciliates *Paramecium* and *Tetrahymena*, in the malaria parasite *Plasmodium* and in the ameboid *Dictyostelium*. In ciliates guanylyl cyclase activity is calcium-regulated suggesting a structural kinship to similarly regulated membrane-bound guanylyl cyclases in vertebrates. Yet, cloning of ciliate guanylyl cyclases revealed a novel combination of known modular building blocks. Two cyclase homology domains are inversely arranged in a topology of mammalian adenylyl cyclases, containing two cassettes of six transmembrane spans. In addition the protozoan guanylyl cyclases contain an N-terminal P-type ATPase-like domain. Sequence comparisons indicate a compromised ATPase function. The adopted novel function remains enigmatic to date. The topology of the guanylyl cyclase domain in all protozoans investigated is identical. A recently identified *Dictyostelium* guanylyl cyclase lacks the N-terminal P-type ATPase domain. The close functional relation of *Paramecium* guanylyl cyclases to mammalian adenylyl cyclases has been established by heterologous expression, respective point mutations and a series of active mammalian adenylyl cyclase/ *Paramecium* guanylyl cyclase chimeras. The unique structure of protozoan guanylyl cyclases suggests that unexpectedly they do not share a common guanylyl cyclase ancestor with their vertebrate congeners but probably originated from an ancestral mammalian-type adenylyl cyclase. (Mol Cell Biochem **230**: 149–158, 2002)

Key words: guanylyl cyclase, adenylyl cyclase, *Paramecium*, *Plasmodium*, *Tetrahymena*, *Dictyostelium*, calcium, calmodulin, evolution, chimera, substrate specificity

Introduction

The cyclic nucleotides cAMP and cGMP are the classical intracellular second messengers in signal transduction processes. Almost innumerable studies during the last four decades demonstrates that cAMP can be regarded as a universal signalling molecule which is present in gram-positive as well as in gram-negative bacteria, in protozoa and in metazoa. The formation of cGMP on the other hand has mainly been investigated in metazoa and research on the regulation and function of cGMP in unicellular organisms, be they pro- or eukaryotes, is still in its infancy. In prokaryotes, the presence of cGMP has only been reported for cyanobacteria. There, a putative cyanobacterial guanylyl cyclase (GC) gene has been detected by gene deletion [1], yet the biochemical activity of the expressed protein as a GC remains to be demonstrated. In protozoa cGMP formation has been studied in the ciliates *Paramecium* and *Tetrahymena*, the malaria parasite *Plasmodium* and in the ameboid eukaryote *Dictyostelium*. The most consistent case for cGMP as a signalling molecule and a GC protein in a unicell has been made for *Paramecium*. Consequently this review will concentrate on the description of the *Paramecium* GC and refer to data concerning other unicells as deemed appropriate.

Guanylyl cyclase in *Paramecium*

Regulation of native Paramecium *GC*

A high GC activity in *Paramecium* was discovered in the early 1980's [2]. Ninety percent of the activity are present in cell bodies whereas 10% localize to the ciliary membranes, in which the specific activity seems to be higher [3]. The GC activity is exclusively membrane-bound and requires treatment with selected powerful detergents to solubilize at least a fraction of its activity [4].

Address for offprints: J.U. Linder, Abteilung Pharmazeutische Biochemie, Pharmazeutisches Institut, Universität Tübingen, Morgenstelle 8, 72076 Tübingen, Germany

In the regulation of enzyme activity Ca^{2+}-ions seem to play a prominent role. Removal of Ca^{2+} by washing membrane preparations with a low concentration of EGTA and suspension in Ca^{2+}-depleted buffers results in an 80% reduction of GC activity [4, 5]. This inhibition is specifically relieved by addition of Ca^{2+}-ions (ED_{50} = 8 μM). Sr^{2+} is less effective, Ba^{2+} does not reactivate GC [5].

The Ca^{2+}-dependence of the GC *in vitro* is in agreement with a boost-like increase of cGMP levels *in vivo* upon a rise of intracellular Ca^{2+}-concentrations. *Paramecium* cell-membranes contain a depolarization-activated, voltage-gated Ca^{2+}-channel. Depolarization is physiologically achieved by a sudden elevation of extracellular K^+. This is always accompanied by a reversal and a frequency increase of the ciliary powerstroke which result in a backswimming response. The K^+-induced depolarization causes only a brief behavioral response. As a consequence of Ca^{2+}-channel inactivation within 5 msec the global cGMP-content of the cells rises only slightly (approximately from 3.8 to 5.5 pmol/mg protein; [6]). A sustained Ca^{2+}-entry can be pharmacologically provoked by a Ba^{2+}-induced depolarization. Concomitantly with an exaggerated and extended behavioral backswimming response this causes a short-lived, up to 6-fold increase in cGMP-levels [6]. It is evident that Ca^{2+}-entry is responsible for the intracellular cGMP increase because (i) Ba^{2+} does not affect GC-activity *in vitro*; (ii) the peak levels of cGMP upon Ba^{2+}-stimulation *in vivo* correlate with the presence of higher extracellular Ca^{2+}; and (iii) the modulation of cGMP-levels parallels the behavioral changes of *Paramecium* mutants with defective depolarization-gated Ca^{2+}-conductance [6]. *Pawn*-mutants which do not possess a depolarization-activated Ca^{2+}-influx, show no backward swimming response [7] and do not display a significant cGMP increase upon Ba^{2+}-stimulation [6]. On the other hand the mutant called *dancer* which has a very slowly inactivating depolarization-activated Ca^{2+}-channel [8], displays prolonged bouts of backward swimming upon addition of Ba^{2+} and has a sustained cGMP-response with maximal cGMP-levels similar to the wild-type [6]. Moreover due to the impaired Ca^{2+}-channel inactivation cGMP levels in *dancer* cells rise 3 fold upon a K^+-evoked depolarization. This is in sharp contrast to the poor effect seen with wild-type cells under the same conditions [6].

The physiological response to hyperpolarization of *Paramecium* is a period of fast forward swimming. Among others, this can be accomplished by addition of 5% ethanol, 20 mM colchicine or 2 mM amiloride. These agents cause a 6–30 fold transient elevation of cGMP within 10–20 sec [9]. In the case of amiloride it has been demonstrated that cGMP formation depends on the presence of extracellular Ca^{2+}. Amiloride itself does not affect GC activity *in vitro*. Thus, Ca^{2+}-ions seem also to be involved in the transduction of a hyperpolarization signal onto GC [9]. Furthermore the cGMP-response to amiloride is unaltered in pawn-mutants showing that Ca^{2+}-influx most likely did not occur through the depolarization-activated Ca^{2+}-channels but through a distinct hyperpolarization-activated Ca^{2+}-conductance [10].

In light of these data a major effort has been undertaken to elucidate the mechanism by which Ca^{2+} stimulates the protozoan GC. Because of the high specificity of the GC-activation for Ca^{2+} it was speculated that a Ca^{2+}-binding protein is involved. The observation that La^{3+}-ions cause an inactivation of GC [11], and that such an inactivated enzyme cannot be reactivated after removal of La^{3+} by Ca^{2+} alone unless a soluble fraction of ciliary protein is added supported this idea. Calmodulin (CaM) appeared as a likely candidate for a Ca^{2+}-dependent activator, because it was shown to be abundant in *Paramecium* [12, 13] and responsible for activation of a Ca^{2+}-dependent K^+-current [14, 15]. Initial data were highly promising as addition of a protozoan CaM to *Paramecium* ciliary membranes or detergent-solubilized enzyme stimulates GC reproducibly by 20% [4] and as Ca^{2+}/CaM is capable to fully reactivate a La^{3+}-treated GC [11]. However, the CaM concentration needed for a half-maximal reactivation of the GC is high (200 nM) compared to the CaM-dependent phosphodiesterase [16, 17]. Furthermore, CaM does not remain associated with the GC upon reactivation, whereas the intrinsic Ca^{2+}-dependence is retained over several purification steps [11]. Finally reactivation of La^{3+}-treated GC is also possible with Ca^{2+}/EDTA or Ca^{2+}/EGTA mixtures [18]. Thus CaM does not appear to be required as the physiological Ca^{2+}-dependent activator and may be replaced by other means. The true intermediary molecule which relays the Ca^{2+}-signal to the GC protein may be a CaM-related factor such as a member of the GCAP-family, but this remains speculation (similar arguments apply to the GC in *Tetrahymena*; see below).

Structure of cloned Paramecium *guanylyl cyclase*

Biochemical analyses of the native *Paramecium* GC did not yield insights into the identity of the enzyme as a purification to homogeneity was impossible due to the highly restricted availability of *Paramecium* starting material. With the advent of molecular biology techniques homology cloning of candidate genes became the method of choice to get such enzymes at hand. The basis of cloning, however, is an initial assumption that a known gene of another organism is highly similar to the gene to be searched for. Strikingly, a GC from *Paramecium* was cloned as a result of an attempt to reach out for a *Paramecium* adenylyl cyclase (AC). This approach was based on the speculation that the *Paramecium* AC is related to mammalian ACs. Starting with a PCR fragment from *Paramecium* DNA which has a significant similarity to mammalian ACs, a 7.2 kb cDNA was finally obtained which codes for a two-domain integral membrane protein of 280 kDa [19]. The N-terminal domain of 155 kDa contains 10 putative transmembrane helices and is similar to P-type ion transport ATPases (see below). A short peptide-linker establishes the connection to the C-terminal cyclase domain which has a predicted membrane topology identical to that of mammalian ACs [20] (Fig. 1A). Two sets of six transmembrane helices, designated as M1 and M2, serve as anchors for two cytosolic catalytic domains, termed C1a and C2 according to mammalian ACs. Both catalytic subdomains are similar to each other and constitute classical cyclase homology domains. Finally the two 'MC' half-molecules are joined by a hydrophilic region termed C1b.

Inspection of the sequence of the catalytic C1a and C2 subdomains in the *Paramecium* cyclase revealed that in the frame of a mammalian topology two major shifts have occurred [19]. First, the C1a and C2-regions are functionally switched (Fig. 1B). The GDCY-motif which contains a metal-cofactor binding Asp [21] and is part of the canonical C1a domain of mammalian ACs is found in C2 of the *Paramecium* enzyme. Similarly, the conserved TYMA-motif which includes a purine-ring binding Met [22], is found in C1a of the *Paramecium* cyclase whereas it is always located in C2 of mammalian ACs. In this manner all the crucial catalytic amino acid residues of the mammalian C1a domains are present in the *Paramecium* C2 domain and the complementary part of the catalytic amino acid residues of mammalian AC C2 domains is localized in the *Paramecium* C1a loop (Fig. 1B). Viewed functionally metal-binding in mammalian ACs is exclusively exerted by the C1a domain and purine-ring binding mainly by the C2 domain. This functional division is switched in the *Paramecium* protein, i.e. primary structure of the cloned *Paramecium* cyclase shows an opposite distribution of tasks.

Second, substrate specificity of the *Paramecium* cyclase has appeared to be shifted from ATP to GTP. In general cyclase homology domains form dimers with two pockets [22] (symbolized in Fig. 1C). In mammalian ACs one pocket binds ATP and catalyzes the cyclization, the second non-catalytic pocket represents the binding site for the diterpene activator forskolin. Adenine ring binding in ACs is specified by two amino acid residues Lys and Asp of C2 with assistance of a Gln residue in C1a [22, 23]. Taking the switched C1a and C2 functions in the *Paramecium* sequence into account the corresponding residues in the protozoan cyclase are Glu1681 and Ser1748 of C1a and Arg2347 of C2, nearly identical to the guanine-binding triad Glu-Cys-Arg present in the mammalian GCs [23]. Therefore, the cloned *Paramecium* cyclase was unmasked as a guanylyl cyclase in the topological disguise of a mammalian AC with switched C1a and C2-functions and an additional N-terminal P-type ATPase-domain.

The similarity of the 155 kDa, N-terminal domain to the large family of P-type ATPases [24] is, on the one hand, evident by the topology of its three membrane domains with a canonical 2+2+6 distribution of transmembrane helices. On the other hand, P-type ATPases contain several blocks of conserved amino acids which are closely linked to functional properties of the transporters. The invariant DKTGT(L/I)T signature-motif at the aspartate-phosphorylation site [25] is also present in the *Paramecium* GC. However, the conserved (TSND)GE(SNT) motif in the transduction domain [26] is altered to ^{240}SGNT^{243} and at an extremely conserved position in the ATP-binding domain where an Asp residue has been identified to be functionally indispensable [27, 28], a Glu is present at position 848 in the *Paramecium* domain. These decisive changes indicate that the N-terminal domain of the *Paramecium* GC is most likely not an active transporter but adopted a new, as yet enigmatic, function [19].

Catalytic properties of the cloned Paramecium *GC*

The results of the analysis of the primary structure of *Paramecium* GC have been verified by heterologous expression of the protein and biochemical assays [19, 29]. The main goals of these studies were to prove that the cloned gene codes for a specific GC, to correlate the cloned *Paramecium* GC with the activity of the native enzyme and to establish the relationship of the *Paramecium* GC to mammalian ACs in terms of topology, switched C1a and C2 functions, substrate-specifying amino acids and catalytic mechanism.

Heterologous expression of *Paramecium* genes is impossible because the universal TAA/TAG stop codons are translated to glutamine in the ciliate. To achieve expression in metazoan cells all 99 TAA/TAG codons of the GC gene had to be mutated to CAA/CAG. Moreover the high A/T-content of the gene (66%) typical for *Paramecium* DNA caused problems in HEK293 cells initially used as an expression system, probably because of instability of the mRNA *in vivo*. Replacement of about 40% of the gene by stretches of synthetic DNA employing the G/C-rich mammalian codon usage was required and sufficient to solve this problem [19].

Expression of the cyclase domain alone (amino acids 1380–2412) produces robust membrane-bound GC activity of up to 150 pmol/(mg*min) in HEK293 cells and up to 1 nmol/(mg*min) in insect Sf9 cells. Expression of the holoenzyme in insect cells yields GC activity of about 500 pmol/(mg*min). Western blot analyses revealed that the holoenzyme expressed in Sf9 cells is cleaved into two large peptides. Probably this proteolytic processing occurred intracellularly, because it is not affected by protease inhibitors added during membrane preparation and because the mRNA in Sf9 cells is not fragmented as evaluated by RT-PCR. The exact site of proteolysis is unknown, but appears to be at the P-type ATPase/GC boundary as judged by the size of the fragments.

No significant differences in the enzymatic properties between the cyclase domain expressed alone and the GC fragment resulting from proteolytic processing of the holoenzyme are observed. GTP is the specific substrate with Mg^{2+} as a cofactor, AC activity is negligible. Addition of ATP in a 10 fold molar excess over GTP reduces GC activity by only 25%. When Mn^{2+} ions are used as cofactor a small AC side-activity of 20% is apparent. This is not unusual for GCs. Consequently under physiological conditions the cloned enzyme is a highly specific GC. The K_M for MgGTP is 50 µM, use of MnGTP increases activity by about 80%. These enzymatic

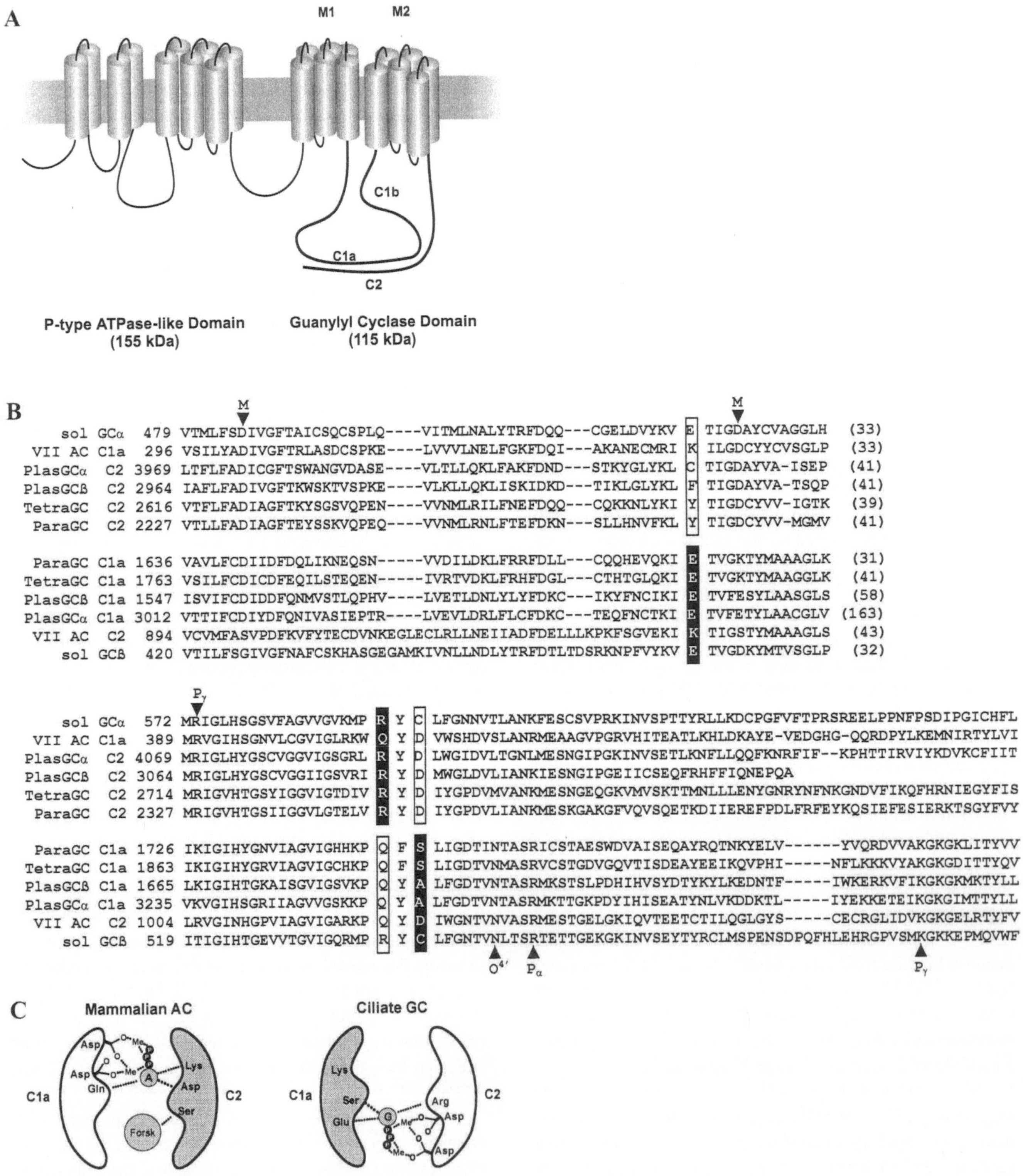

properties of the recombinant protein are almost identical to those of the native *Paramecium* GC. However, the activity of the recombinant GC is completely unresponsive to Ca^{2+} or Ca^{2+}/CaM. Also 50 µM La^{3+} inhibits by only 25% (J. Linder, unpublished) whereas native GC is almost totally turned off by 5 µM La^{3+} [11]. This discrepancy might be explained by the absence of a *Paramecium*-specific Ca^{2+}-dependent regulator in the heterologous expression systems, but other explanations are also possible. (i) The cloned GC may be a minor Ca^{2+}-independent species among several isoforms (see below); (ii) a *Paramecium*-specific post-translational modification may be absent in the engineered protein; (iii) Ca^{2+}-dependence may rely on an intact unfragmented holoenzyme.

Immuno-gold localization of the GC in *Paramecium* tetraurelia shows an exclusive distribution of the cyclase among cell surface membranes, i.e. ciliary membranes, cell membrane and inner alveolar sac membranes [19]. The occurrence in the ciliary and cell membranes is consistent with an *in situ* regulation of native GC by a Ca^{2+}-influx. Alveolar sacs are subplasmalemmal Ca^{2+}-stores and GC localized there may be regulated by an internal Ca^{2+}-release mechanism, although this is currently not supported by physiological or pharmacological data.

Experimental correlation of Paramecium *GC with mammalian ACs*

A first set of experiments to correlate *Paramecium* GC with mammalian ACs showed that the *Paramecium* GC cyclase domain is not only topologically identical to mammalian ACs but that the cyclase homology domains of the *Paramecium* enzyme are biochemically compatible with an authentic mammalian AC frame. The cyclase homology domains C1a and C2 of bovine type VII AC can be simultaneously exchanged by the cyclase homology domains of the *Paramecium* GC [19]. No matter whether the original GC order (C1a-C2) or the mammalian functional order (*Paramecium* GC: C2-C1a) is employed in the membrane attachment, the resulting chimeras are specific GCs. Even a concomitant exchange of the mammalian C1b region for that of the *Paramecium* GC, i.e. *Paramecium* C1-C2 with type VII AC membrane anchors, is well tolerated [19].

To further characterize the structure function relationship of the *Paramecium* GC catalyst the cyclase homology domains have been expressed as soluble proteins in *E. coli* in analogy to successful experiments carried out with mammalian ACs [30–32]. When C1a and C2 are expressed individually, cyclase activity is absent in the single domains, but specific GC activity is reconstituted upon mixing of both [29]. Expression of C1a and C2 linked by a short peptide also produces a specific GC. The K_M-values for both variants are in the range of 60–75 µM, i.e. in the range of the K_M of the membrane-bound cyclase domain. Thus general catalytic properties of the *Paramecium* GC activity do not significantly change upon removal of M1, M2 and C1b.

Such a soluble catalyst is amenable to mutational analyses [23]. To probe the *Paramecium* GC substrate specificity, the Glu-Ser-Arg triad which is corresponding to the substrate-defining Lys-Asp-Gln triad in mammalian ACs, has been targeted. An E1681K/S1748D double mutation introduced into *Paramecium* GC-C1a is sufficient to convert the GC activity into a specific AC activity [29]. This experiment has proven two hypotheses based on the primary structure analysis: the substrate-defining amino acids of *Paramecium* GC are at functionally equivalent positions in the cyclase homology domains to those of the mammalian ACs and *Paramecium* C1a indeed functions as the substrate-defining domain, a task which has been taken over by C2 in mammalian ACs. The third residue in the triad, Arg2347 of *Paramecium* GC-C2 appears to have a minor role. ^{2345}LVR^{2347} has been mutated to the mammalian C1a consensus KWQ. Combining *Paramecium* GC wild-type C1a and the mutant C2 leads to a 10 fold increase of K_M measured with MgGTP, but induces no shift of substrate specificity. Furthermore the mutant *Paramecium* GC-C2 displays no effect on ATP substrate specificity or on the K_M value when tested with the *Paramecium* GC-C1a E1648K/S1748D mutant.

The homology of *Paramecium* GC-C2 with the mammalian AC-C1a has been further substantiated by generation of chimeric enzymes [29]. Isolated mammalian C2 domains from type II and type IX AC which are inactive by themselves gain specific AC activity when *Paramecium* GC-C2 is added. Thus, *Paramecium* GC-C2 fully complements mammalian

←

Fig. 1. Structure of protozoan GCs and comparison to mammalian ACs. (A) Topology of *Paramecium* GC as described in the text (adapted from [19]). (B) Alignment of the C1a/C2 domains of bovine type VII AC (VII AC) with rat soluble GC α_1/β_1 (sol GC) and C2/C1a portions of alveolata GCs from *Paramecium* (ParaGC), *Tetrahymena* (TetraGC) and *Plasmodium* (PlasGCα/β). Amino acid positions are shown on the left. Two blocks of high conservation are separated by the number of amino acid residues shown in brackets. The upper part of each block contains the counterparts of mammalian AC-C1a domains, the lower part the ones of mammalian AC-C2 domains. Arrowheads indicate the amino acids involved in metal-binding (M), binding of phosphate-groups (P_α and P_γ) and of the ribose-ring oxygen atom (O4'). The amino acid residues of the substrate-defining triads in the catalytic pockets are inversed, their counterparts in the non-catalytic pockets are boxed. (C) Model of a cross-section through the C1a/C2 heterodimers of mammalian ACs and ciliate GCs. Left halves symbolize the C1a domains, right halves the C2 domains. The functionally inverted order of *Paramecium* domains is symbolized by a switched shading. Chemical groups are depicted by circles (A – adenine; G – guanine; P – phosphate; Forsk – forskolin). The metal ions (Me) are complexed to the side-chain carboxyl groups of two conserved Asp residues marked in (B) and bind to the triphosphate moiety of the substrates. The amino acid residues of the substrate-defining triads select the purines. In mammalian ACs a prominent Ser residue is involved in binding of forskolin to the non-catalytic pocket which is a Lys in ciliate GCs.

AC-C1a in terms of catalytic activity. The mutation of ^{2345}LVR^{2347} to KWQ in *Paramecium* GC-C2 doubles the specific activity of such chimeras, again pointing to a mere assistant role of Arg2347 in substrate binding.

The *Paramecium* GC has been further compared to mammalian ACs by analysis of additional properties of the catalytic and non-catalytic pockets [29]. Mammalian cyclases are inhibited by purine-nucleotides which interfere with product release via binding to the so-called P-site in the catalytic pocket [33]. Thereby binding of the purine-ring is dependent on the substrate-defining triad. The most potent commercially available P-site inhibitors are 2′-d-3′ AMP for ACs and 2′-d-3′ GMP for GCs [23, 33]. The efficiency of P-site inhibition depends on the activation-state, the more the cyclases are stimulated by activators such as Mn^{2+}, forskolin, G_sa-GTP, the lower the IC_{50} values of the inhibitors. Soluble *Paramecium* GC-C1a/C2 is inhibited by about 60 % by high (1 mM) concentrations of 2′-d-3′ GMP while no significant effect is seen with 2′-d-3′ AMP. P-site inhibition is dependent on Mn^{2+}-activation as the effect of the inhibitor is poor with MgGTP as a substrate. The E1681K/S1748D mutation of C1a shifts P-site inhibition towards 2′-d-3′ AMP and also mammalian AC-C2/*Paramecium* GC-C2 chimeras are preferentially inhibited by 2′-d-3′ AMP. Thus the *Paramecium* GC parallels mammalian cyclases with respect to P-site inhibition.

In the non-catalytic pocket of mammalian ACs the substrate-defining triad Lys-Asp-Gln has been retained during evolution, although no adenine-containing molecule is known to bind to this site. The pocket is the target for the pharmacological AC activator forskolin whose binding largely depends on a Ser-residue at the corresponding position of a metal-binding Asp in the catalytic pocket [22, 34] (Figs 1B, 1C). In the *Paramecium* GC the triad in the non-catalytic pocket is Tyr-Asp-Gln, i.e. somewhat like Lys-Asp-Gln in ACs. This supports the idea that the *Paramecium* GC evolved from a mammalian AC-like ancestor molecule (see below). In the position of the Ser needed for forskolin binding in mammalian ACs a Lys is carried by the *Paramecium* GC and, expectedly, even 100 μM forskolin do not stimulate the *Paramecium* GC. However, as in the case of mammalian ACs it remains unknown whether a physiological factor may act on the non-catalytic pocket of the *Paramecium* GC.

The second classical activator of mammalian ACs is G_sa-GTP which binds laterally to the C1a/C2 catalytic dimer attaching to both cyclase homology domains [22]. *Paramecium* seems to lack canonical heterotrimeric G-proteins as there is neither any biochemical evidence for their existence nor do data from a genome survey project (P. Dessen *et al.*, in press) reveal any homologue of such G-proteins in the ciliate. Consistently, neither soluble *Paramecium* GC-C1a/C2 nor any mutant thereof nor any *Paramecium* GC-C2/mammalian AC-C2 chimera has been found to be affected by mammalian G_sa-GTP [29].

In conclusion the properties of the *Paramecium* GC suggest that there is room for GC regulation and activation as implied by the moderate P-site inhibition and suggested by the presence of a non-catalytical pocket. This brings Ca^{2+}-activation back to stage. The sole known physiological activator of *Paramecium* GC is still to be explored.

Guanylyl cyclase isoforms in Alveolata

Guanylyl cyclase in Tetrahymena

The ciliate *Tetrahymena* is related to *Paramecium* by genetics, morphology and behavior. In this protozoan native GC appears to have very similar biochemical properties as *Paramecium* GC, but the regulation of *Tetrahymena* GC and cGMP-levels *in vivo* have been investigated to a lesser extent. The GC is particulate and abundant in ciliary membranes. It is stimulated by Ca^{2+} *in vitro*, mediated by a soluble activator [35]. The Ca^{2+}-stimulation is not tightly associated with the cyclase; after centrifugation of *Tetrahymena* homogenates the GC is pelleted whereas the activating factor remains in the supernatant. *Tetrahymena* GC can be inactivated by La^{3+} or EGTA in a similar fashion as *Paramecium* GC; reactivation is possible by high concentrations of Ca^{2+}/CaM [16, 36]. Exactly as for the *Paramecium* GC CaM is able to mediate the Ca^{2+}-stimulation of *Tetrahymena* GC *in vitro*, but appears not to be the physiological activator. *Tetrahymena* GC has not been purified and its identity and structure remained unknown until the discovery of a *Paramecium* GC gene facilitated the homology cloning of a *Tetrahymena* GC [19]. The predicted protein has a two domain structure consisting of an N-terminal P-type ATPase and a C-terminal cyclase domain with mammalian AC topology, but inversed C1a and C2 functions (Fig. 1B), features identical to the *Paramecium* GC. The substrate-defining triad is Glu-Ser-Arg, indicative of a GC and identical to the *Paramecium* GC. Patterns of the cyclase domain of *Tetrahymena* GC in general parallel the *Paramecium* enzyme, but the sequence of the P-type ATPase-like domain yields additional information. Whereas the DKTGT(L/I)T signature is retained in *Paramecium* GC, it is altered to SKSGTLM in the *Tetrahymena* GC virtually excluding any active membrane-transport activity of this domain. Like in the *Paramecium* GC significant mutations are found in the (TSND)GE(SNT) block in the transduction domain and at the conserved Asp residue in the hinge domain, confirming a novel role of P-type ATPase structures in cGMP-mediated signal transduction. Comparative analysis of the transmembrane helices of *Paramecium* GC and *Tetrahymena* GC shows that residues thought to be involved in cation binding of classical P-type ATPases [28, 37] have been randomly altered, but that novel consensus sequences specific for the cyclase-associated P-type ATPase-like domains are present in the fourth and fifth transmembrane helix (Fig. 2). For this reason it seems probable that these helices play an intricate role in the function of the P-type-ATPase-like domains which may include binding of molecules within the membrane, i.e. a function of these domains as a novel type of receptors is an attractive possibility.

Multiple GC isoforms in Paramecium *and* Tetrahymena

To search for novel GC-isoforms homology PCR has been carried out with degenerate primers which have targeted a GC-C1a domain which comprises the serine defining GTP as a substrate [19]. Three more cyclase fragments have been amplified from *Paramecium* DNA; five novel fragments have been obtained from *Tetrahymena* DNA. In all eight amplification products a Ser residue is located at the substrate-defining position, thus they all represent fragments of ciliate GC isoforms. The open reading frames of the fragments code for 60–75 amino acids of which a stretch of 24 amino acids surrounding the serine residue is highly conserved (87–100% identity). The remainder corresponds to a region which in analogy to mammalian cyclases appears to carry mainly structural information. In this region the ciliate GCs display a considerable degree of divergence with 21–93% identity among isoforms. These features are highly similar to those in mammals which possess nine membrane-bound AC isoforms with a similar degree of divergence in this structural region (24–83% identity; [20]).

One of the *Tetrahymena* GC fragments is 95% identical to the fully cloned *Paramecium* GC. Such a remarkable interspecies conservation, in view of the otherwise considerable intraspecies divergence, is a strong indication of a specific task of at least these two orthologs and argues against a mere redundancy of ciliate GC genes.

```
ParaGC    383 QLMN YSGLLPLYFYFLIDLLYFTQ MLYN
TetraGC   398 YFLE YMIFLPCIVYPYLDIMQLII VYKS      TM 4
PlasGCα   386 YTLL YSNIIPISILISVDLISILQ SILI
PlasGCß   336 FVGL YILVLPIMMYSEKSLIYIIQ SLRI

ParaGC   1116 YFER ILIFALFRSFGIIYLLCVTQLI RDEI
TetraGC  1234 QKED LVIMSFYRSFLIGIILFFWNTY QSSY    TM 5
PlasGCα  2083 TISI ILYWNFFKNILLILPIFFYQAY ASWS
PlasGCß   985 SLCK VFLWIIYLKITVVSFYFFHNFD NYFS
```

Fig. 2. Alignment of conserved transmembrane spans in the P-type ATPase-like domains. Abbreviations are as in the legend to Fig. 1B. Arrowheads indicate positions of significant conservation in the fourth (TM4) and fifth (TM5) transmembrane span. A conserved motif in TM5 of ciliate enzymes is boxed. Four amino acid residues N- and C-terminal of each transmembrane span are included.

Plasmodium *guanylyl cyclases*

In the malaria parasite Plasmodium neither a guanylyl cyclase enzyme activity nor defined cGMP-levels have been unequivocally determined so far. In the few studies concerning cGMP-mediated signal transduction, pharmacological tools validated for research on mammalian GCs have been applied to Plasmodium without any verification and thus the results obtained are interpretable to a very limited extent only [38, 39]. Again identification of GCs was possible by molecular biology techniques. Independent analysis of genomic data bases [19] and homology cloning [40] revealed the existence of guanylyl cyclases in Plasmodium falciparum which have a topology identical to *Paramecium* and *Tetrahymena* GCs. The homology of GC genes of ciliates and this apicomplexan parasite confirms the assignment of both to the parvkingdom of alveolata which originally has been based predominantly on morphological characteristics. Two GC isoforms termed GCα and GCβ are present in Plasmodium [40].

In both, the C1a and C2 cyclase homology domains are in the same functional order as in the ciliate GCs, the substrate-defining triads of both are Glu-Ala-Arg (Fig. 1B). The unexpected replacement of the ciliate Ser residue by Ala, lacking the potential for a hydrogen bond, indicates that substrate selection in Plasmodium GCs is largely left to the canonical Glu residue in C1a.

As in ciliates the P-type ATPase-like domains display marked deviations from the general P-type ATPase consensus; the DKTGT(L/I)T signature is only present in GCα but completely marred in GCβ [40]. Therefore these domains do not appear to be involved in conventional active membrane transport processes. Interestingly, the fourth and fifth transmembrane-helices of Plasmodium GCα show significant similarity to the respective helices of P-type-ATPase-like domains in ciliate GCs (Fig. 2), which points to a common function of these domains.

Heterologous expression of Plasmodium GC was successful with GCβ only [40]. A mixture of C1a and C2, produced in *E. coli*, yields GC activity of only about 4 pmol/(mg*min). Forskolin and Ca^{2+}/CaM do not activate Plasmodium GCβ as is the case with recombinant *Paramecium* GC.

Clues to the function of Plasmodium GCs *in vivo* have been obtained by expression and localization studies [40]. mRNA of both isoforms is detected in sexual blood stages, but absent in asexual forms. Thus a role of Plasmodium GCs in signal-transduction of gametocytes may be envisaged. The GCα isoform has been localized to the parasitophorous vacuole and/or plasma membrane of intraerythrocytic gametocytes whereas no GCα could be detected in extracellular gametocytes, suggesting a specific role of GCα in the pre-emergence state [40]. Whether Plasmodium GCs are related to ciliate GCs by Ca^{2+}-regulation is one of the most interesting questions to be addressed in the future.

Guanylyl cyclase with mammalian AC topology in *Dictyostelium*?

In the ameboid eukaryote *Dictyostelium* the native GC activity is reported to be particulate [41]. It is transiently activated *in vivo* by chemoattractants such as cAMP or folate and upon application of osmotic stress. The chemotactic signals are reported to be coupled to GC regulation via heterotrimeric G-proteins [42]. Ca^{2+} in submicromolar concentrations is inhibitory [43]. As a third component of GC-regulation a cGMP-binding protein is invoked which in the cGMP-free state supposedly stimulates [44]. Neither of the three GC-regulatory pathways is particularly characterized in molecular terms. Most recently a cDNA has been published which purportedly codes for a *Dictyostelium* GC [45]. The predicted GC protein resembles that in the protozoans outlined above, a P-type ATPase module is absent. As a substrate-binding triad a **Glu**-His-His arrangement is predicted. Functional analysis of the cyclase overexpressed in *Dictyostelium* is as yet limited because the enzymatic reaction is initiated by the addition of the cyclizable GTP analogue GTPγS at the time of lysis and an addition of a 10 fold excess of GTP barely has any further effect. A knock-out of the proposed *Dictyostelium* GC did not result in a reduction of total cellular GC activity compared to the wild-type, nor in a defect of development or chemotaxis [45]. At this point the function of this type of GC remains to be established.

Evolutionary aspects

For two reasons alveolata-type GCs are much more distant to mammalian GCs than to mammalian ACs. First, mammalian membrane-bound GCs contain a single cyclase homology domain and are attached to the plasma membrane by a single transmembrane helix [46]. So far no metazoan GC with mammalian AC topology has been detected. Second, although mammalian soluble GCα and GCβ subunits functionally correspond to the mammalian AC C1a and C2 domains, the substrate-defining triad of mammalian soluble GCs is Glu-Cys-Arg in the catalytic as well as in the non-catalytic pocket. In the alveolata-type GCs the triad in the non-catalytical pocket is akin to that in mammalian ACs. Therefore, we tend to exclude the possibility that mammalian and protozoan GCs were derived from a common ancestral GC. Rather, the alveolata-type GCs probably evolved from a precursor AC of mammalian topology [19]. The fact that cyclases exist with C1a and C2 in either order sheds some light on the evolution of mammalian-type AC structures. It is conceivable that initially membrane-anchored monomers formed homodimeric cyclases. Such a progenitor protein is still present in the Rv1625c gene of Mycobacterium tuberculosis which encodes a homodimeric AC of the eukaryotic type (Y. Guo *et al.*, submitted). Following gene duplication heterodimeric ACs may have evolved before monomers were fused into a single polypeptide chain. Again the amplification of monomeric AC genes is still evident in M. tuberculosis where three membrane-bound myxobacterial-type ACs of high similarity are clustered in genes Rv1318c to Rv1320c. The fusion of mammalian-type AC monomers obviously occurred either way, resulting in the canonical mammalian AC structure on the one hand and an ancestral protozoan AC on the other hand, from which the alveolata-type GCs emerged by point mutations in the catalytic pocket to switch to GTP as a substrate. Subsequently a P-type ATPase domain may have been acquired by a primordial GC in alveolates. Future genome projects will reveal whether ACs with mammalian AC topology, but switched C1a and C2 functions are still in use. Also a variety of alveolata-type GCs are to be discovered.

Conclusion

Native protozoan GC activity is membrane-bound and at least in some cases Ca^{2+}-regulated, probably via a Ca^{2+}-binding factor. These properties initially suggested a relation to mammalian GCs, namely the membrane-bound Ca^{2+}-regulated retinal rod outer segment GC.

Molecular cloning however, unravelled a new highly specialized species of mammalian AC-like GCs with an additional P-type ATPase-like domain. The close relation to mammalian ACs concerning topology and basic catalytic properties does not extent to the action of mammalian AC stimulators like forskolin or heterotrimeric G-proteins on the protozoan GCs. Thus pharmacological tools for research on protozoan GCs have yet to be developed. These will be needed to investigate GC and cGMP functions in parasites like Plasmodium falciparum which are difficult to access biochemically. Tools used for research on mammalian GCs or on mammalian ACs are not a priori applicable. Genome projects combined with heterologous expression will facilitate characterization of a whole variety of protozoan GCs and help to investigate the connection of these enzymes to their surrounding signal transduction network.

Acknowledgements

We thank Dr. Christian Beyer for help with creating the figures.

References

1. Ochoa De Alda JA, Ajlani G, Houmard J: Synechocystis strain PCC 6803 cya2, a prokaryotic gene that encodes a guanylyl cyclase. J Bacteriol 182: 3839–3842, 2000
2. Schultz JE, Klumpp S: Guanylate cyclase in the excitable ciliary membrane of *Paramecium*. FEBS Lett 122: 64–66, 1980
3. Schultz JE, Klumpp S: Calcium-regulated guanylyl cyclases from *Paramecium* and *Tetrahymena*. Meth Enzymol 195: 466–474, 1991
4. Klumpp S, Gierlich D, Schultz JE: Adenylate cyclase and guanylate cyclase in the excitable ciliary membrane from *Paramecium*: Separation and regulation. FEBS Lett 171: 95–99, 1984
5. Klumpp S, Schultz JE: Characterization of a Ca^{2+}-dependent guanylate cyclase in the excitable ciliary membrane from *Paramecium*. Eur J Biochem 124: 317–324, 1982
6. Schultz JE, Pohl T, Klumpp S: Voltage-gated Ca^{2+} entry into *Paramecium* linked to intraciliary increase in cyclic GMP. Nature 322: 271–273, 1986
7. Oertel D, Schein SJ, Kung C: Separation of membrane currents using a *Paramecium* mutant. Nature 268: 120–124, 1977
8. Hinrichsen RD, Saimi Y: A mutation that alters properties of the calcium channel in *Paramecium tetraurelia*. J Physiol 351: 397–410, 1984
9. Schultz JE, Guo Y, Kleefeld G, Volkel H: Hyperpolarization- and depolarization-activated Ca^{2+} currents in *Paramecium* trigger behavioral changes and cGMP formation independently. J Membr Biol 156: 251–259, 1997
10. Preston RR, Saimi Y, Kung C: Calcium current activated upon hyperpolarization of *Paramecium tetraurelia*. J Gen Physiol 100: 233–251, 1992
11. Klumpp S, Kleefeld G, Schultz JE: Calcium/calmodulin-regulated guanylate cyclase of the excitable ciliary membrane from *Paramecium*. Dissociation of calmodulin by La^{3+}: Calmodulin specificity and properties of the reconstituted guanylate cyclase. J Biol Chem 258: 12455–12459, 1983
12. Walter MF, Schultz JE: Calcium receptor protein calmodulin isolated from cilia and cells of *Paramecium tetraurelia*. Eur J Cell Biol 24: 97–100, 1981
13. Schaefer WH, Lukas TJ, Blair IA, Schultz JE, Watterson DM: Amino acid sequence of a novel calmodulin from *Paramecium tetraurelia* that contains dimethyllysine in the first domain. J Biol Chem 262: 1025–1029, 1987
14. Hinrichsen RD, Burgess-Cassler A, Soltvedt BC, Hennessey T, Kung C: Restoration by calmodulin of a Ca^{2+}-dependent K^+ current missing in a mutant of *Paramecium*. Science 232: 503–506, 1986
15. Kink JA, Maley ME, Preston RR, Ling KY, Wallen-Friedman MA, Saimi Y, Kung C: Mutations in *paramecium* calmodulin indicate functional differences between the C-terminal and N-terminal lobes *in vivo*. Cell 62: 165–174, 1990
16. Kakiuchi S, Sobue K, Yamazaki R, Nagao S, Umeki S, Nozawa Y, Yazawa M, Yagi K: Ca^{2+}-dependent modulator proteins from *Tetrahymena pyriformis*, sea anemone, and scallop and guanylate cyclase activation. J Biol Chem 256: 19–22, 1981
17. Kudo S, Muto Y, Nagao S, Naka M, Hidaka H, Sano M, Nozawa Y: Specificity of *Tetrahymena* calmodulin in activation of calmodulin-regulated enzymes. FEBS Lett 149: 271–276, 1982
18. Schultz JE, Klumpp S: Cyclic GMP in lower forms. Adv Pharmacol 26: 285–303, 1994
19. Linder JU, Engel P, Reimer A, Kruger T, Plattner H, Schultz A, Schultz JE: Guanylyl cyclases with the topology of mammalian adenylyl cyclases and an N-terminal P-type ATPase-like domain in *Paramecium*, *Tetrahymena* and *Plasmodium*. EMBO J 18: 4222–4232, 1999
20. Sunahara RK, Dessauer CW, Gilman AG: Complexity and diversity of mammalian adenylyl cyclases. Annu Rev Pharmacol Toxicol 36: 461–480, 1996
21. Tesmer JJ, Sunahara RK, Johnson RA, Gosselin G, Gilman AG, Sprang SR: Two-metal-Ion catalysis in adenylyl cyclase. Science 285: 756–760, 1999
22. Tesmer JJ, Sunahara RK, Gilman AG, Sprang SR: Crystal structure of the catalytic domains of adenylyl cyclase in a complex with Gsα GTPγS. Science 278: 1907–1916, 1997
23. Sunahara RK, Beuve A, Tesmer JJ, Sprang SR, Garbers DL, Gilman AG: Exchange of substrate and inhibitor specificities between adenylyl and guanylyl cyclases. J Biol Chem 273: 16332–16338, 1998
24. Fagan MJ, Saier MH: P-type ATPases of eukaryotes and bacteria: Sequence analyses and construction of phylogenetic trees. J Mol Evol 38: 57–99, 1994
25. Allen G, Green NM: A 31-residue tryptic peptide from the active site of the [Ca^{2+}]- transporting adenosine triphosphatase of rabbit sarcoplasmic reticulum. FEBS Lett 63: 188–192, 1976
26. Clarke DM, Loo TW, MacLennan DH: Functional consequences of mutations of conserved amino acids in the beta-strand domain of the Ca^{2+}-ATPase of sarcoplasmic reticulum. J Biol Chem 265: 14088–14092, 1990
27. Clarke DM, Loo TW, MacLennan DH: Functional consequences of alterations to amino acids located in the nucleotide binding domain of the Ca^{2+}-ATPase of sarcoplasmic reticulum. J Biol Chem 265: 22223–22227, 1990
28. Toyoshima C, Nakasako M, Nomura H, Ogawa H: Crystal structure of the calcium pump of sarcoplasmic reticulum at 2.6 A resolution. Nature 405: 647–655, 2000
29. Linder JU, Hoffmann T, Kurz U, Schultz JE: A guanylyl cyclase from *Paramecium* with 22 transmembrane spans. Expression of the catalytic domains and formation of chimeras with the catalytic domains of mammalian adenylyl cyclases. J Biol Chem 275: 11235–11240, 2000
30. Tang WJ, Gilman AG: Construction of a soluble adenylyl cyclase activated by Gs alpha and forskolin. Science 268: 1769–1772, 1995
31. Whisnant RE, Gilman AG, Dessauer CW: Interaction of the two cytosolic domains of mammalian adenylyl cyclase. Proc Natl Acad Sci USA 93: 6621–6625, 1996
32. Dessauer CW, Gilman AG: Purification and characterization of a soluble form of mammalian adenylyl cyclase. J Biol Chem 271: 16967–16974, 1996
33. Johnson RA, Yeung SM, Stubner D, Bushfield M, Shoshani I: Cation and structural requirements for P site-mediated inhibition of adenylate cyclase. Mol Pharmacol 35: 681–688, 1989
34. Zhang G, Liu Y, Ruoho AE, Hurley JH: Structure of the adenylyl cyclase catalytic core. Nature 386: 247–253, 1997
35. Nakazawa K, Shimonaka H, Nagao S, Kudo S, Nozawa Y: Magnesium-sensitive guanylate cyclase and its endogenous activating factor in *Tetrahymena pyriformis*. J Biochem (Tokyo) 86: 321–324, 1979
36. Schultz JE, Schonefeld U, Klumpp S: Calcium/calmodulin-regulated guanylate cyclase and calcium-permeability in the ciliary membrane from *Tetrahymena*. Eur J Biochem 137: 89–94, 1983
37. Zhang P, Toyoshima C, Yonekura K, Green NM, Stokes DL: Structure of the calcium pump from sarcoplasmic reticulum at 8-A resolution. Nature 392: 835–839, 1998
38. Kawamoto F, Alejo-Blanco R, Fleck SL, Kawamoto Y, Sinden RE: Possible roles of Ca^{2+} and cGMP as mediators of the exflagellation of *Plasmodium berghei* and *Plasmodium falciparum*. Mol Biochem Parasitol 42: 101–108, 1990
39. Kawamoto F, Fujioka H, Murakami R, Syafruddin, Hagiwara M, Ishikawa T, Hidaka H: The roles of Ca^{2+}/calmodulin- and cGMP-dependent pathways in gametogenesis of a rodent malaria parasite, *Plasmodium berghei*. Eur J Cell Biol 60: 101–107, 1993

40. Carucci DJ, Witney AA, Muhia DK, Warhurst DC, Schaap P, Meima M, Li JL, Taylor MC, Kelly JM, Baker DA: Guanylyl cyclase activity associated with putative bifunctional integral membrane proteins in *Plasmodium falciparum*. J Biol Chem 275: 22147–22156, 2000
41. Janssens PM, De Jong CC, Vink AA, Van Haastert PJ: Regulatory properties of magnesium-dependent guanylate cyclase in *Dictyostelium discoideum* membranes. J Biol Chem 264: 4329–4335, 1989
42. Wu L, Valkema R, Van Haastert PJ, Devreotes PN: The G protein beta subunit is essential for multiple responses to chemoattractants in *Dictyostelium*. J Cell Biol 129: 1667–1675, 1995
43. Valkema R, Van Haastert PJ: Inhibition of receptor-stimulated guanylyl cyclase by intracellular calcium ions in *Dictyostelium* cells. Biochem Biophys Res Commun 186: 263–268, 1992
44. Kuwayama H, Van Haastert PJ: Regulation of guanylyl cyclase by a cGMP-binding protein during chemotaxis in *Dictyostelium discoideum*. J Biol Chem 271: 23718–23724, 1996
45. Roelofs J, Snippe H, Kleineidam RG, Van Haastert PJ: Guanylate cyclase in *Dictyostelium discoideum* with the topology of mammalian adenylate cyclase. Biochem J 354: 697–706, 2001
46. Wedel B, Garbers D: The Guanylyl Cyclase Family at Y2k. Annu Rev Physiol 63: 215–233, 2001

Molecular and Cellular Biochemistry **230:** 159–164, 2002.

Isoforms of NO-sensitive guanylyl cyclase

Michael Russwurm and Doris Koesling
Pharmakologie und Toxikologie, Medizinische Fakultät MA N1, Ruhr-Universität Bochum, 44780 Bochum, Germany

Abstract

By the formation of cGMP the NO-sensitive guanylyl cyclase plays a key role within the NO/cGMP signaling cascade involved in vascular regulation and neurotransmission. The prosthetic heme group of the enzyme acts as the NO sensor, and binding of NO induces conformational changes leading to an up to 200-fold activation of the enzyme. The unexpected fast dissociation half-life of NO of a few seconds is fast enough to account for the deactivation of the enzyme in biological systems. YC-1 and its analogues acting as NO sensitizers uncovered a new pharmacologically and conceivably physiologically relevant regulatory principle of the enzyme.

Two existing isoforms of the heterodimeric guanylyl cyclase ($\alpha_1\beta_1$, $\alpha_2\beta_1$) are known that are functionally indistinguishable. Up to now, the NO-sensitive guanylyl cyclase has been considered as a soluble enzyme. However, recent evidence about the $\alpha_2\beta_1$ isoform interacting with a PDZ domain of the postsynaptic scaffold protein PSD-95 suggests that the α_2 subunit directs a membrane association of this isoform. The interaction with PSD-95 locates the $\alpha_2\beta_1$ isoform in close proximity to the NO-generating NO synthase thereby enabling the NO sensor to respond to locally raised NO concentrations. (Mol Cell Biochem **230**: 159–164, 2002)

Key words: nitric oxide, NO-sensitive guanylyl cyclase, PDZ domain, cGMP

Introduction

cGMP, like cAMP, acts as an intracellular signal molecule in the regulation of various cellular events. Although the precise mechanisms of cGMP action are not yet fully understood, cGMP is now an established second messenger molecule involved in smooth muscle relaxation, platelet aggregation, and neurotransmission. cGMP exerts its effects through cGMP-dependent protein kinases, cGMP-regulated phosphodiesterases and cGMP-regulated ion channels. Two major types of cGMP-forming enzymes have been identified: the membrane-spanning peptide-regulated guanylyl cyclases and the NO-sensitive guanylyl cyclases (NO-sensitive GCs) found mainly in the cytosol. This review, however, will focus solely on the NO-sensitive GC. The NO-sensitive GC represents the best characterized receptor for the signal molecule NO produced by the NO synthases. Similarly, NO donors used in the treatment of coronary heart disease cause vasodilatation by stimulation of the enzyme. Relatively high levels of the NO-sensitive GC are found in lung, brain, and platelets, and already 20 years ago, the enzyme has been purified to apparent homogeneity in several laboratories. Subsequently, mainly by the progress in molecular biology, the subunit composition of the NO-sensitive GC was clarified.

Structure

NO-sensitive GC is a heterodimeric enzyme composed of two different subunits, α and β. The $\alpha_1\beta_1$ heterodimer represents the enzyme purified from lung; both subunits have been cloned and sequenced in the late 1980's [1–4]. By homology screening, two other subunits have been identified, designated α_2 and β_2 [5, 6]. The subunits initially termed α_3 and β_3 represent the human orthologs of the α_1 and β_1 subunits, respectively [7]. Expression experiments revealed that one α and one β subunit are required for the formation of a catalytically active enzyme [8, 9]. However, only one report for the functional expression of a β_2 subunit exists [10], whereas most laboratories, including our own, were not able to detect catalytic activity of the β_2 subunit upon coexpression with any other subunit.

Address for offprints: D. Koesling, Pharmakologie und Toxikologie, Medizinische Fakultät MA N1, Ruhr-Universität Bochum, 44780 Bochum, Germany (E-mail: koesling@iname.com)

Each subunit contains a C-terminal cyclase catalytic domain (Fig. 1) which is also conserved in the peptide receptor guanylyl cyclases and the adenylyl cyclases. Analysis of truncated mutants revealed that these catalytic parts are sufficient for cyclic nucleotide-forming activity while stimulation by NO requires the intact N termini [11]. The NO-sensitive GC as well as the adenylyl cyclases contain two different catalytic domains, whereas the peptide-regulated guanylyl cyclases contain two identical catalytic domains. Structure analysis of the adenylyl cyclase catalytic domains suggests a 'head-to-tail'-like fashion of the catalytic parts of the guanylyl cyclase subunits, with both α and β subunit forming the catalytic core [12, 13]. Mutation analysis identified three amino acids that are responsible for the specificity of an adenylyl or a guanylyl cyclase [14]. Furthermore the crystal structure suggested a second 'catalytic-like' site that, in the adenylyl cyclase, binds the activator forskolin [13].

The conserved central part of the subunits are thought to be involved in dimerization which is concluded from studies performed with the peptide-regulated guanylyl cyclases [15]; however, direct experimental evidence for the NO-sensitive GC is still lacking. The N-termini of the subunits are referred to as the regulatory domains; expression of the respective region of the β_1 subunit yielded a homodimer and revealed that this region is sufficient for heme binding [16]. One important residue, the histidine 105 of the β subunit, that acts as the proximal heme ligand ([17], see below) is located within this region. The N-terminal putative regulatory domains of the a subunits differ considerably in the primary structure (see Fig. 1), however the $\alpha_1\beta_1$ and $\alpha_2\beta_1$ isoforms share the same regulatory properties ([18], see below).

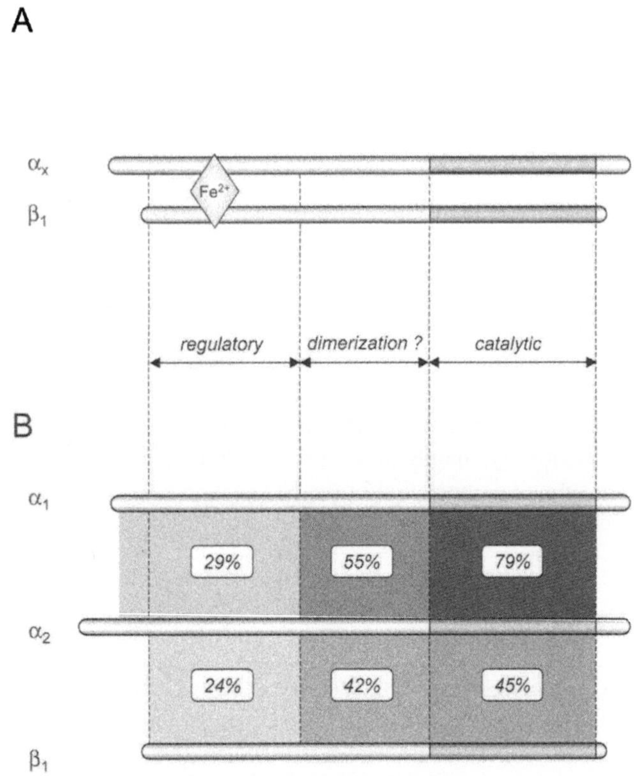

Fig. 1. Isoforms of NO-sensitive GC. A schematic representation of the two isoforms of NO-sensitive GC, composed of an α subunit (either α_1 or α_2), the β_1 subunit and a prosthetic heme group is shown in the upper section. The catalytic regions are shaded. The bottom part depicts a comparison of the primary structures of the three subunits. The relative numbers of identical amino acids shared between the respective subunits within a certain region (vertical lines) are given in percent. The C termini of the subunits of NO-sensitive GC are unrelated. The regions chosen for comparison are: amino acids 1–280 of the α_1, 1–321 of the α_2, and 1–211 of the β_1 subunit, corresponding to the regulatory domains; amino acids 281–443 of the α_1, 322–483 of the α_2, and 212–383 of the β_1 subunit, corresponding to the putative dimerization domains; amino acids 444–657 of the α_1, 484–697 of the α_2, and 384–604 of the β_1 subunit, corresponding to the catalytic regions, respectively.

Regulation by nitric oxide

NO-sensitive GC contains a prosthetic heme group that mediates the up to 200-fold activation of the enzyme by NO [19–21]. The heme content of NO-sensitive GC has been a matter of debate for some years. Whereas we and others [20, 22, 23] found one or less than one mole of heme per mole of heterodimer, there was one report of 1.5 moles heme per mole heterodimer. The higher heme content was due to the use of a correction factor of 1.66 for the protein determination [21]. However, in a recent paper by this group [24] the heme stoichiometry was corrected to be 1 mole heme per mole heterodimer.

The absorbance spectrum of NO-sensitive GC with a peak at 430 nm, called Soret band, is indicative of a five-coordinated ferrous heme with a histidine as axial ligand [25]. Binding of NO to the heme iron results in breakage of the histidine-iron bond. The resulting five-coordinated nitrosyl-heme complex features a shifted absorption maximum at 398 nm. The iron-free heme precursor protoporphyrin IX stimulates NO-sensitive GC independently of NO [26] by mimicking the NO-heme complex in the enzyme. The breakage of the histidine-iron bond is required for activation of the enzyme as CO, that binds to the heme iron without disruption of the histidine-iron bond (thereby forming a six-coordinated heme-iron with an absorption maximum at 424 nm) causes an only 4–6 fold activation of the enzyme [25, 27].

The histidine 105 of the β_1 subunit was identified as the axial heme-binding ligand. Substitution of this residue with phenylalanine produced an NO-insensitive, heme-deficient enzyme with unaffected basal activity [17]. Although the heme-deficient mutant was still able to bind heme in reconstitution experiments, the heme-reconstituted enzyme remained insensitive to NO [28]. The non-responsiveness to NO of the reconstituted mutant suggests that, besides heme-binding, histidine 105 is involved in the transduction of the NO-induced, heme-mediated stimulus, but not in the catalytic reaction. With a somewhat different approach Zhao *et al.* [29] confirmed the histidine 105 as the heme proximal ligand in a truncated homomeric β_1 subunit.

From other hemoproteins with a five coordinated heme iron the association of NO is known to be diffusion-controlled whereas the dissociation of NO is generally rather slow resulting in very high affinities of hemoproteins to NO [30]. Studies on the dissociation of NO and the resulting deactivation of the enzyme are contradictory and confusing. It should be noted, that the reported desensitization differs from deactivation as it probably involves other regulatory factors only present in intact cells and is not yet fully understood [31]. In contrast, dissociation and deactivation describe the behavior of the enzyme upon addition of an NO-scavenger monitored spectrophotometrically or in enzymatic assays, respectively. The addition of a scavenger molecule is a prerequisite in studies on the dissociation of gaseous ligands from hemoproteins, as otherwise the dissociation of the ligand is immediately followed by re-association and the resulting data describe the decay of the ligand in solution rather than the dissociation from the hemoproteins [30].

By spectrophotometric analysis, we found a half-life of the NO-GC complex of about 2 min at 37°C using oxyhemoglobin as NO scavenger [30]. Discrepancies to the deactivation of the purified enzyme determined in catalytic assays upon addition of an NO-scavenger prompted us to include the substrate MgGTP in the spectrophotometric experiments, and we obtained a half-life of approximately 5 sec under those conditions [32]. However, because of the relatively constant MgGTP levels in the cell, this effect of the substrate on the dissociation can be assumed not to be regulatory; rather, one should keep in mind that handling the enzyme in the absence of the substrate is artificial and the resulting data do possibly not describe the enzyme in its native conformation.

In a somewhat different approach Brandish *et al.* [24] confirmed the dissociation half-life of NO from NO-sensitive GC to be in the range of a few minutes. However, they did not find a much faster dissociation in the presence of the substrate MgGTP, possibly because the enzyme was handled shortly in the absence of MgGTP in the NO-bound state.

Recently, an attempt to determine the deactivation rate of the enzyme was made. Margulis and Sitaramayya [33] measured deactivation of NO-sensitive GC in cytosolic preparations of bovine retina, and reported a half-life of about 18 sec at 20°C, corresponding to about 5 sec at 37°C.

Besides the *in vitro* measurements, Bellamy *et al.* [31] reported a rapid desensitization of the NO-sensitive GC with a time constant of 6.9 sec upon treatment of cerebellar cells and platelets with NO [34]. Furthermore, this group described deactivation of NO-sensitive GC upon NO removal in cerebellar cells with a half-life of 0.2 sec [34], which is 25-fold faster than the value reported by Margulis and Sitaramayya [33] for the deactivation of the NO-sensitive GC *in vitro*. However, the underlying mechanism of this divergent behavior of NO-sensitive GC *in vivo* remains to be elucidated.

Inhibition of NO-sensitive GC by ODQ

In 1995, the quinoxalin derivative 1H-[1,2,4]oxadiazolo[4,3-a]-quinoxalin-1-one (ODQ) has been shown to be a potent and selective inhibitor of NO-sensitive GC in brain slices, thus providing an important tool to discriminate between cGMP-dependent and cGMP-independent NO signaling [35]. Subsequently, the inhibitory effects of ODQ on NO-sensitive GC have been confirmed in a variety of other cells and tissues [36, 37]. The compound binds to NO-sensitive GC in an NO-competitive manner and inhibits NO-stimulated activity irreversibly, thereby leaving basal activity almost unchanged. Spectrophotometric analysis suggests that the inhibitory effect of ODQ is due to oxidation of the heme iron [38]. Further evidence for this concept came from the observation that ODQ does not inhibit the stimulation of NO-sensitive GC by the iron free heme precursor protoporphyrin IX [39].

Modulators/NO-independent activators

The benzyl indazole derivative YC-1 has been published as an inhibitor of platelet aggregation which led to an increase in the intracellular cGMP concentration [40]. Subsequently, the substance has been shown to inhibit platelet-rich thrombosis *in vivo* [41] and to induce a concentration-dependent relaxation of endothelial-free rat aortic rings precontracted with phenylephrine [42, 43].

We [27] and others [42] have demonstrated that YC-1 stimulates NO-sensitive GC about 10-fold. As the activation was not blocked by the NO scavenger oxyhemoglobin, the YC-1 effect is referred to as NO-independent. In addition to the small stimulation of the enzyme by YC-1 in the absence of NO, YC-1 exerts tremendous effects on the NO-stimulated enzyme. YC-1 increased maximal NO-induced catalytic rate of NO-sensitive GC by about 40% and shifted the NO-concentration response curve by one order of magnitude to the left, indicating that YC-1 sensitizes NO-sensitive GC towards NO. Furthermore, YC-1 sensitized the enzyme towards CO, leading to a stimulation similar to that of NO. Interestingly, YC-1 stimulation requires the presence of the prosthetic heme group as YC-1 did not stimulate the heme-depleted enzyme, although it was still able to bind to NO-sensitive GC [44]. As the deactivation in the presence of YC-1 upon addition of an NO scavenger to the NO-stimulated enzyme was slowed down [44] it is tempting to assume that YC-1 exerts its effects by decreasing the dissociation rates of the gaseous ligands from the heme moiety. The immense effects on the cGMP levels in platelets are not only caused by an increased synthesis of cGMP but at least in part by a reduced degradation of cGMP through inhibition of phosphodiesterases by YC-1 [45, 46].

Recently, BAY 41-2272 has been published as a YC-1 analog that stimulates NO-sensitive GC 30-fold in the absence of NO [47]. More interestingly, it features an approx. 100-fold higher potency (EC_{50} ~ 0.5 μM) than YC-1 and sensitizes the enzyme towards physiologically occurring NO concentrations. In contrast to YC-1, BAY 41-2272 was reported not to inhibit phosphodiesterase-5 activity. Based on photoaffinity labeling, two cysteines in the α_1 subunit (Cys 238 and Cys 243) were proposed to be involved in YC-1/BAY 41-2272 binding. However, these cysteines are not conserved in the α_2 subunit, although the $\alpha_2\beta_1$ isoform exhibited the same YC-1 effects as the $\alpha_1\beta_1$ isoform. In doses of up to 3 mg per kg body wt BAY 41-2272 increased the tail-bleeding time in rats; doses of 10 mg per kg body wt effectively decreased the arterial blood pressure of spontaneously hypertensive rats, and increased the survival time of rats treated with the NO synthase inhibitor L-NAME.

YC-1 analogs represent a novel class of drugs that sensitize the NO receptor towards its natural activator NO. This type of action may have broad implications in the future therapy of coronary heart disease, hypertension and erectile dysfunction; by exerting their effects selectively on the sites of endogenous NO release, YC-1 analogs avoid unwanted effects of NO releasing compounds. On the other hand, an allosteric site of NO-sensitive GC suggests the existence of an endogenous ligand that modulates the sensitivity towards NO and – assuming there is a CO signaling cascade – turns NO-sensitive GC in an efficient CO receptor.

Tissue distribution and isoforms

Studies concerning the subunit expression of NO-sensitive GC in different tissues are rare. In Northern blots, Budworth and coworkers [48] demonstrated a widespread distribution of the α_1 and β_1 subunits in accordance with the assumption of the $\alpha_1\beta_1$ heterodimer as the predominant, ubiquitously expressed isoform. They found the α_2 subunit to be expressed solely in brain, placenta and uterus. Recently, Gibb and Garthwaite [49] studied the expression of the NO-sensitive GC subunits in rat brain using RT-PCR and *in-situ* hybridization. They found a widespread distribution of the α_1, α_2, and β_1 subunits, with some regions predominantly expressing either the $\alpha_1\beta_1$ isoform or the $\alpha_2\beta_1$ isoform. Hence, the mismatches between the α_1 and β_1 subunit expression found in some brain areas earlier [50] can be explained by the expression pattern of the α_2 subunit.

With the existence of two α and two β subunits, in theory four αβ heterodimers may exist. However, to date only two isoforms have been demonstrated to exist on the protein level: the $\alpha_1\beta_1$ isoform corresponding to the enzyme initially purified from lung, and the $\alpha_2\beta_1$ isoform that has been shown exist in human placenta [18]. Because of the low number of identical amino acids on the N-terminal regulatory domains of the α subunits (see Fig. 1), both isoforms purified from Sf9 cells were characterized extensively. Surprisingly, no differences were detected with respect to heme conformation, substrate affinity, catalytic rates and sensitivity towards NO. Also, the new NO-sensitive GC modulator YC-1 sensitized both enzymes towards NO, and turned CO into a potent activator [18]. This raised the question of the physiological relevance of two isoforms.

Recently, we have demonstrated the occurrence of the $\alpha_2\beta_1$ isoform in rat brain synaptosomes and its association with the postsynaptic density protein-95 (PSD-95) [51]. The interaction with PSD-95 results in the membrane-association of the so-far considered soluble enzyme. PSD-95 was the first member of a family of synaptic scaffold proteins containing three PDZ domains besides other typical protein interaction motifs (for reviews see [52–54]). PDZ domains are adapter domains of about 100 amino acids which apparently mainly occur in proteins of the cytoskeleton. They bind to other proteins by interacting with the C-termini of their binding partners in most cases. PSD-95 is anchored to the postsynaptic membrane and has been implicated in NMDA-receptor clustering and formation of 'transducisomes', in which the downstream signaling machinery is located in a close neighborhood. Interestingly, the neuronal NO synthase has been shown to be associated with PSD-95. PSD-95's first and second PDZ domains can bind to the NMDA receptor NR2 subunit [55, 56], the second PDZ domain binds to an internal stretch of the neuronal NO synthase [57, 58]. The $\alpha_2\beta_1$GC-PSD-95 interaction occurs directly between the C-terminal amino acids of the α_2 subunit and the third PDZ domain of PSD-95 [51]. Those multiple PSD-95 interactions would give raise to the formation of a quaternary complex with the NMDA receptor, the neuronal NO synthase and the $\alpha_2\beta_1$GC, thereby forming a 'transducisome' as depicted in Fig. 2.

In sum, it appears that the isoforms do not represent differently regulated NO-receptors; rather, different subcellular targeting directed by the α subunit requires the existence two α subunits of NO-sensitive GC. Furthermore, the membrane association of the $\alpha_2\beta_1$GC by interaction with PSD-95 contradicts the up to now used categorization of guanylyl cyclases in soluble and membrane-bound isoforms, rather the guanylyl cyclases should be classified according to their regulatory properties, i.e. NO-sensitive guanylyl cyclases and peptide-regulated guanylyl cyclases.

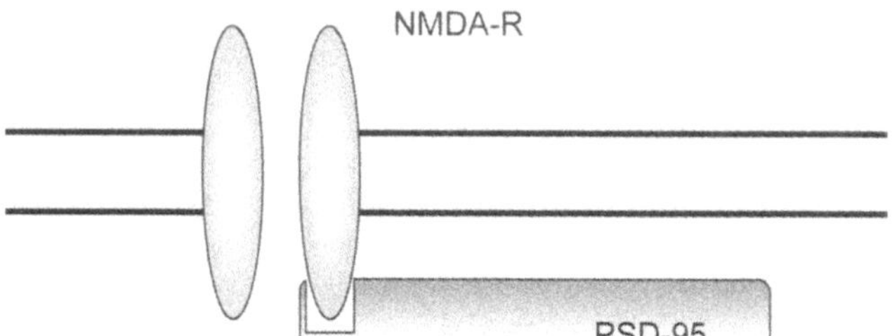

Fig. 2. Signaling complex of NMDA receptor, NO synthase and NO-sensitive GC. The NMDA-receptor NR2 subunit, the neuronal NO synthase and the $\alpha_2\beta_1$GC bind to the first second and third PDZ domains of PSD-95, respectively, thereby forming a 'transducisome' for fast and efficient signaling without raising second messenger concentrations throughout the cell.

References

1. Koesling D, Herz J, Gausepohl H, Niroomand F, Hinsch KD, Mulsch A, Bohme E, Schultz G, Frank R: The primary structure of the 70 kDa subunit of bovine soluble guanylate cyclase. FEBS Lett 239: 29–34, 1988
2. Nakane M, Saheki S, Kuno T, Ishii K, Murad F: Molecular cloning of a cDNA coding for 70 kilodalton subunit of soluble guanylate cyclase from rat lung. Biochem Biophys Res Commun 157: 1139–1147, 1988
3. Koesling D, Harteneck C, Humbert P, Bosserhoff A, Frank R, Schultz G, Bohme E: The primary structure of the larger subunit of soluble guanylyl cyclase from bovine lung. Homology between the two subunits of the enzyme. FEBS Lett 266: 128–132, 1990
4. Nakane M, Arai K, Saheki S, Kuno T, Buechler W, Murad F: Molecular cloning and expression of cDNAs coding for soluble guanylate cyclase from rat lung. J Biol Chem 265: 16841–16845, 1990
5. Yuen PS, Potter LR, Garbers DL: A new form of guanylyl cyclase is preferentially expressed in rat kidney. Biochemistry 29: 10872–10878, 1990
6. Harteneck C, Wedel B, Koesling D, Malkewitz J, Bohme E, Schultz G: Molecular cloning and expression of a new α-subunit of soluble guanylyl cyclase. Interchangeability of the α-subunits of the enzyme. FEBS Lett 292: 217–222, 1991
7. Zabel U, Weeger M, La M, Schmidt HH: Human soluble guanylate cyclase: Functional expression and revised isoenzyme family. Biochem J 335: 51–57, 1998
8. Harteneck C, Koesling D, Soling A, Schultz G, Bohme E: Expression of soluble guanylyl cyclase. Catalytic activity requires two enzyme subunits. FEBS Lett 272: 221–223, 1990
9. Buechler WA, Nakane M, Murad F: Expression of soluble guanylate cyclase activity requires both enzyme subunits. Biochem Biophys Res Commun 174: 351–357, 1991
10. Gupta G, Azam M, Yang L, Danziger RS: The β_2 subunit inhibits stimulation of the α_1/β_1 form of soluble guanylyl cyclase by nitric oxide. Potential relevance to regulation of blood pressure. J Clin Invest 100: 1488–1492, 1997
11. Wedel B, Harteneck C, Foerster J, Friebe A, Schultz G, Koesling D: Functional domains of soluble guanylyl cyclase. J Biol Chem 270: 24871–24875, 1995
12. Zhang G, Liu Y, Ruoho AE, Hurley JH: Structure of the adenylyl cyclase catalytic core. Nature 386: 247–253, 1997
13. Tesmer JJ, Sunahara RK, Gilman AG, Sprang SR: Crystal structure of the catalytic domains of adenylyl cyclase in a complex with $G_s\alpha$.GTPγS. Science 278: 1907–1916, 1997
14. Sunahara RK, Beuve A, Tesmer JJ, Sprang SR, Garbers DL, Gilman AG: Exchange of substrate and inhibitor specificities between adenylyl and guanylyl cyclases. J Biol Chem 273: 16332–16338, 1998
15. Wilson EM, Chinkers M: Identification of sequences mediating guanylyl cyclase dimerization. Biochemistry 34: 4696–4701, 1995
16. Schelvis JP, Zhao Y, Marletta MA, Babcock GT: Resonance raman characterization of the heme domain of soluble guanylate cyclase. Biochemistry 37: 16289–16297, 1998
17. Wedel B, Humbert P, Harteneck C, Foerster J, Malkewitz J, Bohme E, Schultz G, Koesling D: Mutation of His-105 in the β_1 subunit yields a nitric oxide-insensitive form of soluble guanylyl cyclase. Proc Natl Acad Sci USA 91: 2592–2596, 1994
18. Russwurm M, Behrends S, Harteneck C, Koesling D: Functional properties of a naturally occurring isoform of soluble guanylyl cyclase. Biochem J 335: 125–130, 1998
19. Gerzer R, Hofmann F, Schultz G: Purification of a soluble, sodium-nitroprusside-stimulated guanylate cyclase from bovine lung. Eur J Biochem 116: 479–486, 1981
20. Humbert P, Niroomand F, Fischer G, Mayer B, Koesling D, Hinsch KD, Gausepohl H, Frank R, Schultz G, Bohme E: Purification of soluble guanylyl cyclase from bovine lung by a new immunoaffinity chromatographic method. Eur J Biochem 190: 273–278, 1990
21. Stone JR, Marletta MA: Heme stoichiometry of heterodimeric soluble guanylate cyclase. Biochemistry 34: 14668–14674, 1995
22. Gerzer R, Bohme E, Hofmann F, Schultz G: Soluble guanylate cyclase purified from bovine lung contains heme and copper. FEBS Lett 132: 71–74, 1981
23. Tomita T, Tsuyama S, Imai Y, Kitagawa T: Purification of bovine soluble guanylate cyclase and ADP-ribosylation on its small subunit by bacterial toxins. J Biochem (Tokyo) 122: 531–536, 1997
24. Brandish PE, Buechler W, Marletta MA: Regeneration of the ferrous heme of soluble guanylate cyclase from the nitric oxide complex: Acceleration by thiols and oxyhemoglobin. Biochemistry 37: 16898–16907, 1998
25. Stone JR, Marletta MA: Soluble guanylate cyclase from bovine lung: Activation with nitric oxide and carbon monoxide and spectral characterization of the ferrous and ferric states. Biochemistry 33: 5636–5640, 1994
26. Ignarro LJ, Wood KS, Wolin MS: Activation of purified soluble guanylate cyclase by protoporphyrin IX. Proc Natl Acad Sci USA 79: 2870–2873, 1982

27. Friebe A, Schultz G, Koesling D: Sensitizing soluble guanylyl cyclase to become a highly CO-sensitive enzyme. Embo J 15: 6863–6868, 1996
28. Foerster J, Harteneck C, Malkewitz J, Schultz G, Koesling D: A functional heme-binding site of soluble guanylyl cyclase requires intact N-termini of α_1 and β_1 subunits. Eur J Biochem 240: 380–386, 1996
29. Zhao Y, Schelvis JP, Babcock GT, Marletta MA: Identification of histidine 105 in the β_1 subunit of soluble guanylate cyclase as the heme proximal ligand. Biochemistry 37: 4502–4509, 1998
30. Kharitonov VG, Sharma VS, Magde D, Koesling D: Kinetics of nitric oxide dissociation from five- and six-coordinate nitrosyl hemes and heme proteins, including soluble guanylate cyclase. Biochemistry 36: 6814–6818, 1997
31. Bellamy TC, Wood J, Goodwin DA, Garthwaite J: Rapid desensitization of the nitric oxide receptor, soluble guanylyl cyclase, underlies diversity of cellular cGMP responses. Proc Natl Acad Sci USA 97: 2928–2933, 2000
32. Kharitonov VG, Russwurm M, Magde D, Sharma VS, Koesling D: Dissociation of nitric oxide from soluble guanylate cyclase. Biochem Biophys Res Commun 239: 284–286, 1997
33. Margulis A, Sitaramayya A: Rate of deactivation of nitric oxide-stimulated soluble guanylate cyclase: Influence of nitric oxide scavengers and calcium. Biochemistry 39: 1034–1039, 2000
34. Bellamy TC, Garthwaite J: Sub-second kinetics of the nitric oxide receptor, soluble guanylyl cyclase, in intact cerebellar cells. J Biol Chem 9: 4287–4292, 2000
35. Garthwaite J, Southam E, Boulton CL, Nielsen EB, Schmidt K, Mayer B: Potent and selective inhibition of nitric oxide-sensitive guanylyl cyclase by 1H-[1,2,4]oxadiazolo[4,3-a]quinoxalin-1-one. Mol Pharmacol 48: 184–188, 1995
36. Brunner F, Schmidt K, Nielsen EB, Mayer B: Novel guanylyl cyclase inhibitor potently inhibits cyclic GMP accumulation in endothelial cells and relaxation of bovine pulmonary artery. J Pharmacol Exp Ther 277: 48–53, 1996
37. Abi-Gerges N, Hove-Madsen L, Fischmeister R, Mery PF: A comparative study of the effects of three guanylyl cyclase inhibitors on the L-type Ca^{2+} and muscarinic K^+ currents in frog cardiac myocytes. Br J Pharmacol 121: 1369–1377, 1997
38. Schrammel A, Behrends S, Schmidt K, Koesling D, Mayer B: Characterization of 1H-[1,2,4]oxadiazolo[4,3-a]quinoxalin-1-one as a heme-site inhibitor of nitric oxide-sensitive guanylyl cyclase. Mol Pharmacol 50: 1–5, 1996
39. Koesling D, Friebe A: Soluble guanylyl cyclase: Structure and regulation. Rev Physiol Biochem Pharmacol 135: 41–65, 1999
40. Wu CC, Ko FN, Kuo SC, Lee FY, Teng CM: YC-1 inhibited human platelet aggregation through NO-independent activation of soluble guanylate cyclase. Br J Pharmacol 116: 1973–1978, 1995
41. Teng CM, Wu CC, Ko FN, Lee FY, Kuo SC: YC-1, a nitric oxide-independent activator of soluble guanylate cyclase, inhibits platelet-rich thrombosis in mice. Eur J Pharmacol 320: 161–166, 1997
42. Mulsch A, Bauersachs J, Schafer A, Stasch JP, Kast R, Busse R: Effect of YC-1, an NO-independent, superoxide-sensitive stimulator of soluble guanylyl cyclase, on smooth muscle responsiveness to nitrovasodilators. Br J Pharmacol 120: 681–689, 1997
43. Wegener JW, Gath I, Forstermann U, Nawrath H: Activation of soluble guanylyl cyclase by YC-1 in aortic smooth muscle but not in ventricular myocardium from rat. Br J Pharmacol 122: 1523–1529, 1997
44. Friebe A, Koesling D: Mechanism of YC-1-induced activation of soluble guanylyl cyclase. Mol Pharmacol 53: 123–127, 1998
45. Friebe A, Mullershausen F, Smolenski A, Walter U, Schultz G, Koesling D: YC-1 potentiates nitric oxide- and carbon monoxide-induced cyclic GMP effects in human platelets. Mol Pharmacol 54: 962–967, 1998
46. Galle J, Zabel U, Hubner U, Hatzelmann A, Wagner B, Wanner C, Schmidt HH: Effects of the soluble guanylyl cyclase activator, YC-1, on vascular tone, cyclic GMP levels and phosphodiesterase activity. Br J Pharmacol 127: 195–203, 1999
47. Stasch JP, Becker EM, Alonso-Alija C, Apeler H, Dembowsky K, Feurer A, Gerzer R, Minuth T, Perzborn E, Pleiss U, Schroder H, Schroeder W, Stahl E, Steinke W, Straub A, Schramm M: NO-independent regulatory site on soluble guanylate cyclase. Nature 410: 212–215, 2001
48. Budworth J, Meillerais S, Charles I, Powell K: Tissue distribution of the human soluble guanylate cyclases. Biochem Biophys Res Commun 263: 696–701, 1999
49. Gibb BJ, Garthwaite J: Subunits of the nitric oxide receptor, soluble guanylyl cyclase, expressed in rat brain. Eur J Neurosci 13: 539–544, 2001
50. Furuyama T, Inagaki S, Takagi, H: Localizations of α_1 and β_1 subunits of soluble guanylate cyclase in the rat brain. Brain Res Mol Brain Res 20: 335–344, 1993
51. Russwurm M, Wittau N, Koesling D: Guanylyl cyclase/PSD-95 interaction: Targeting of the NO-sensitive $\alpha_2\beta_1$ guanylyl cyclase to synaptic membranes. J Biol Chem 276: 44647–44652, 2001
52. Craven SE, Bredt DS: PDZ proteins organize synaptic signaling pathways. Cell 93: 495–498, 1998
53. Garner CC, Nash J, Huganir RL: PDZ domains in synapse assembly and signalling. Trends Cell Biol 10: 274–280, 2000
54. Sheng M, Pak DT: Ligand-gated ion channel interactions with cytoskeletal and signaling proteins. Annu Rev Physiol 62: 755–778, 2000
55. Kornau HC, Schenker LT, Kennedy MB, Seeburg PH: Domain interaction between NMDA receptor subunits and the postsynaptic density protein PSD-95. Science 269: 1737–1740, 1995
56. Niethammer M, Kim E, Sheng M: Interaction between the C terminus of NMDA receptor subunits and multiple members of the PSD-95 family of membrane-associated guanylate kinases. J Neurosci 16: 2157–2163, 1996
57. Brenman JE, Chao DS, Gee SH, McGee AW, Craven SE, Santillano DR, Wu Z, Huang F, Xia H, Peters MF, Froehner SC, Bredt DS: Interaction of nitric oxide synthase with the postsynaptic density protein PSD-95 and a1-syntrophin mediated by PDZ domains. Cell 84: 757–767, 1996
58. Christopherson KS, Hillier BJ, Lim WA, Bredt DS: PSD-95 assembles a ternary complex with the N-methyl-D-aspartic acid receptor and a bivalent neuronal NO synthase PDZ domain. J Biol Chem 274: 27467–27473, 1999

Molecular and Cellular Biochemistry **230**: 165–176, 2002.

The receptor-like properties of nitric oxide-activated soluble guanylyl cyclase in intact cells

Tomas C. Bellamy and John Garthwaite
Wolfson Institute for Biomedical Research, University College London, London, UK

Abstract

Soluble guanylyl cyclase (sGC) is the main receptor for nitric oxide (NO), and so mediates a wide range of effects (e.g. vasodilatation, platelet disaggregation and neural signalling) through the accumulation of cGMP and the engagement of various downstream targets, such as protein kinases and ion channels. Until recently, our understanding of sGC functioning has been derived exclusively from studies of the enzyme in tissue homogenates or in its purified form. Here, NO binds to the haem prosthetic group of sGC, triggering a conformational change and a large increase in catalytic activity. The potency (EC_{50}) of NO appears to be about 100–200 nM. The rate of activation of sGC by NO is rapid (milliseconds) and, in the presence of excess substrate, cGMP is formed at a constant rate; on removal of NO, sGC deactivates slowly (seconds–minutes). Recent investigation of the way that sGC behaves in its natural environment, within cells, has revealed several key differences. For example, the enzyme exhibits a rapidly desensitizing profile of activity; the potency of NO is 45 nM for the minimally-desensitized enzyme but becomes higher with time; deactivation of sGC on removal of NO is 25-fold faster than the fastest estimate for purified sGC. Overall, within cells, sGC behaves in a way that is analogous to the way that classical neurotransmitter receptors operate. The properties of cellular sGC have important implications for the understanding of NO-cGMP signalling. For example, the dynamics of the enzyme means that fluctuations in the rate of NO formation, even on subsecond time scale, will result in closely synchronized sGC activity in neighbouring cells; desensitization of sGC provides an economical way of generating a cellular cGMP signal and, in concert with phosphodiesterases, provides the basis for cGMP signal diversity, allowing different targets (outputs) to be selected from a common input (NO). Thus, despite exhibiting only limited molecular heterogeneity, cellular sGC functions in a way that introduces speed, complexity, and versatility into NO-cGMP signalling pathways. (Mol Cell Biochem **230**: 165–176, 2002)

Key words: soluble guanylyl cyclase, nitric oxide, kinetics, cyclic GMP, desensitization

Introduction

Soluble guanylyl cyclase (sGC) is the principal physiological target for the freely diffusible intercellular messenger, nitric oxide (NO), in cells. NO mediates a wide range of physiological effects, from vasodilatation and platelet disaggregation in the cardiovascular system, to neural signalling in both the central and peripheral nervous systems [1–3]. Through sGC, the NO signals are coded into cytosolic cGMP accumulation [1]. In turn, cGMP targets several classes of downstream effectors, principally cGMP-dependent kinase (cGK), cyclic nucleotide-gated ion channels (CNG channels), and cGMP-stimulated and -inhibited phosphodiesterases (PDE 2 and 3 respectively), to bring about the physiological effects [4–6]. These cGMP responses are terminated by phosphodiesterases, which catalyse the hydrolysis of cGMP to 5′ GMP [7]. The NO signal transduction pathway is summarized in Fig. 1.

The successful purification of sGC from bovine lung was achieved in the late 1970s [8], and considerable progress has since been made in the molecular characterization of the enzyme. sGC is an obligate heterodimer and, at present, four subunits are known: α1, α2, β1, and β2. The molecular biology of the isoforms of sGC is reviewed elsewhere in this volume and so will only be touched upon here. Instead, the aim is to summarize the present state of understanding of sGC

Address for offprints: T.C. Bellamy, Division of Neurophysiology, National Institute for Medical Research, The Ridgeway, Mill Hill, London NW7 1AA, UK (E-mail: tbellam@nimr.mrc.ac.uk)

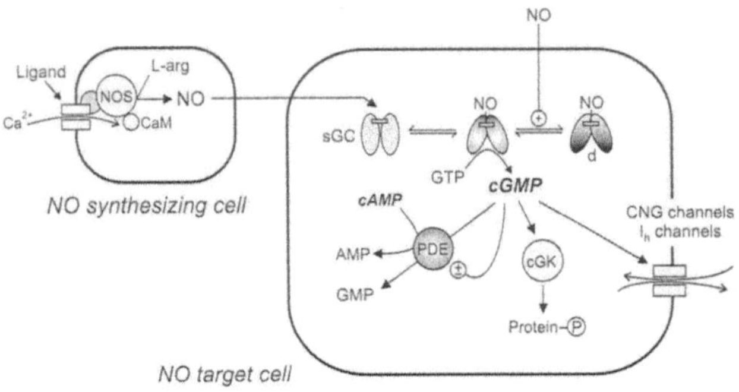

Fig. 1. The NO signal transduction pathway. A NOS-containing cell synthesizes NO in response to, for example, Ca^{2+} influx through a ligand-gated Ca^{2+} channel. NO diffuses to a target cell and activates sGC, catalysing the synthesis of cGMP. After activation sGC undergoes progressive desensitization (d), a process accelerated at high NO levels. cGMP acts on several cellular targets, including cyclic nucleotide-gated (CNG) ion channels and hyperpolarization-activated (I_h) channels, activation and inhibition of phosphodiesterases (PDE families 2 and 3 respectively) which modulate the levels of both cGMP and cAMP, and cGMP-dependent protein kinases (cGK) which phosphorylate a wide range of intracellular proteins, modifying their function (see [4]). PDEs terminate the cGMP signal by catalysing hydrolysis to GMP.

function in cells as the keystone for translation of NO signals into cGMP signals. In particular we focus on recent progress in understanding how the NO receptor, sGC, functions within intact cells in terms of the mechanism and kinetics of sGC activation by NO, deactivation following removal of NO, and catalysis of cGMP synthesis. This perspective is an attempt to bridge the conceptual gap between the enzymological features of the purified components of the pathway, and the physiological consequences of cGMP accumulation in intact tissues. To achieve this it is necessary to directly monitor the operation of the signalling enzymes within cells, and to understand how the routes for cGMP synthesis and degradation interact to code the cGMP signal following exposure of cells to NO. The majority of experiments undertaken so far have been conducted with cell suspensions derived from the postnatal rat cerebellum, a brain region involved in motor learning.

Initially, the properties of purified sGC are discussed, and subsequently these are compared and contrasted with the properties of cellular sGC. Thereafter, the applicability of classical models of receptor function to cellular sGC is highlighted, and the functional implications for NO signal transduction are discussed.

The activation of purified sGC by NO

Early attempts at purification of sGC were hampered by large losses in response to NO-releasing compounds [9]. This problem was resolved by the discovery that, when maximally active, the enzyme contained an associated haem moiety, and that earlier purification protocols had resulted in the loss of this prosthetic group [10, 11]. It was thus deduced that the binding site for NO on sGC was the haem group, a finding reconcilable with the well-known interaction of NO with the archetypal haemoproteins, haemoglobin (Hb) and myoglobin (Mb).

Each sGC heterodimer incorporates a single haem moiety, with the haem-binding site located to the β-subunit, where histidine residue 105 forms a covalent link to the Fe^{2+} centre of the haem group [12–15]. It is thought, by analogy with Hb and Mb, that other non-covalent (presumably hydrophobic) interactions at the periphery of the porphyrin ring help to stabilize haem binding to sGC. This proposal is supported by the N-terminal regions of both the α1 and β1 subunits being required for the incorporation of haem into sGC [16]. The β2 subunit lacks the first 63 N-terminal amino acid residues relative to the β1 subunit, and some groups have reportedly failed to find functional sGC of the α1β2 form [17]. This is consistent with the finding that deletion of the first 64 residues of β1 leads to formation of a α1β1 heterodimer that cannot bind haem [16].

The prevailing model [18] for the haem-based activation of sGC by NO is illustrated in Fig. 2. The initial step is the binding of NO to the Fe^{2+} centre of the haem moiety, resulting in the formation of 6-coordinated Fe^{2+}. The increase in catalytic activity of sGC is widely conceived to follow a second step: cleavage of the proximal histidine 105 bond [19–21], forming a pentavalent nitrosyl-Fe^{2+} structure, and resulting in a conformational change, presumably propagated to the enzyme's active site, which causes a several hundred-fold increase in the rate of cGMP synthesis [10, 22–24]. A feature of this haem-based activation mechanism of sGC that has greatly facilitated mechanistic studies is the characteristic UV-visible absorbance spectra for the three different haem species (i.e. 5-coordinate his-ligated sGC, 6 coordinate his- and NO-ligated, and 5 coordinate NO-ligated sGC). The absorbance maximum (known as the Soret band)

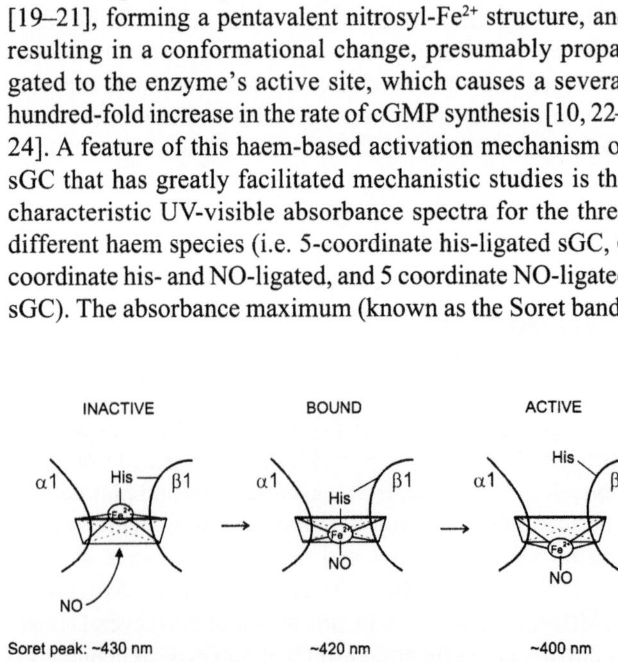

Fig. 2. Two-step activation mechanism for sGC. Inactive sGC exists as a 5-coordinated haem, bound by his105 of the protein β subunit. NO initially binds to form a 6-coordinate complex, and then triggers the cleavage of the his105 bond forming the active 5-coordinate nitrosyl complex of sGC.

shifts from ~430 to ~420, and finally ~400 nm, as sGC moves through the binding and activation steps after exposure to NO [11, 25, 26]. As a result, through the use of stopped-flow, and laser photolysis techniques, the kinetics of sGC activation can be followed on a sub-second time scale.

This kinetics of NO association with the haem group and the subsequent conformational change are presently matters of some controversy. It is agreed that the initial rate of association is very rapid (bimolecular rate constant of 10^7–10^8 $M^{-1}sec^{-1}$), approaching that of a diffusion-limited reaction. Different research groups, however, have reported different kinetics for the activation step. In a recent study by Zhao *et al.* [26], it was found that the binding of NO occurred with a rate constant, $k > 1.4 \times 10^8$ $M^{-1}sec^{-1}$ at 4°C. The subsequent conformational change step was rate-limiting for activation. It was reported that this second step proceeded with second order kinetics ($k = 2.4 \times 10^5$ $M^{-1}sec^{-1}$ at 4°C) in a manner that depended on NO concentration. This was interpreted as evidence for a second (non-haem) NO binding site, an hypothesis with significant mechanistic implications. In comparison, an alternative study by Makino *et al.* [25] found that NO binding was so rapid as to be complete in the dead-time (~2.5 msec) of the stopped flow apparatus at 15°C, which is consistent with Zhao *et al.*, except that the conformational change step proceeded in a straightforward first-order manner (k = 38 sec^{-1}), that remained unchanged over a 6-fold NO concentration range. In these experiments, however, the concentration of NO was apparently in excess of that of sGC by a minimum of approximately 6-fold.

These differences are potentially reconcilable, the confusion stemming from a misinterpretation of the kinetics of the two-step activation model. As will be discussed in more detail later, the evolving model for sGC function is closely analogous to classical models for neurotransmitter receptor activation (e.g. del Castillo and Katz; see Fig. 3 and [27]). In this model, the receptor exists in three states: unbound, agonist-bound, and activated. On application of a fixed concentration of agonist to unbound receptor, the system will proceed to equilibrium in two exponential phases (equation 1):

$$R_t/R_o = Ae^{-\lambda_1 t} + Be^{-\lambda_2 t} + C \qquad (1)$$

where R_t is the concentration of unbound receptor at time = t and R_o = concentration of unbound receptor before addition of agonist. A, B and C are constants related to the microscopic rate constants for agonist binding and conformational steps (i.e. k_1, k_{-1}, α and β in Fig. 3). $A + B + C = 1$, that is, the fraction of free receptor before application of agonist. After addition of agonist, the two exponential terms will tend to zero over time (the rates being determined by the values of λ_1 and λ_2), so that at equilibrium (i.e. t = ∞) the fraction of free receptor is equal to C.

This mechanistic model has been applied to sGC [25, 26].

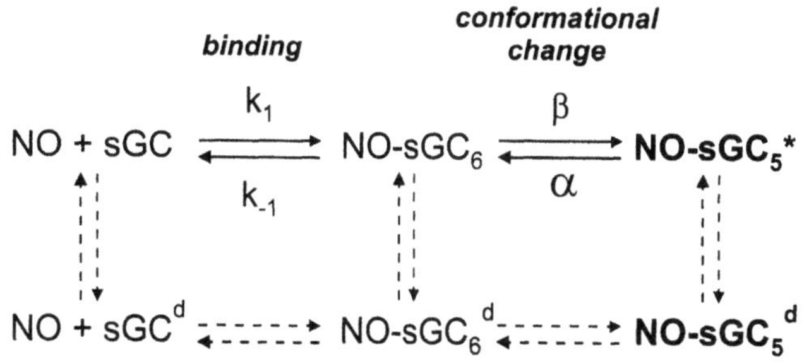

Fig. 3. Receptor-like model for sGC function in cells. A two-step model analogous to del Castillo–Katz constitutes the activation/deactivation pathway. The desensitized form of sGC may be reached by several routes (dashed lines). Key: sGC = free receptor; $NO\text{-}sGC_6$ = 6-coordinate bound receptor; $\mathbf{NO\text{-}sGC_5^*}$ = 5-coordinate active receptor; sGC^d = desensitized receptor.

The rate constants for the two phases (λ_1 and λ_2) have both been reported to vary with NO concentration, particularly when NO levels were substoichiometric [26]. This was considered as evidence that the conformational change which leads to sGC activation is dependent on NO binding at a previously unrecognised non-haem binding site. This conclusion is not necessarily warranted by the data. In measuring the disappearance of free sGC as two exponential phases, the measured constants correspond to λ_1 and λ_2 in equation 1. These observed constants do not correspond to the constants of the forward steps of receptor activation (k_1 and β; see Fig. 3), but are related to all of the microscopic rate constants for forward and backward steps. Consequently, both λ_1 and λ_2 will be dependent on NO concentration, especially at low levels of NO. There is therefore no requirement *per se* to infer a second NO binding event to trigger the transition to the active state of sGC. The prevailing model for sGC activation is therefore not directly challenged by the findings of Zhao *et al.*, and so a simple three-state receptor model still appears to be a good description of purified sGC activation.

A recent study by Kharitonov *et al.* [28] examined the rate of association of carbon monoxide (CO) with sGC by a different technique: laser photolysis and recombination of haem–ligand interactions. This method allows finer temporal resolution than stopped-flow methods, and demonstrated slower binding of CO ($k \sim 9 \times 10^5$ $M^{-1}sec^{-1}$) than NO ($k = 1.8 \times 10^8$ $M^{-1}sec^{-1}$) to sGC. It is interesting to compare CO with NO on sGC, as the carboxyl-sGC species is principally 6-coordinate, and exhibits a negligible (~4-fold in a CO-saturated solution) increase in sGC catalytic activity. It is not clear whether this small activation is due to a steady-state equilibrium with a minor population of 5 coordinate CO-sGC, or whether the strain on the proximal his bond in the 6 coordinate state is sufficient to enhance catalysis by a small amount [29]. More interesting still would be the results from photolysis and recombination of NO-sGC, and this work is awaited with anticipation.

A recent crystallographic study by Lawson *et al.* [30] on microbial cytochrome *c′* has revealed an unexpected result of potential importance for sGC activation. This group found that the NO-bound crystal structure of the cytochrome *c′* of *Alicaligenes xylosoxidans* contained NO bound to the *proximal* side of the haem-iron. This proximal rather than distal binding had not been observed (or perhaps even considered) previously. Cytochrome *c′* has closely similar ligand-binding properties to sGC: exclusion of O_2, a 5-coordinate ferrous resting haem, and a 5-coordinate nitrosyl complex but 6-coordinate carboxyl complex. Furthermore, it was found that a transient 6-coordinate species was formed on binding NO, which converted to a 5-coordinate (NO proximal) state in an NO concentration-dependent way. A mechanism was proposed, based on this finding, in which NO initially binds to the (cytochrome *c′* and sGC) haem in the distal position. The subsequent step, required for formation of the stable 5-coordinate species, is the displacement of the proximal histidine and the distal NO by a second NO molecule, which forms a bond at the proximal side. Again, this hypothesis may have arisen from confusion over the activation kinetics (see earlier), but nevertheless, a proximal binding position for NO in a similar haemoprotein requires a re-examination of some of the assumptions regarding the sequence of events involved in NO activation of sGC.

Deactivation of purified sGC

Deactivation refers to the rate at which active sGC disappears following an abrupt removal of free agonist. From the model, this will correspond to the rate at which sGC reverts from the active state to the bound state, and subsequently releases NO. For the majority of haem-containing proteins, the rate of spontaneous dissociation of NO from an NO-Fe^{2+} bond is very slow, even at physiological temperatures (half-time = hours–days [31]). Furthermore, under aerobic conditions, those haemoproteins which bind oxygen, such as haemoglobin (Hb) and myoglobin (Mb), will react with NO to form nitrate and ferric (Fe^{3+}) haem. This reaction is sufficiently rapid for mixing time to be the limiting factor (rate constant, $k > 3.4 \times 10^7$ $M^{-1}sec^{-1}$ [32]).

The rate of NO dissociation from purified sGC is considerably faster than the archetypal haemoproteins (half-time = minutes–hours [31, 33]), although this remains slow by comparison with other physiological receptors. As sGC-haem does not bind O_2, NO is not turned-over by sGC, and can associate and dissociate freely. In an experimental context this is advantageous, as oxyHb can be used as a tool to rapidly remove free NO from an aerobic sGC reaction milieu, allowing the rate of dissociation of NO from the haem group of sGC to be monitored. Using such a method, Kharitonov *et al.* [34] discovered that the dissociation rate of NO from sGC was greatly accelerated in the presence of co-factor (Mg^{2+}) and substrate (GTP). Under these conditions, they report a half-life of 2–5 sec, which brings sGC deactivation into a more reasonable time frame for responding dynamically to changes in the concentration of a physiological NO stimulus. These findings have been refuted however, in a study by Brandish *et al.* [33], who found that co-factor and substrate had little impact on their NO dissociation half-time of 2 min. Instead these authors highlighted the ability of low molecular weight thiols to accelerate dissociation (from hours down to minutes).

These studies have examined the rate of NO dissociation by monitoring changes in the absorbance of the haem group associated with the ligand-binding state of sGC. In comparison, a more recent study by Margulis and Sitaramayya [35] examined the rate at which cGMP synthesis declined to zero after removal of free NO by addition of Hb. Thus, deactivation of sGC was measured as a decrease in catalytic activity. As with the findings of Kharitonov *et al.* [34], these authors found that deactivation proceeded with a half-time of ~ 20 sec at 20°C, which would presumably reduce to a few seconds at 37°C. Furthermore, these authors report that the concentration of free Ca^{2+} is a determinant of the rate of deactivation [35].

The reasons for the variability of experimental findings for sGC deactivation are not clear. Perhaps different purification protocols yield proteins with differing haem environments. Clearly more work is needed to account for the present uncertainty in the rate of deactivation of purified sGC.

Catalysis of cGMP synthesis

The stereochemical course of the sGC catalysed conversion of GTP to cGMP was elucidated by Senter *et al.* [36] as an inversion of configuration at the proximal phosphate of GTP, indicating a simple displacement reaction without the formation of a covalently bound enzyme-substrate intermediate, a mechanism common to the related adenylyl cyclases. For catalysis, a nucleophilic amino acid residue accepts a proton from the hydroxyl group at position 5 of the ribose moiety of GTP, allowing displacement of pyrophosphate and formation of cGMP. The identity of this basic residue has not been determined, but point-mutational studies are reportedly underway [18]. It has been speculated that the release of histidine 105 from bonding with the haem group (the step proposed to result in enzyme activation) could provide the necessary base for the catalytic mechanism.

The kinetics of sGC catalysis has been studied using both tissue homogenates [37, 38] and the purified form of the enzyme [10, 39]. On activation, under conditions of surplus substrate (GTP) and activator (NO), the enzyme exhibits straightforward Michaelis–Menten-type kinetics: cGMP accumulates at a constant rate with time (a graph of cGMP concentration against time being linear). This indicates that the enzyme is not subject to feedback, or 'end-product', inhibition.

The Michaelis constant (K_m) of sGC purified from bovine lung for Mg^{2+}-GTP has been determined as 40–150 μM in the absence of NO [40, 41–43], and 20–40 μM in the presence of NO [40, 41–43]. The observed maximum rate of synthesis (quantified as specific activity) has been determined as 10–100 nmol cGMP.mg^{-1}.min^{-1} in the absence, and 10–40 μmol cGMP.mg^{-1}.min^{-1} in the presence, of added NO [17, 40, 42–44].

The evolving model of the activity of purified sGC is therefore of a rapidly activated ($k \sim 10^5$ $M^{-1}sec^{-1}$), slowly deactivating ($k \sim 0.04$ to 0.005 sec^{-1}), and kinetically straightforward enzyme. This picture has largely developed from studies on NO-haem interactions based on analogy with archetypal haemoproteins such as Hb and Mb, and enzymological studies on both tissue homogenates and purified preparations of sGC. In contrast, very little work has been undertaken to examine the behaviour of sGC within cells. To that end, we have recently examined the validity of the predictions arising from this biophysical data for the biochemical, kinetic and pharmacological properties of sGC in intact cells. The cells were mainly derived from the immature rat cerebellum. These studies have revealed that within its natural, physiological, environment the NO receptor behaves substantially differently from the way it does in cell-free or purified preparations.

Deactivation of sGC within cells: A strategy for measuring cellular sGC activity

The first major divergence between sGC when purified and the receptor in operation within cells is the rate at which NO dissociates from the haem site, deactivating cGMP synthesis. Addition of Hb to a cell suspension will result in the effectively immediate removal of free NO from the medium bathing the cells (see above). Hb will not penetrate cells, however, and so by measuring the rate at which cGMP synthesis declines after removal of free NO, one can measure the rate at which active sGC is lost, i.e. the rate at which NO dissociates from sGC, diffuses from the cell, and is trapped by Hb. In cerebellar cell suspensions it was found that complete deactivation occurred within a few seconds or less [45]. This finding had valuable implications, as it facilitated a novel approach to the kinetic analysis of cGMP accumulation within the cells. This is not a simple task, as the intracellular cGMP concentration at any given time is governed by the rate at which it is being synthesized by sGC, and the rate at which it is degraded by phosphodiesterases [45]. Unravelling the contributions of these two enzymes to cGMP levels is not straightforward. Complete inhibition of PDE cannot confidently be achieved. For example incubation with 1 mM IBMX, which is frequently used as a method to inhibit PDEs, is clearly insufficient [46, 47]. A direct method for measurement of one or other activity is required.

Following the rapid arrest of sGC activity by Hb addition, cGMP levels decline progressively, due to hydrolysis by PDEs. By fitting the cGMP decline with the integrated form of the Michaelis–Menten equation, a quantitative measurement of PDE activity can be achieved [48]. Consequently, with an explicit description of the contribution of cGMP degradation to the observed cGMP accumulation over time, the profile can be deconvoluted to reveal the rate of cGMP synthesis.

When this strategy was applied to both cerebellar astrocytes and human platelets, it revealed that, in stark contrast to the simple kinetics of the purified enzyme, sGC within intact cells exhibits a desensitizing response to NO [45]. After exposure to NO donors, the rate of cGMP synthesis declines progressively with time, to reach a steady-state rate approximately 10% of the maximal rate within around 30 sec. This property of sGC would not have been predicted from studies of the purified enzyme. Of course, there are alternative explanations for a falling-off in sGC activity with time, other than a functional desensitization: depletion of substrate (GTP), feedback inhibition of sGC by cGMP or pyrophosphate (PPi; [49]), or active export of cGMP from cells [50]. None of these alternatives can, however, account for desensitization as described [45]. These deductions can be made from two key observations: (a) Preincubation of cells with a sGC inhibitor (ODQ) and activator (YC-1) varied the amplitude of the cGMP response to NO over a 10-fold range, but the kinetics of the response was unchanged, indicating that a 10-fold range in GTP consumed and cGMP/PPi produced had no influence on sGC desensitization kinetics; (b) in cerebellar cells, the measured export of cGMP was negligible (2% of total measured cGMP).

It appears, therefore, that sGC within cells undergoes functional desensitization – that is, it can exist in an inactive form with agonist (NO) still bound. Desensitization is a property common to numerous neurotransmitter receptors, but its role in neurotransmission remains somewhat obscure [51]. In an experimental context, it is often considered something of a nuisance, as a continually changing response impacts on the measurement of most properties of an ion channel. This applies equally to a receptor-enzyme such as sGC, and desensitization complicates the experimental measurement of key kinetic parameters for sGC activity.

Kinetics of sGC activation, deactivation, desensitization and resensitization in cells

One of the major limitations in investigating the function of sGC within intact cells is the rapidity of the NO receptor's response to its agonist. Thus, the rates of activation and deactivation cannot be resolved with conventional methods and, when mixing times are slow, the data will be vulnerable to problems associated with desynchronization of receptor activation and desensitization within a population. In an attempt to address this problem in part, and to obtain a quantitative measurement of cellular sGC activity, a novel method for the rapid quenching of cell suspensions was recently devised [52].

When applied to cerebellar cells, it was found that the kinetics for sGC deactivation was complex. Deactivation did not appear to proceed in a simple exponential manner, as would be predicted for a simple decline in the levels of active sGC. Instead, the most straightforward operational description of deactivation was as an exponential decline (rate constant of 3.7 sec^{-1}) following an immediate 0.4 fractional loss of activity. Even the value of 3.7 sec^{-1} is approximately 25-fold greater than the fastest estimate for deactivation of purified sGC [34].

Using the same method, it was found that the rate of activation of cellular sGC following photolysis of a caged derivative of NO was too rapid to be measured even with a 20 msec sampling interval. Consequently, whether measurements for purified sGC apply to the receptor within cells could not be directly determined. It should be noted, however, that estimates for the potency (EC_{50}) of NO for activating purified sGC range from ~80 nM [53, 54] to 250 nM [23]. In contrast, the potency of NO for activating sGC in cells (measured 100 msec after photolysis of caged NO) is an estimated 45 nM [52]. This indicates that an enhanced rate of deactivation of cellular sGC does not compromise the potency of NO for activation, meaning that the rate of activation must also be equally higher (as the potency of NO is determined by the rate constant for deactivation over the rate constant for activation).

Progressive desensitization could be satisfactorily described by a single exponential, with a time constant of 6.7 sec at ~70 nM NO. This was reduced to 5.4 sec at a 10-fold higher concentration, indicating that NO accelerates desensitization. The consequence of this effect is that the concentration-response relationship of cellular sGC to NO becomes increasingly bell-shaped with time, as maximal sGC activity becomes truncated by desensitization. This presents complications in assessing the potency of NO for activating sGC and for desensitizing sGC. A concentration-response curve obtained 100 msec after release of NO from a photolabile compound gave an estimated EC_{50} of 45 nM, but the value progressively declined with time. In effect, the potency of NO increases with duration of NO exposure. Thus, measurement of tissue cGMP levels on a minute time scale will return an EC_{50} value lower than estimates based on the activation of purified sGC.

The impact of desensitization on agonist potency is clear, but this kinetic device may also account for the apparent complexity of sGC deactivation in cells. Deactivation has been measured by monitoring the fall in rate of cGMP formation after removal of free NO, which will be proportional to the concentration of sGC in the active state. However, reversal of sGC from the agonist bound states to the unbound state after loss of free NO is potentially complicated by the persistence of a population of sGC in the desensitized state. It is unknown how the desensitized state of sGC is reached (i.e. isomerization from the unbound, bound, or active states), but it is possible that following removal of free NO, desensitized sGC may revert to the unbound state with a sojourn(s) into the active state. Thus, the potential exists for the prolongation of a low level of sGC activity after removal of NO. This may account for the immediate fractional loss of active sGC, followed by a slower decline over a few hundred milliseconds, observed when Hb is added to the cell suspension [52]. An analogous scenario is well illustrated by GABA receptors, which apparently shuttle between desensitized and active states after a pulse of agonist [55].

Recovery of cellular sGC from desensitization was much slower than the onset of desensitization. The rate was measured by exposing cells to a supra-maximal concentration of NO to provoke desensitization, sGC was then deactivated by addition of Hb, and then a second dose of NO sufficient to saturate the Hb and fully stimulate sGC was re-applied after different times. The half-time for recovery of full activity was found to be approximately 1.5 min [45].

Cellular sGC as a cytosolic receptor

With the developing understanding of sGC function within cells, it becomes increasingly apparent that, as a transducer of NO signals, sGC more closely behaves like a neurotransmitter receptor than either a kinetically simple 'housekeeping' enzyme, or the archetypal NO-binding haemoproteins.

A model based on an extension of that of del Castillo-Katz for receptor activation is therefore useful for considering the function of sGC in cells, and is illustrated in Fig. 3. It seems fair to consider sGC as a cytosolic NO receptor, directly analogous to a membrane-bound ionotropic neurotransmitter receptor, which conducts a cGMP flux analogous to the ion-fluxes of conventional receptors. Although the route to the desensitized form of sGC is unknown, the profile of catalytic activity can be described most simply by an exponentially declining rate, reflecting progressive desensitization of sGC.

This model fits experimental data obtained with cerebellar cells satisfactorily (Fig. 4a). In fact, a formal model would predict desensitization kinetics to be more complex, with the sum of 3 exponentials describing the change in concentration of active sGC with time (assuming a single desensitized state exists), but given the precision of experimental sampling of this type, adopting an explicit model is unlikely to add much to our understanding at this stage. Testing of these formal models will have to await more sophisticated methods for the measurement of the kinetics of intracellular cGMP.

Functional implications of sGC as a desensitizing receptor

The properties of sGC in cells are different from those of the purified enzyme, and the question obviously arises as to the functional consequences for NO signal transduction in cells.

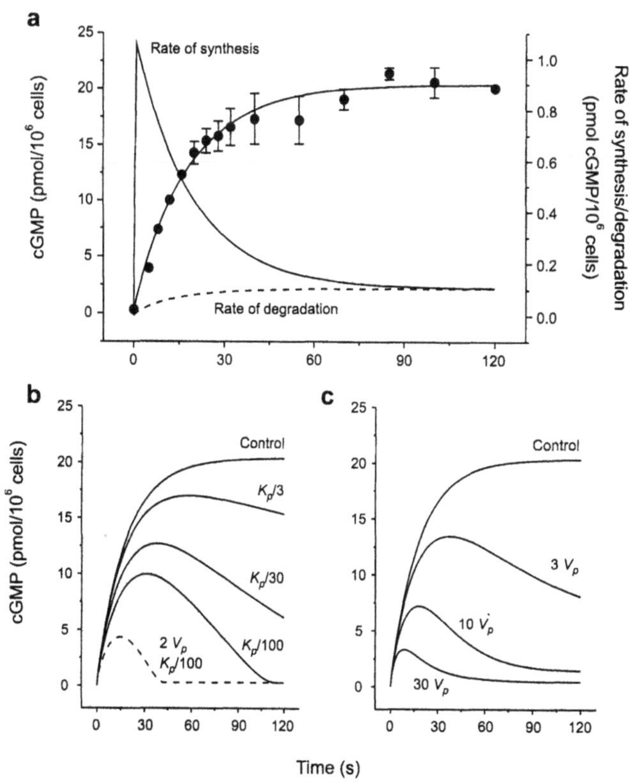

Fig. 4. Desensitization underlies diversity in intracellular cGMP profile. (a) A simple model of exponentially-decreasing rate of cGMP synthesis (solid line) is overlaid on experimentally measured cGMP accumulation (●). Constants for rate of degradation are experimentally-determined (dashed line). With a desensitizing rate of cGMP synthesis, diversity is introduced into the cGMP profile when the rate of cGMP degradation is augmented by decreasing the substrate concentration giving half-maximal rate (K_p; (b)) or by increasing maximal rate of degradation (V_p; (c)).

Put succinctly, the receptor-like properties of sGC within cells will dictate how a particular NO signal will be translated into a particular cGMP signal and ultimately into a physiological response.

The first discrepancy between purified and cellular sGC with functional implications is the dynamics of sGC response to fluctuations in NO concentration. As previously mentioned, sGC is activated sufficiently rapidly to exhibit no detectable delay in onset of catalytic activity on exposure to NO. This means that the receptor will respond, in effect, contemporaneously with the onset of NO synthesis, as the rate of diffusion for NO from a source to a nearby target is rapid (approximately two orders of magnitude faster than glutamate for example) [56]. Furthermore, as the rate of deactivation of sGC in cells is rapid ($k = 3.7\ sec^{-1}$), sGC will 'switch off' within a few hundred milliseconds of agonist removal. This is in contrast to the second–minute timescale predicted by the data on purified sGC [33, 34]. Therefore, rather than a sluggish sGC response exhibiting a persistent 'memory' for an NO signal, cellular sGC will respond dynamically to fluctuations in NO level, offering a translated signal of impressive fidelity. Hence, the properties of cellular sGC provide a means for a target cell to receive and translate information about nearby NO synthesis in close synchrony with NOS activity itself.

Another important feature of cellular sGC is agonist-induced desensitization, occurring over a time scale of seconds. This kinetic device has several consequences for cell function and signalling, which will be considered in turn. In considering the roles of desensitization in NO signal transduction, it is first useful to explain the character of the cGMP response to maximal NO stimulation observed in cerebellar cells, where most experiments to date have been carried out. Here, exposure of the cells to NO leads to a hyperbolic increase in cGMP culminating in the establishment of a steady-state cGMP level after around 1 min (Fig. 4a; [45, 57]).

In contrast to ionotropic neurotransmitter receptors, which evoke a change in membrane potential by opening ion-conducting pores, the signal evoked by sGC activation is an accumulation of intracellular cGMP. To achieve this end, sGC must consume GTP – a molecule which constitutes both an intermediate in other metabolic pathways (e.g. RNA synthesis), and important cellular 'currency' in energy utilization (GTP is maintained in equilibrium with ATP). Therefore to sustain a cGMP steady-state signal, it is desirable for the cell to match a low level of cGMP synthesis to a low level of degradation. The problem with this scenario kinetically, is that it would take considerable time to attain an elevated steady-state if sGC and PDE were operating with matched constant low rates of activity. The alternative scenario, where a steady-state is rapidly attained by constant high rates of synthesis and degradation is very costly in terms of GTP consumption. In contrast, given a desensitizing profile of sGC activity, an initially high rate of synthesis will ensure that the cGMP steady-state is reached rapidly, but desensitization means considerably less GTP consumption is required to sustain the steady-state level. These alternative scenarios are illustrated in Fig. 5.

In addition to this energetic benefit, desensitization has important consequences for the diversification of the cGMP response resulting from a given NO signal. This is of particular interest given the apparently limited molecular heterogeneity of the NO receptor. Only 4 possible subunits combinations are so far known, and only two isoforms have been found at the protein level: $\alpha 1\beta 1$ which is widespread and $\alpha 2\beta 1$ which is found in human placenta [58]. Diversity in NO signalling may be provided by the kinetics of the cGMP response.

Diversity is introduced at two levels: firstly desensitization leads to a gradual increase in NO potency with time, and secondly it provides cells with the capacity for varying the profile of longer-term cGMP responses. The former effect is enabled by the enhancement of sGC desensitization by high NO levels. This means that the agonist effects of NO on sGC will be in competition with the inhibitory effects of NO through desensitization of sGC. Therefore, the peak cGMP level will be determined by the NO concentration where activation and desensitization are most closely balanced, corresponding to the peak of a bell-shaped concentration-response curve. This finding can account for the 5- to 10-fold greater potency of NO in mediating both cGMP increases [45] and vasodilatation [59, 60] over time courses of minutes, than its potency for activating the non-desensitizing purified sGC [23, 53]. This mechanism is the first way in which diversity results from sGC desensitization: the sensitivity of sGC to NO will change with time. A short-term (msec) exposure to NO will result in a simple sGC response of increasing magnitude with increasing NO concentration (with a maximum at around 240 nM NO), whereas a long-term exposure (seconds–minutes) will result in an sGC response over a narrower concentration range (max. < 80 nM) and with a lower maximal activity [52].

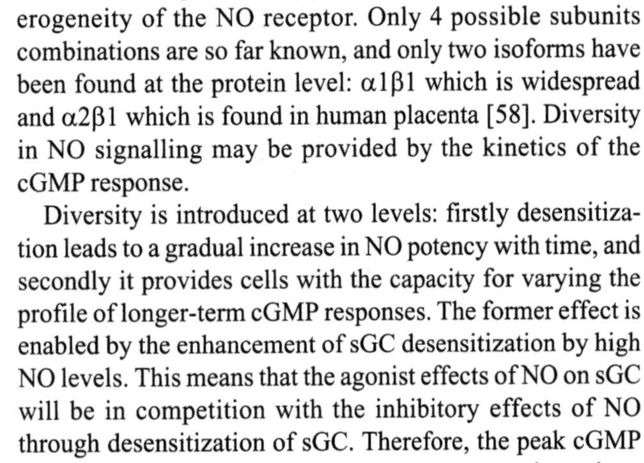

Rate of synthesis	Approx. time to steady-state	GTP consumption at steady state
Constant high	1.5 min	1 pmol/10^6 cells/s
Constant low	15 min	0.1 pmol/10^6 cells/s
Desensitizing	1.5 min	0.1 pmol/10^6 cells/s

Fig. 5. Establishing a cGMP steady-state with a non-desensitizing sGC. With a constant high rate of synthesis (2.75 pmol/10^6 cells/sec), a steady-state as measured experimentally (O) is rapidly attained (a), but energetically costly. In contrast a constant low rate of synthesis (0.275 pmol/10^6 cells/sec) is slow to reach steady-state, but more economical (b). Desensitization combines both benefits.

The second mechanism for introducing diversity results from sGC activity dwindling with time due to desensitization. With a desensitizing profile of cGMP synthesis, variation in the rate of cGMP degradation will alter both the amplitude and temporal characteristics of the resultant cGMP accumulation. This can be demonstrated by modelling an exponentially declining rate of synthesis with variation in the (substrate-linked) PDE activity of cells. This concept is illustrated in Figs 4a–c, where a cGMP accumulation obtained from cerebellar cells is overlaid with measured values for cGMP degradation. By manipulating the maximal rate, or substrate affinity of the cellular PDE, the potential diversity in the shape of the cGMP accumulation is revealed. Depending on the PDE activity of the cell, the kinetic profile of intracellular cGMP will range from low amplitude transients (where high PDE activity overwhelms desensitized sGC activity) to high amplitude plateaux (where low PDE activity matches desensitized sGC activity and cGMP reaches steady-state).

What are the likely consequences of this diversity? It is anticipated that the differential sensitivity of the downstream targets of cGMP (cGK, PDE and CNG channels) offers a means for the cGMP response to select a particular target. This means that diversity in the magnitude of the cGMP response evoked by a common signal (NO) will selectively engage downstream effector mechanisms of varying affinity, provoking different physiological outcomes. In addition to simple amplitude coding, however, desensitization allows versatility in the temporal coding of cGMP signals. This means that particular cGMP targets can be activated for differing periods, and so the duration of a physiological response to prolonged NO exposure can be graded.

The manner of cGMP coding will therefore have profound effects on the 'output' from an NO input, and this translation is largely under the control of sGC activity. A desensitizing profile of activity offers cells the optimum capacity for variety in selection of output from a common input, and so this relatively simple kinetic device empowers the cell to carefully direct physiological events resulting from NO signalling.

The significance of the temporal and amplitude coding of intracellular signals is clear in analogous signalling pathways. In particular the numerous cellular targets for cytosolic Ca^{2+} increases are largely selected based on the spatiotemporal and amplitude characteristics of the Ca^{2+} influx into the cytosol. The variety of Ca^{2+} signal kinetics includes sparks or puffs, oscillations, and intra- and intercellular waves [61]. Elementary events (i.e. sparks) are sufficient to trigger some effects such as neurotransmitter release, whereas oscillations and waves mediate longer-term effects such as activation of transcription factors and control of cell cycle progression.

Mechanisms for modulating cellular sGC

In addition to appreciating the functional consequences of the complex kinetics of cellular sGC, it is desirable to understand the mechanism by which accelerated deactivation and desensitization are achieved. The observation that lysis of cells leads to loss of desensitization implies the existence of a cellular factor, the action of which is lost, possibly by dilution [45].

The literature on modulation of sGC activity in cells by other signalling pathways is not extensive, but a few studies are of particular interest. A protein inhibitor of sGC has been partially purified from bovine lung, but has yet to be characterized in detail [63]. This has potential significance for desensitization, as the peak rate of cGMP synthesis in intact cells is approximately 4-fold less than the rate in lysed cells. The desensitizing factor may thus be an inhibitory factor. Phosphorylation of sGC by PKA and PKC has been reported in homogenate preparations and PC12 cells, to lead to a modest increase in activity [64–66], but the influence of phosphorylation on NO-activated sGC was not apparently investigated. Similarly, ADP-ribosylation [43], and direct interaction with G-proteins linked to β-adrenergic receptors in pinealocytes [67] have also been postulated as regulatory mechanisms. Unfortunately, no evidence of these mechanisms being engaged and modulating sGC function in response to a physiological stimulus has yet been reported, making their relevance in NO signalling difficult to ascertain.

A recent study by Ferraro *et al.* [68] of the NO-cGMP pathway in chromaffin cells reported that sGC exists in a phosphorylated state under normal resting conditions. Elevation of cGMP levels, and activation of cGMP-dependent protein kinase (cGK) led to activation of a phosphatase which dephosphorylated sGC, resulting in a pronounced decrease in NO-evoked cGMP accumulation. This mechanism is of particular interest, as an analogous mechanism (cGK-stimulated dephosphorylation) operates to desensitize membrane-bound isoforms of guanylyl cyclase [69, 70]. Unfortunately the time course of cGMP accumulation in the presence and absence of phosphorylation was not presented, and so the direct relevance to sGC desensitization cannot be easily assessed. Nevertheless, this feedback inhibition is clearly of importance in NO signal transduction, and it will be of interest to know if this mechanism is widespread.

Uncompetitive inhibition of sGC by Ca^{2+} has been reported recently as a mechanism for regulating sGC [71], revisiting a debate that has been persisting since the early investigations of NO signal transduction [72]. The Ca^{2+} concentrations required to elicit half-maximal inhibition of purified sGC are, however, at least an order of magnitude greater than those commonly observed in cells, even when responding to Ca^{2+}-elevating stimuli. A demonstration of Ca^{2+}-influx leading to decreased cGMP accumulation in NO-stimulated HEK-293 cells [71], apparently failed to address the possible involvement of a cGMP-hydrolysing Ca^{2+}-calmodulin sensitive phosphodiesterase (PDE 1) which has a sensitivity to Ca^{2+} in the appropriate range for physiological Ca^{2+} signalling [46].

Long term down-regulation of sGC expression: nitrate tolerance

Prolonged exposure (of the order of hours) of many tissues to NO has been shown to provoke a loss in tissue responsiveness to subsequent NO treatment [73–75]. This phenomenon is of clinical importance as it accounts for so-called 'nitrate tolerance', where nitrovasodilators become progressively less efficacious with time. This has been linked to a decline in sGC levels with long-term exposure to NO donors, apparently due to destabilization of subunit mRNA and/or accelerated turnover of sGC protein [74, 76]. In a similar study, exposure of cultured cerebellar cells to bacterial lipopolysaccharide (LPS) was shown to cause a progressive down-regulation of sGC levels. This effect was NO-independent however, suggesting that expression of sGC is potentially under the control of several transcriptional factors [77].

These effects are sometimes also referred to as desensitization, but should not be confused with sGC desensitization as earlier described. In this circumstance, rather than a functional decrease in activity as a property of the receptor in operation, the absolute levels of the receptor in cells are diminished by long-term NO or LPS treatment. This is an adaptive response to persistent (and potentially deleterious) stimulation rather than a property inherent in the kinetics of cellular sGC. An analogy (though imperfect perhaps) would be the difference between a neurotransmitter receptor entering a desensitized state within milliseconds of agonist binding, and being internalized after prolonged activation.

A shorter-term apparent change in sGC sensitivity has been demonstrated by brief (min) exposure of isolated vessels to NOS inhibitors [78]. This inhibition of basal NO 'tone' resulted in a marked increase in the potency of subsequent NO donor application for provoking relaxation of the vessel. More recent studies with eNOS knockout mice [79] have corroborated this finding (i.e. NO donors have higher potency in knockout compared to wild type mice). No mechanism has yet been proposed for these effects but, again, the findings of Ferraro *et al.* [68] are of possible relevance.

The activity of sGC within a physiological environment is thus under numerous regulatory mechanisms, which operate over a range of time scales. The response of sGC to NO is dynamic and kinetically complex, diversifying cGMP kinetics. Short term alterations in catalytic activity are observed due to basal levels of NO synthesis and phosphorylation state of the receptor. Long term adaptation to several stimuli reflects modulation of sGC gene expression, and/or protein stability, modifying the tissue levels of the receptor. In all these cases, our understanding of the mechanisms underlying these properties is in the early stages. Progress in these areas will be critical to our appreciation of how the enzymological features of purified sGC relate to the observed physiological outcomes of NO signalling.

Acknowledgements

We thank John Wood for valuable discussions on classical receptor models. Work in this laboratory referred to in this review was supported by an E.C. grant (B104-CT98-0034) and The Wellcome Trust.

References

1. Moncada S, Palmer RM, Higgs EA: Nitric oxide: Physiology, pathophysiology, and pharmacology. Pharmacol Rev 43: 109–142, 1991
2. Garthwaite J, Boulton CL: Nitric oxide signaling in the central nervous system. Annu Rev Physiol 57: 683–706, 1995
3. Sanders KM, Ward SM: Nitric oxide as a mediator of nonadrenergic noncholinergic neurotransmission. Am J Physiol 262: G379–G392, 1992
4. Francis SH, Corbin JD: Progress in understanding the mechanism and function of cyclic GMP-dependent protein kinase. Adv Pharmacol 26: 115–170, 1994
5. Zagotta WN, Siegelbaum SA: Structure and function of cyclic nucleotide-gated channels. Annu Rev Neurosci 19: 235–263, 1996
6. Juilfs DM, Soderling S, Burns F, Beavo JA: Cyclic GMP as substrate and regulator of cyclic nucleotide phosphodiesterases (PDEs). Rev Physiol Biochem Pharmacol 135: 67–104, 1999
7. Beavo JA: Cyclic nucleotide phosphodiesterases: Functional implications of multiple isoforms. Physiol Rev 75: 725–748, 1995
8. Garbers DL: Purification of soluble guanylate cyclase from rat lung. J Biol Chem 254: 240–243, 1979
9. Gerzer R, Hofmann F, Bohme E, Ivanova K, Spies C, Schultz G: Purification of soluble guanylate cyclase without loss of stimulation by sodium nitroprusside. Adv Cyclic Nucleotide Res 14: 255–261, 1981
10. Gerzer R, Hofmann F, Schultz G: Purification of a soluble, sodium-nitroprusside-stimulated guanylate cyclase from bovine lung. Eur J Biochem 116: 479–486, 1981
11. Gerzer R, Bohme E, Hofmann F, Schultz G: Soluble guanylate cyclase purified from bovine lung contains heme and copper. FEBS Lett 132: 71–74, 1981
12. Wedel B, Harteneck C, Foerster J, Friebe A, Schultz G, Koesling D: Functional domains of soluble guanylyl cyclase. J Biol Chem 270: 24871–24875, 1995
13. Wedel B, Humbert P, Harteneck C, Foerster J, Malkewitz J, Bohme E, Schultz G, Koesling D: Mutation of His-105 in the beta 1 subunit yields a nitric oxide-insensitive form of soluble guanylyl cyclase. Proc Natl Acad Sci USA 91: 2592–2596, 1994
14. Zhao Y, Marletta MA: Localization of the heme binding region in soluble guanylate cyclase. Biochemistry 36: 15959–15964, 1997
15. Zhao Y, Schelvis JP, Babcock GT, Marletta MA: Identification of histidine 105 in the beta1 subunit of soluble guanylate cyclase as the heme proximal ligand. Biochemistry 37: 4502–4509, 1998
16. Foerster J, Harteneck C, Malkewitz J, Schultz G, Koesling D: A functional heme-binding site of soluble guanylyl cyclase requires intact N-termini of alpha 1 and beta 1 subunits. Eur J Biochem 240: 380–386, 1996
17. Koesling D: Studying the structure and regulation of soluble guanylyl cyclase. Methods 19: 485–493, 1999
18. Sharma VS, Magde D: Activation of soluble guanylate cyclase by carbon monoxide and nitric oxide: A mechanistic model. Methods 19: 494–505, 1999
19. Wolin MS, Wood KS, Ignarro LJ: Guanylate cyclase from bovine lung. A kinetic analysis of the regulation of the purified soluble enzyme by protoporphyrin IX, heme, and nitrosyl-heme. J Biol Chem 257: 13312–13320, 1982
20. Traylor TG, Sharma VS: Why NO? Biochemistry 31: 2847–2849, 1992
21. Dierks EA, Hu S, Vogel KM, Yu AE, Spiro TG, Burstyn JN: Demonstration of the role of scission of the proximal histidine-iron bond in the activation of soluble guanylyl cyclase through metalloporphyrin substitution studies. J Am Chem Soc 119: 7316–7323, 1997
22. Humbert P, Niroomand F, Fischer G, Mayer B, Koesling D, Hinsch KD, Gausepohl H, Frank R, Schultz G, Bohme E: Purification of soluble guanylyl cyclase from bovine lung by a new immunoaffinity chromatographic method. Eur J Biochem 190: 273–278, 1990
23. Stone JR, Marletta MA: Spectral and kinetic studies on the activation of soluble guanylate cyclase by nitric oxide. Biochemistry 35: 1093–1099, 1996
24. Stone JR, Marletta MA: The ferrous heme of soluble guanylate cyclase: Formation of hexacoordinate complexes with carbon monoxide and nitrosomethane. Biochemistry 34: 16397–16403, 1995
25. Makino R, Matsuda H, Obayashi E, Shiro Y, Iizuka T, Hori H: EPR characterization of axial bond in metal center of native and cobalt-substituted guanylate cyclase. J Biol Chem 274: 7714–7723, 1999
26. Zhao Y, Brandish PE, Ballou DP, Marletta MA: A molecular basis for nitric oxide sensing by soluble guanylate cyclase. Proc Natl Acad Sci USA 96: 14753–14758, 1999
27. Colquhoun D: Binding, gating, affinity and efficacy: The interpretation of structure-activity relationships for agonists and of the effects of mutating receptors. Br J Pharmacol 125: 924–947, 1998
28. Kharitonov VG, Sharma VS, Magde D, Koesling D: Kinetics and equilibria of soluble guanylate cyclase ligation by CO: Effect of YC-1. Biochemistry 38: 10699–10706, 1999
29. Kharitonov VG, Sharma VS, Pilz RB, Magde D, Koesling D: Basis of guanylate cyclase activation by carbon monoxide. Proc Natl Acad Sci USA 92: 2568–2571, 1995

30. Lawson DM, Stevenson CE, Andrew CR, Eady RR: Unprecedented proximal binding of nitric oxide to heme: Implications for guanylate cyclase. EMBO J 19: 5661–5671, 2000
31. Kharitonov VG, Sharma VS, Magde D, Koesling D: Kinetics of nitric oxide dissociation from five- and six-coordinate nitrosyl hemes and heme proteins, including soluble guanylate cyclase. Biochemistry 36: 6814–6818, 1997
32. Eich RF, Li T, Lemon DD, Doherty DH, Curry SR, Aitken JF, Mathews AJ, Johnson KA, Smith RD, Phillips GN Jr, Olson JS: Mechanism of NO-induced oxidation of myoglobin and hemoglobin. Biochemistry 35: 6976–6983, 1996
33. Brandish PE, Buechler W, Marletta MA: Regeneration of the ferrous heme of soluble guanylate cyclase from the nitric oxide complex: Acceleration by thiols and oxyhemoglobin. Biochemistry 37: 16898–16907, 1998
34. Kharitonov VG, Russwurm M, Magde D, Sharma VS, Koesling D: Dissociation of nitric oxide from soluble guanylate cyclase. Biochem Biophys Res Commun 239: 284–286, 1997
35. Margulis A, Sitaramayya A: Rate of deactivation of nitric oxide-stimulated soluble guanylate cyclase: Influence of nitric oxide scavengers and calcium. Biochemistry 39: 1034–1039, 2000
36. Senter PD, Eckstein F, Mulsch A, Bohme E: The stereochemical course of the reaction catalyzed by soluble bovine lung guanylate cyclase. J Biol Chem 258: 6741–6745, 1983
37. Katsuki S, Arnold W, Mittal C, Murad F: Stimulation of guanylate cyclase by sodium nitroprusside, nitroglycerin and nitric oxide in various tissue preparations and comparison to the effects of sodium azide and hydroxylamine. J Cyclic Nucleotide Res 3: 23–35, 1977
38. Troyer EW, Hall IA, Ferrendelli JA: Guanylate cyclases in CNS: Enzymatic characteristics of soluble and particulate enzymes from mouse cerebellum and retina. J Neurochem 31: 825–833, 1978
39. Stone JR, Marletta MA: Soluble guanylate cyclase from bovine lung: Activation with nitric oxide and carbon monoxide and spectral characterization of the ferrous and ferric states. Biochemistry 33: 5636–5640, 1994
40. Lewicki JA, Brandwein HJ, Mittal CK, Arnold WP, Murad F: Properties of purified soluble guanylate cyclase activated by nitric oxide and sodium nitroprusside. J Cyclic Nucleotide Res 8: 17–25, 1982
41. Denninger JW, Schelvis JP, Brandish PE, Zhao Y, Babcock GT, Marletta MA: Interaction of soluble guanylate cyclase with YC-1: Kinetic and resonance Raman studies. Biochemistry 39: 4191–4198, 2000
42. Tomita T, Ogura T, Tsuyama S, Imai Y, Kitagawa T: Effects of GTP on bound nitric oxide of soluble guanylate cyclase probed by resonance Raman spectroscopy. Biochemistry 36: 10155–10160, 1997
43. Tomita T, Tsuyama S, Imai Y, Kitagawa T: Purification of bovine soluble guanylate cyclase and ADP-ribosylation on its small subunit by bacterial toxins. J Biochem Tokyo 122: 531–536, 1997
44. Stone JR, Marletta MA: Heme stoichiometry of heterodimeric soluble guanylate cyclase. Biochemistry 34: 14668–14674, 1995
45. Bellamy TC, Wood J, Goodwin DA, Garthwaite J: Rapid desensitization of the nitric oxide receptor, soluble guanylyl cyclase, underlies diversity of cellular cGMP responses. Proc Natl Acad Sci USA 97: 2928–2933, 2000
46. Mayer B, Klatt P, Bohme E, Schmidt K: Regulation of neuronal nitric oxide and cyclic GMP formation by Ca^{2+}. J Neurochem 59: 2024–2029, 1992
47. Beavo JA, Houslay MD: Cyclic Nucleotide Phosphodiesterases: Structure, Regulation, and Drug Action. Wiley, New York, 1990
48. Bellamy TC, Garthwaite J: 'cAMP-specific' phosphodiesterase contributes to cGMP degradation in cerebellar cells exposed to nitric oxide. Mol Pharmacol 59: 54–61, 2001
49. Lee YC, Martin E, Murad F: Human recombinant soluble guanylyl cyclase: Expression, purification, and regulation. Proc Natl Acad Sci USA 97: 10763–10768, 2000
50. Jedlitschky G, Burchell B, Keppler D: The multidrug resistance protein 5 functions as an ATP-dependent export pump for cyclic nucleotides. J Biol Chem 275: 30069–30074, 2000
51. Jones MV, Westbrook GL: The impact of receptor desensitization on fast synaptic transmission. Trends Neurosci 19: 96–101, 1996
52. Bellamy TC, Garthwaite J: Sub-second kinetics of the nitric oxide receptor, soluble guanylyl cyclase, in intact cerebellar cells. J Biol Chem 276: 4287–4292, 2001
53. Schrammel A, Behrends S, Schmidt K, Koesling D, Mayer B: Characterization of 1H-[1,2,4]oxadiazolo[4,3-a]quinoxalin-1-one as a heme-site inhibitor of nitric oxide-sensitive guanylyl cyclase. Mol Pharmacol 50: 1–5, 1996
54. Schmidt K, Desch W, Klatt P, Kukovetz WR, Mayer B: Release of nitric oxide from donors with known half-life: A mathematical model for calculating nitric oxide concentrations in aerobic solutions. Naunyn Schmiedebergs Arch Pharmacol 355: 457–462, 1997
55. Jones MV, Westbrook GL: Desensitized states prolong GABAA channel responses to brief agonist pulses. Neuron 15: 181–191, 1995
56. Wood J, Garthwaite J: Models of the diffusional spread of nitric oxide: Implications for neural nitric oxide signalling and its pharmacological properties. Neuropharmacology 33: 1235–1244, 1994
57. Garthwaite J, Charles SL, Chess Williams R: Endothelium-derived relaxing factor release on activation of NMDA receptors suggests role as intercellular messenger in the brain. Nature 336: 385–388, 1988
58. Russwurm M, Behrends S, Harteneck C, Koesling D: Functional properties of a naturally occurring isoform of soluble guanylyl cyclase. Biochem J 335: 125–130, 1998
59. Carter TD, Bettache N, Ogden D: Potency and kinetics of nitric oxide-mediated vascular smooth muscle relaxation determined with flash photolysis of ruthenium nitrosyl chlorides. Br J Pharmacol 122: 971–973, 1997
60. Kelm M, Feelisch M, Spahr R, Piper H-M, Noack E, Schrader J: Quantitative and kinetic characterization of nitric oxide and EDRF released from cultured endothelial cells. Biochem Biophys Res Commun 154: 236–244, 1988
61. Berridge MJ, Bootman MD, Lipp P: Calcium – a life and death signal. Nature 395: 645–648, 1998
62. Hollmann M, Heinemann S: Cloned glutamate receptors. Annu Rev Neurosci 17: 31–108, 1994
63. Kim TD, Burstyn JN: Identification and partial purification of an endogenous inhibitor of soluble guanylyl cyclase from bovine lung. J Biol Chem 269: 15540–15545, 1994
64. Zwiller J, Revel MO, Malviya AN: Protein kinase C catalyzes phosphorylation of guanylate cyclase *in vitro*. J Biol Chem 260: 1350–1353, 1985
65. Zwiller J, Revel MO, Basset P: Evidence for phosphorylation of rat brain guanylate cyclase by cyclic AMP-dependent protein kinase. Biochem Biophys Res Commun 101: 1381–1387, 1981
66. Louis JC, Revel MO, Zwiller J: Activation of soluble guanylate cyclase through phosphorylation by protein kinase C in intact PC12 cells. Biochim Biophys Acta 1177: 299–306, 1993
67. White BH, Klein DC: Stimulation of cyclic GMP accumulation by sodium nitroprusside is potentiated via a Gs mechanism in intact pinealocytes. J Neurochem 64: 711–717, 1995
68. Ferrero R, Rodriguez-Pascual F, Miras-Portugal MT, Torres M: Nitric oxide-sensitive guanylyl cyclase activity inhibition through

cyclic GMP-dependent dephosphorylation. J Neurochem 75: 2029–2039, 2000

69. Potter LR, Garbers DL: Protein kinase C-dependent desensitization of the atrial natriuretic peptide receptor is mediated by dephosphorylation. J Biol Chem 269: 14636–14642, 1994
70. Potter LR: Phosphorylation-dependent regulation of the guanylyl cyclase-linked natriuretic peptide receptor B: Dephosphorylation is a mechanism of desensitization. Biochemistry 37: 2422–2429, 1998
71. Parkinson SJ, Jovanovic A, Jovanovic S, Wagner F, Terzic A, Waldman SA: Regulation of nitric oxide-responsive recombinant soluble guanylyl cyclase by calcium. Biochemistry 38: 6441–6448, 1999
72. Knowles RG, Palacios M, Palmer RM, Moncada S: Formation of nitric oxide from L-arginine in the central nervous system: A transduction mechanism for stimulation of the soluble guanylate cyclase. Proc Natl Acad Sci USA 86: 5159–5162, 1989
73. Waldman SA, Rapoport RM, Ginsburg R, Murad F: Desensitization to nitroglycerin in vascular smooth muscle from rat and human. Biochem Pharmacol 35: 3525–3531, 1986
74. Filippov G, Bloch DB, Bloch KD: Nitric oxide decreases stability of mRNAs encoding soluble guanylate cyclase subunits in rat pulmonary artery smooth muscle cells. J Clin Invest 100: 942–948, 1997
75. Ujiie K, Hogarth L, Danziger R, Drewett JG, Yuen PS, Pang IH, Star RA: Homologous and heterologous desensitization of a guanylyl cyclase-linked nitric oxide receptor in cultured rat medullary interstitial cells. J Pharmacol Exp Ther 270: 761–767, 1994
76. Scott WS, Nakayama DK: Sustained nitric oxide exposure decreases soluble guanylate cyclase mRNA and enzyme activity in pulmonary artery smooth muscle. J Surg Res 79: 66–70, 1998
77. Baltrons MA, Garcia A: Nitric oxide-independent down-regulation of soluble guanylyl cyclase by bacterial endotoxin in astroglial cells. J Neurochem 73: 2149–2157, 1999
78. Moncada S, Rees DD, Schultz R, Palmer RMJ: Development and mechanism of a specific supersensitivity to nitrovasodilators after inhibition of vascular nitric oxide synthesis *in vivo*. Proc Natl Acad Sci USA 88: 2166–2170, 1991
79. Hussain MB, Hobbs AJ, MacAllister RJ: Autoregulation of nitric oxide-soluble guanylate cyclase-cyclic GMP signalling in mouse thoracic aorta. Br J Pharmacol 128: 1082–1088, 1999

Molecular and Cellular Biochemistry **230:** 177–186, 2002.

Soluble guanylate cyclases in the retina

Ari Sitaramayya
Eye Research Institute, Oakland University, Rochester, MI, USA

Abstract

Soluble guanylate cyclase catalyzes the formation of cyclic GMP using GTP as substrate. It is now well established that soluble guanylate cyclase is highly activated by nitric oxide, and that many of the effects of nitric oxide on various cells and tissues are mediated through increased production of cyclic GMP. This review discusses the evidence for the presence of soluble guanylate cyclases in different classes of cells in vertebrate retina and the role of these enzymes in retinal physiology. It is concluded that the enzyme is present in nearly every class of cells in the retina and that it may be involved in signal transmission between some cells and in the modulation of signal transmission between others. (Mol Cell Biochem **230**: 177–186, 2002)

Key words: retina, soluble guanylate cyclase, photoreceptor cells, horizontal cells, bipolar cells, amacrine cells, Müller cells, rods, cones, inner plexiform layer, outer plexiform layer, GABA receptor, nitric oxide, cyclic GMP, ON-bipolar cell, OFF-bipolar cell, ON-ganglion cell, OFF-ganglion cell, gap junctions, connexins, glutamate receptors, adaptation, retinal network, glutamate toxicity

Guanylate (or guanylyl) cyclases catalyze the formation of cyclic GMP using GTP as substrate. Two types of guanylate cyclases are known: the membrane-associated forms regulated by peptide hormones or calcium-binding proteins, and the soluble forms activated by nitric oxide [1–5]. Over the last three decades investigations on the guanylate cyclases of retina have focused mainly on the membrane-associated enzymes, particularly those of the photoreceptors [4, 5]. These studies have revealed that membrane-associated guanylate cyclases play a pivotal role in phototransduction, the process in which photoreceptor cells detect changes in light intensity and transmit the information to second order neurons. Evidence emerging from more recent studies suggests that soluble guanylate cyclases are also present in retina, in many cell types, where they may play important roles in signal transduction pathways and in the adaptation of retina to ambient light.

Soluble guanylate cyclases are heterodimeric enzymes consisting of an alpha and a beta subunit [6]. Both subunits are essential for the catalytic activity [7]. The beta subunit contains a heme moiety to which nitric oxide binds and stimulates the enzyme [8, 9]. Multiple forms of both alpha and beta subunits are known, and the nature of the beta subunit appears to determine the level to which the enzyme is activated by nitric oxide [10, 11]. In addition, since nitric oxide is membrane permeant, it can diffuse from the site of production to distant cells and activate cyclase, though the distance traveled may be limited in some cases by unknown factors [12–15]. Soluble guanylate cyclase remains activated as long as nitric oxide is bound, returning to basal activity following its dissociation [16–20].

Evidence for the presence of soluble guanylate cyclases in retina comes from six different types of studies: biochemical measurement of enzyme activity in the soluble fraction of retinal homogenate or the homogenate of subretinal preparations; histochemical detection of larger amounts of cyclic GMP in sections of retina treated with nitric oxide-releasing chemicals; detection of soluble guanylate cyclase subunits in retinal sections using specific antibodies; observation of the expression of soluble guanylate cyclase mRNA in retinal sections with the aid of specific probes; studies on the effects of cyclic GMP and nitric oxide on the spread of intracellularly injected tracers to neighboring cells coupled by gap junctions; and electrophysiological studies in which ion channel activity in a given cell or intercellular coupling between cells was shown to be influenced similarly by cyclic GMP and nitric oxide-releasing chemicals. Though the effects of nitric oxide are not always mediated by cyclic GMP, simulation of its effects on a cell by cyclic GMP is indicative of the presence in that cell of nitric oxide-

Address for offprints: A. Sitaramayya, Eye Research Institute, Oakland University, Rochester, MI 48309-4480, USA stimulated soluble guanylate cyclase. Successful blockade of the effect of nitric oxide by an inhibitor of soluble guanylate cyclase provides additional support for the expression of soluble guanylate cyclase. Though it may be argued that a nitric oxide-inhibited phosphodiesterase could also elevate the concentration of cyclic GMP and produce effects similar to those of nitric oxide, there is no evidence that any of the known phosphodiesterases is affected by nitric oxide. For a comprehensive review on nitric oxide in retina, see Eldred [21].

The purpose of this article is to review the evidence for the presence of soluble guanylate cyclases in the different classes of cells in vertebrate retina and for their potential roles in retinal physiology.

Figure 1 shows a highly simplified depiction of the various classes of retinal neurons: the photoreceptor, horizontal, bipolar, amacrine and ganglion cells. Several different types of cells are known in each class based on their structure and or function of which only a few are shown here (for a detailed description of the different types of retinal cells, see [22, 23]). The outermost layer of retina contains rods and cones, the photoreceptor cells that detect light. Rods, being highly light sensitive, respond to changes in light intensity in dim ambient light (starlight to twilight) and cones, which are less sensitive, function in brighter ambient light (twilight to daylight) [24]. Both types of photoreceptor cells communicate with second order neurons in the outer plexiform layer: cones with the OFF- and ON-bipolar cells, and rods with the rod bipolar cells. Cone bipolar cells form synapses with ganglion cells in the inner plexiform layer; OFF- and ON-bipolar cells communicate with OFF- and ON-ganglion cells, respectively. For rod bipolar cells in mammalian retina, however, the third order neurons are amacrine cells, with which they communicate in the inner plexiform layer [25–27]. Rod amacrine cells, also called AII amacrine cells, in turn network with each other and pass on the visual information to the cone OFF- and ON-bipolar cells, also in the inner plexiform layer [28]. The AII amacrine cells communicate with the ON-

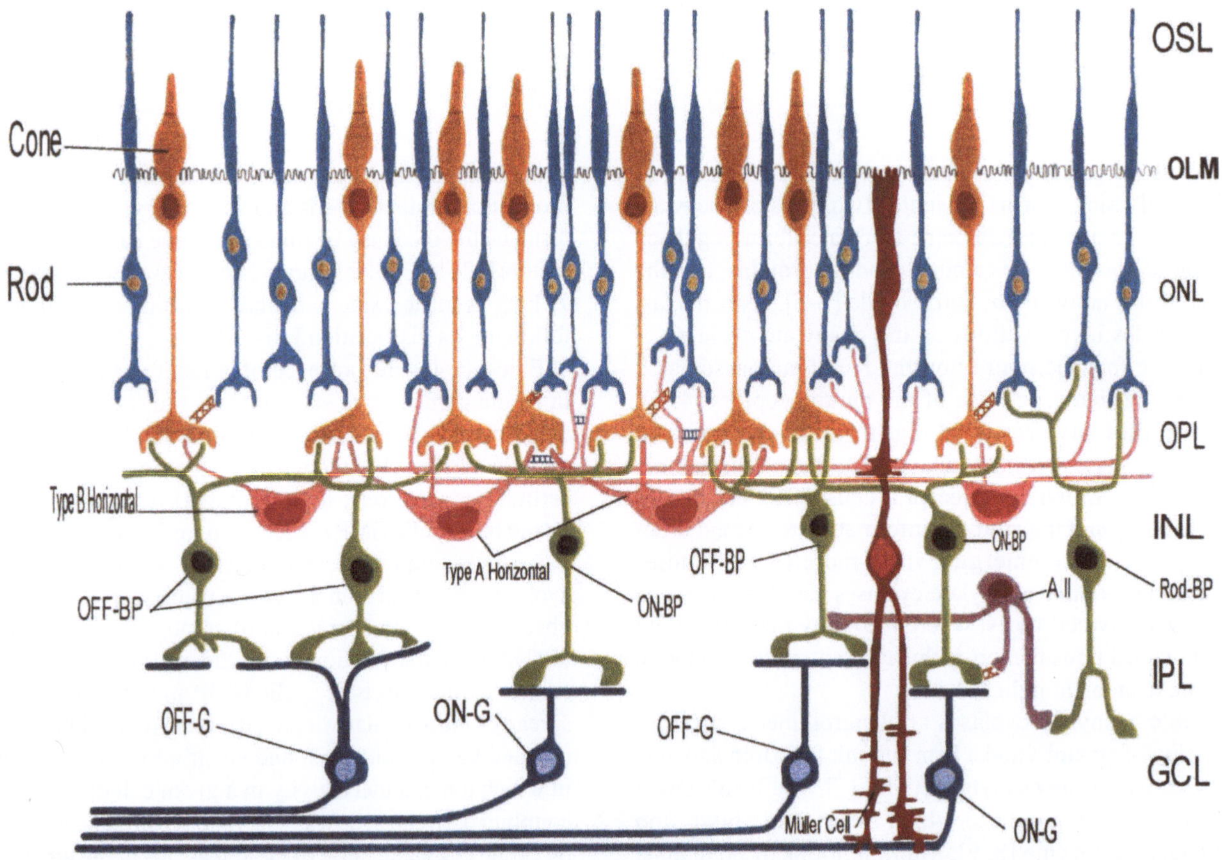

Fig. 1. Schematic representation of some of the retinal cells. AII – AII amacrine cell; ON-BP – ON-bipolar cell; OFF-BP – OFF-bipolar cell; Rod-BP – rod bipolar cell; ON-G – ON-ganglion cell; OFF-G – OFF-ganglion cell; OSL – outer segment layer; OLM – outer limiting membrane; ONL – outer nuclear layer; OPL – outer plexiform layer; INL – inner nuclear layer; IPL – inner plexiform layer; GCL – ganglion cell layer. Homotypic and heterotypic gap junctions are shown by two lines connected by parallel or crossed hatches, respectively.

bipolar cells through gap junctions and with the OFF-bipolar cells through inhibitory chemical synapses [25, 26, 29–31]. In starlight, visual information from rods is carried via AII amacrine cells to the cone pathway [24, 28]. Gap junctions between the amacrine cells permit a wider receptive area and a reliable collection of scant visual information. Under twilight conditions when more rods are visually stimulated, the AII amacrine cell-cone bipolar cell gap junctions are disabled, and the gap junctions between rods and cones permit transmission of visual information through the cone pathway [24, 32]. In addition to the above neurons, retina also contains horizontal cells that lie proximal to the outer plexiform layer. These cells receive synaptic inputs from the photoreceptor cells as well as provide them with feedback inhibition. Horizontal cell feedback increases the range of light intensity to which the photoreceptors can respond [33]. Retina also contains Müller cells, glial cells that traverse nearly the entire width of the retina.

Photoreceptor cells

The soluble guanylate cyclases account for nearly 20% of the total guanylate cyclase activity of the mouse retina and about 15% of the total activity of the bovine retina [34, 35]. So far it has not been possible to isolate a pure population of any cell type from any retina in quantities amenable to biochemical analysis. The quantitative distribution of soluble guanylate cyclase activity between the different cell types, therefore, remains undetermined. However, certain genetic diseases leave the retina essentially devoid of photoreceptor cells, thus permitting a comparison of cyclase activity of the whole retina with that of the photoreceptor-less retina. Such a comparison of mouse retinas suggests a polarized distribution of membrane and soluble cyclases: 78% of the total retinal guanylate cyclase activity is in the photoreceptor cells of which 92% is due to the membrane-associated enzyme; in contrast, of the 22% of total cyclase activity in the inner retina, 82% is due to the soluble guanylate cyclase [34].

Although the soluble enzyme accounts for about 8% of the total guanylate cyclase activity of the photoreceptor cells according to the measurements cited above, it accounts for only about 3% of the total activity of purified preparations of the outer segments [35], indicating that a higher proportion must be present in other parts of the photoreceptor cell.

Subunits of soluble guanylate cyclase have been detected in the cell bodies of the photoreceptor cells [36–38]. However, nitric oxide-stimulated cyclic GMP synthesis has not been observed in the photoreceptor cell bodies in the turtle and rat retinas [39, 40]. Quite to the contrary, inhibition of nitric oxide synthesis in bovine retina enhanced the cyclic GMP immunoreactivity in the cell bodies, inner segments and the synapses of the photoreceptor cells [41]. The reason for this unexpected observation remains unknown.

What is the role of the soluble guanylate cyclase in the photoreceptor cells? Schmidt and colleagues observed that intracellular injection of L-arginine accelerated the recovery phase of the light response in frog photoreceptor outer segments [42, 43]. Since L-arginine is a substrate in the synthesis of nitric oxide and the recovery phase of light response depends, in part, upon the synthesis of cyclic GMP, the above observation suggests the presence of nitric oxide synthase and nitric oxide-stimulated soluble guanylate cyclase in the outer segments. In keeping with this observation, application of L-arginine analogs L-NMMA and L-NNA, which are competitive inhibitors of nitric oxide synthase, produced the opposite result. The physiological significance of this observation is, however, unclear. It is well established that calcium concentration goes down in the outer segments in response to light, which is unfavorable to the calcium-dependent synthesis of nitric oxide [44–48]. In addition, it is generally thought that membrane-associated cyclases, not the soluble ones, participate in the light response, an argument strongly supported by their high concentration in the photoreceptor outer segments and their activation as calcium concentration goes down following light activation [34, 49, 50]. Further, Miyachi *et al.* did not observe any influence of sodium nitroprusside, a nitric oxide donor, or L-arginine on the light response of frog rods [51]. The role of soluble cyclases in the outer segments, if any, is thus, yet to be determined.

In contrast, evidence has been accumulating in support of a role for soluble cyclases in the modulation of neurotransmitter release from the synapses of cone photoreceptors. S-nitrosocysteine, a nitric oxide donor, and pCPT-cGMP, a cyclic GMP analog, were shown to cause a similar increase in glutamate release from the cone synapses [52]. Also, LY83583, an inhibitor of soluble guanylate cyclase, blocked the influence of S-nitrosocysteine. Cones communicate with bipolar and horizontal cells, and horizontal cells, in turn, are known to exert feedback inhibition on the cone terminals (reviewed in [33]). Such inhibition could be mediated by the release of nitric oxide from the horizontal cell [15], which would activate soluble guanylate cyclase in the cone terminal, increase the cyclic GMP concentration and accelerate the release of the neurotransmitter. Since light activation of cones results in a decrease of the transmitter release, increased transmitter release is equivalent to inhibition. In this process the role of cyclic GMP appears to be in the gating of non-specific cation channels permitting calcium influx. Rieke and Schwartz have shown earlier that such a calcium influx via cyclic GMP-gated channels supports exocytosis of the neurotransmitter from the cone terminals [53].

Horizontal cells

Horizontal cells receive synaptic input from photoreceptor cells. In mammals horizontal cells are usually of two types: the axon-less type A (HII in primates), and the type B (HI in primates) with slender axons, about 300 μm long, which end in highly branched processes. The dendrites of both type A and type B cells receive synaptic inputs from cones while the axonal processes of type B cells receive inputs from rods. Within each cell type, many cells are connected to each other via gap junctions; gap junctions also join the arborized axonal processes of many type B cells [54–57].

In both mammals and lower vertebrates, the gap junction-coupled network of horizontal cells is much wider in dim light, but greatly reduced in total darkness and bright ambient light [58, 59]. This network is known to be regulated by dopamine through the mediation of cyclic AMP and cyclic AMP-dependent protein kinase [60–65]. Soluble guanylate cyclases also have a role in regulating the gap junction coupling between horizontal cells. De Vries and Schwartz observed that ionic currents passing through gap junction channels of isolated catfish horizontal cells were greatly reduced upon intracellular injection of cyclic GMP or superfusion with sodium nitroprusside, suggesting that the cells contain nitric oxide-stimulated soluble guanylate cyclase [64]. Miyachi and colleagues made a similar observation on horizontal cells of the turtle retina [51, 66]. In addition, they noted that injection of L-arginine into one cell of a pair also reduced the coupling between the cells, an indication that the cells produce nitric oxide [15], as well as harbor nitric oxide-sensitive soluble cyclase. More recent studies have demonstrated that cyclic GMP controls gap junction communication in both type A and type B horizontal cells [67]. Injection of L-arginine into a type A cell of rabbit retina reduced the coupling with neighboring cells, suggesting that these cells produce nitric oxide from L-arginine and contain nitric oxide-stimulated soluble guanylate cyclase. In contrast, the type B cells did not respond reliably to L-arginine. They did, however, uncouple in response to sodium nitroprusside indicating that they also contain soluble guanylate cyclase, though they may not contain nitric oxide synthase.

Regulation of gap junctions between horizontal cells may be important in the adaptation of retina to ambient light. In dim light, the wide receptive field of the gap junction-coupled cone horizontal cells permits collection of visual information from many cones. In bright light, nitric oxide- and cyclic GMP-mediated uncoupling of gap junctions would reduce the area of the receptive field and favor visual acuity, a feature desirable in bright light. Activation of soluble guanylate cyclase is, thus, likely to be a part of the adaptation of retinal network to bright light [68–70]. Gap junctions between the axonal processes of type B horizontal cells, which make contacts with rods, are not modulated by nitric oxide or cyclic GMP [71], possibly because these processes lack soluble guanylate cyclase, though, as mentioned above, dendrites of the type B cells do contain the enzyme.

The precise mechanism by which cyclic GMP controls gap junction communication between horizontal cells is unknown. However, in other tissues such as the rat heart, cyclic GMP regulates gap junction channels by stimulating the phosphorylation of gap junction proteins, called connexins [72–74]. Such connexins, subject to phosphorylation in a cyclic GMP-dependent manner (or cyclic AMP-dependent manner), have not yet been identified in the retina of any species, though candidate proteins have been investigated in horizontal cell-enriched preparations [75, 76].

In spite of the suggestion in the electrophysiological studies, immunohistochemical investigations have generally not found evidence for soluble cyclase in horizontal cells; nitric oxide-stimulated increase in cyclic GMP immunoreactivity was not observed in the horizontal cells of bovine, rat or turtle retina [39–41]. Neither was immunoreactivity found when tested with an antibody against the $\alpha 1$ subunit of soluble cyclase [37]. Mills *et al.*, however, noted cyclic GMP immunoreactivity in type A horizontal cells of rabbit retina incubated with sodium nitroprusside [67]. Given the evidence for soluble cyclase in the electrophysiological studies and the uncertainty of its presence as gleaned from the immunohistochemical studies, it is likely that the enzyme is present at too low a concentration to be reliably detected with the use of antibodies.

In addition to regulating horizontal cell network, soluble cyclases may also play a role in influencing the responsiveness of horizontal cells to stimulation by glutamate. McMahon and Ponomareva reported that application of nitric oxide or 8-bromo-cyclic GMP reduced by 50% the responsiveness of isolated fish horizontal cells to glutamate or its agonist kainate [77]. Guanylate cyclase inhibitor LY83583 blocked this effect. Inhibition of cyclic GMP-dependent protein kinase with a peptide inhibitor also blocked the effect of cyclic GMP indicating possible receptor phosphorylation. It was suggested that the reduced responsiveness of the receptors, which are continually exposed to glutamate released by the photoreceptors in the dark, might protect the horizontal cells from excitotoxicity. However, Pottek *et al.*, recording from horizontal cells in perfused whole retina, observed that nitric oxide and cyclic GMP actually increased the light responsiveness of horizontal cells [69]. Though this appears contrary to the results of McMahon and Ponomareva, Pottek *et al.* concluded that the increased responsiveness was likely due to effects on the cone synaptic output rather than changes in the properties of the receptors on the horizontal cells. More recent observations by McMahon and Schmidt suggest that nitric oxide has dual effects on glutamate receptors in isolated fish horizontal cells: a reduction in the affinity for the neurotransmitter mediated by cyclic GMP, and an increase in the maximum current carried by the activated receptors [78]. The latter effect was not mimicked by cyclic GMP, and therefore not mediated by soluble guanylate cyclase. From these observations, it appears that the positive effect of cyclic GMP on horizontal cell response in the whole retina might be due to effects on the cone synaptic output.

Bipolar cells

There are three main types of bipolar cells (reviewed in [23, 33]): the OFF- and ON-cone bipolar cells, which receive inputs from cones, and the rod bipolar cells, which receive inputs from rods. OFF-bipolar cells hyperpolarize in response to an increase in the intensity of light falling on the cone they synapse with, while the ON-bipolar and the rod-bipolar cells depolarize. In the inner plexiform layer, the OFF- and ON-bipolar cells make contact with OFF- and ON-ganglion cells, respectively, while the rod bipolar cells contact the rod amacrine cells (AII amacrine cells). In turn, the AII amacrine cells make gap junction contacts with ON-bipolar cells and inhibitory synaptic contacts with OFF-bipolar cells.

There is electrophysiological evidence for the presence of soluble guanylate cyclase in the ON-bipolar cells of fish retina [79]. These cells are activated in the dark by glutamate released from the cone terminals [80, 81]. Metabotropic glutamate receptors on the ON-bipolar cells are linked to a G-protein mediated cascade of reactions leading to a reduction in cyclic GMP-gated cation channel activity [82–88]. Shiells and Falk showed that infusion of the ON-bipolar cells with sodium nitroprusside increased the inward current, that direct injection of cyclic GMP into the cell mimicked this response, and that inhibition of soluble guanylate cyclase by infusion of methylene blue had the opposite effect [79]. These observations indicate that soluble guanylate cyclase produces cyclic GMP in the ON-bipolar cells and that it may be involved in transducing the signal from the photoreceptor cells.

There is also immunohistochemical evidence for the presence of soluble cyclase in the ON-bipolar cells. Incubation of turtle, rat and bovine retinas with nitric oxide donors revealed cyclic GMP immunoreactivity in bipolar cells and in their processes terminating in the inner part of the inner plexiform layer where ON-bipolar cells make contacts with ganglion cells [39–41]. Immunoreactivity was also observed in bipolar cells whose processes migrated to the outer portion of the inner plexiform layer, the area where OFF-bipolar cells usually make synaptic contacts [39, 41].

Rod bipolar cells were also cyclic GMP positive after incubation of rat retina with a nitric oxide donor [40]. Since rod bipolar cells are similar to ON-cone bipolar cells in that both have metabotropic glutamate receptors, both depolarize in response to an increase in the light intensity, and hyperpolarize in response to a decrease [89–94], it is likely that soluble guanylate cyclase serves in a similar role in both cells. It should be noted, however, that in two other studies on rabbit and rat retinas, treatment with nitric oxide donors did not increase cyclic GMP immunoreactivity in the rod bipolar cells [41, 95].

In addition to the role in glutamate receptor-mediated signaling, soluble cyclase may also be involved in the regulation of information flow between ON-cone bipolar and AII amacrine cells. It appears that these two cells are coupled by heterotypic gap junctions, i.e. the connexins making up the half channel from the side of the bipolar cell are different from the connexins from the side of the amacrine cell and, thus, subject to potential, differential regulation. In the mammalian retina, neurobiotin, injected into an AII amacrine cell, diffuses into cone bipolar cells through these gap junctions [96]. Nitric oxide and cyclic GMP can block this flow, apparently from the side of the bipolar cell. Such a block would prevent dissipation of signals from the cone pathway, active in daylight, by preventing diversion into the inactive rod pathway. However, control of information flow between the cone bipolar and AII amacrine cells appears to be more complex: Xin and Bloomfield noted that currents do flow from the cone bipolar cell into the AII amacrine cell in a few minutes after light adaptation of rabbit retina, i.e. gap junctions become permeable in the direction of ON-cone bipolar to AII amacrine cell [97]. Nitric oxide and cyclic GMP, however, inhibited the ion flow through the junctions. It remains to be determined what precise sequence of events triggers the stimulation of the soluble cyclase in the bipolar cells leading to a reduction in the gap junction communication with the AII amacrine cells.

Amacrine cells

Immunohistochemical investigations revealed that soluble cyclase is present in some amacrine cells of bovine, rabbit, rat and turtle retinas [37, 39–41, 98]. Based on the morphology, the area of the inner plexiform layer where their processes migrate, and comparison with previous reports, some of the cyclase-positive cells were identified as putative cholinergic, glycinergic and GABAergic amacrine cells [39].

In electrophysiological studies on amacrine cells isolated by papain digestion of rat retina, nitric oxide and cyclic GMP were found to depress GABA receptor function indicating that some, if not all, of the amacrine cells responding to GABA have soluble guanylate cyclase [99]. The physiological relevance of this effect is unknown.

AII amacrine cells, which pass the rod bipolar cell signals on to the cone bipolar cells, also couple to each other via gap junctions, similar to horizontal cells in the outer retina [25, 26]. However, unlike the horizontal cell network, which is regulated by both cyclic GMP and cyclic AMP, the AII cell network is uncoupled only by cyclic AMP [60, 62, 63, 68, 96, 97, 100, 101]. Light adaptation does regulate the coupling between AII amacrine cells [102], but it is apparently not mediated by soluble guanylate cyclase.

Ganglion cells

Like the amacrine cells there are many types of morphologically distinct ganglion cells in the vertebrate retina. Immunohistochemical evidence shows that at least some of the ganglion cells contain soluble guanylate cyclase [37, 39, 41].

There appear to be species-related differences in the role of soluble cyclases in the ganglion cells. In electrophysiological studies on ganglion cells of rat retina, Ahmad *et al.* and Kawai and Sterling observed non-specific cation channels gated by cyclic GMP [103, 104]. The activity of the channels increased in response to activation of soluble guanylate cyclase by nitric oxide donors and it was suggested that the neighboring amacrine cells could be the source of nitric oxide stimulating the ganglion cell soluble cyclase *in vivo*. Interestingly such cyclic GMP-gated cation channels were not found in the isolated ganglion cells of salamander retina, which, however, contained voltage-gated calcium channels whose activation is also facilitated by nitric oxide and cyclic GMP, thus, indicating a role for soluble guanylate cyclase [105].

Müller cells

There is electrophysiological and immunohistochemical evidence for the presence of soluble guanylate cyclase in the Müller cells. Kusaka *et al.* observed that perfusion of cyclic GMP into isolated bovine and human Müller cells induced an inward current, which was blocked by cadmium just like the current carried by non-selective cyclic GMP-gated cation channels [106]. Perfusion of cells with membrane permeant cyclic GMP analog 8-bromo-cGMP or with the nitric oxide donor sodium nitroprusside produced results similar to those of internally perfused cyclic GMP, suggesting that Müller cells contain soluble guanylate cyclase. There is also evidence that Müller cells contain nitric oxide synthase [15, 107–109]. Endogenously produced nitric oxide could, therefore, stimulate soluble cyclase and elevate cyclic GMP concentration. Photomicrographs of cyclic GMP immunoreactivity in rat retina following sodium nitroprusside treatment show prominently stained processes traversing the whole length of the retina, most likely attributable to Müller cells [110].

Kusaka *et al.* suggest that nitric oxide- and cyclic GMP-activated cation channels in Müller cells might have a role in K^+ homeostasis in the retina [106]. Since the cells extend nearly through the whole length of the retina, and the cation channels are non-selective, Müller cells can move K^+ from areas of high concentration to areas of low concentration. In addition to permitting K^+ to pass through, the non-specific channels also carry calcium, which, in turn, could activate K^+ channels.

Inner plexiform layer

Though the discussion hitherto was on individual cell types in retina, the inner plexiform layer where the bipolar, amacrine and ganglion cells make synaptic contacts deserves special attention because soluble cyclase is highly concentrated in that region and the significance of the concentration remains unknown. Nearly 20 years ago Berger *et al.* measured guanylate cyclase activity in microdissected sections of frozen rabbit retina and discovered that the highest specific activity of the enzyme is in the inner plexiform layer, nearly 6 times higher than that of the outer plexiform layer and 2–3 times higher than that of any other retinal layer [111]. Recent biochemical investigations on isolated synaptosomal fractions of bovine retina showed similar results: the specific activity of the enzyme in the synaptosomal fraction derived from the inner plexiform layer was about 4-fold higher than that of the fraction from the outer plexiform layer or the cytosol [112]. It should be noted that nearly 90% of the retinal soluble cyclase activity was in the cytosol, though the specific activity was much higher in the fraction derived from the inner plexiform layer. Studies by Blute *et al.* also revealed intense cyclic GMP immunoreactivity in the inner plexiform layer of turtle retina following stimulation by nitric oxide donors or glutamate agonists [39, 113].

In both turtle and rat retina, cyclic GMP synthesized in response to nitric oxide was found stratified in distinct layers in the ON- (inner two-thirds) and OFF- (outer third) sublamina of the inner plexiform layer [39, 40]. The processes forming these strata appeared to come mainly from a variety of amacrine cells in the turtle retina, and from bipolar cells in the rat retina, though cyclic GMP-positive processes of bipolar, amacrine and ganglion cells were found in both. In the turtle retina, cyclic GMP immunoreactivity was more intense in the OFF-sublamina, and was probably contributed by both amacrine cells whose processes arborized here as well as OFF-bipolar cells. However, a role for cyclic GMP in the OFF-pathway remains to be established. In the rat retina, processes of rod bipolar cells migrating to the vitreal border of the inner plexiform layer were found to produce cyclic GMP in response to nitric oxide. Here again, a role for cyclic GMP in the synaptic terminal of rod bipolar cells has yet to be determined.

In both rat and turtle retina, cyclic GMP was stratified in 2–3 layers in the ON-sublamina. While it has been shown that cyclic GMP regulates gap junctions in ON-bipolar cells, it still doesn't explain the presence of cyclic GMP in multiple layers. Evidently, cyclic GMP is involved in multiple signaling pathways in the inner plexiform layer that are yet to be identified.

Summary and future investigations

As indicated at the beginning of this review, a similar influence of nitric oxide and cyclic GMP on a cellular process is an indication of the presence of soluble guanylate cyclase in that cell. By that criterion, soluble guanylate cyclase is present in most, if not all the cell types of the vertebrate retina and there is good evidence that it is involved in transmitter release from the cone synapses, in the regulation of the horizontal cell network, in glutamate receptor-mediated signaling in the ON-bipolar cells, in the regulation of communication between ON-bipolar and AII amacrine cells, and in the control of ion channel activity in ganglion and Müller cells. It is, however, fair to say that much more work is required to definitively establish relationships between the observed effects of cyclic GMP and their physiological relevance. For example, there is good electrophysiological evidence that nitric oxide and cyclic GMP inactivate gap junction communication between ON-bipolar and amacrine cells, thereby preventing the flow of information from bipolar to amacrine cells in light-adapted retina [96, 97]. However, light and dark adaptation of retina do not produce discernable changes in the concentration of cyclic GMP in any cell type [39, 113]. What then is the physiological significance of the observed effect of cyclic GMP? It is possible that cyclic GMP concentration does increase in the ON-bipolar cells in response to light adaptation and it does play an important role in regulating gap junctions, but the increase might be too small or localized to be reliably detected by the methods used. It is also possible that cyclic GMP concentration increases transiently with some delay following light adaptation, which could easily go undetected.

Further investigation is also required to elucidate the role of soluble cyclase in the various bipolar and amacrine cells that apparently have a high concentration of the enzyme in their synaptic processes in the inner plexiform layer [39, 40].

Another area that requires investigation is the role of soluble cyclase in glutamate toxicity. Prolonged elevation of glutamate concentration caused by ischemia or a rise in the intraocular pressure results in the death of retinal neurons, in particular, the ganglion cells [114–117]. Stimulation of ionotropic receptors by the accumulated glutamate, elevation of intracellular calcium concentration, and calcium/calmodulin-activated production of nitric oxide, are thought to be the sequence of events leading to ganglion cell loss [118]. Among other effects, elevation of nitric oxide concentration should result in the stimulation of soluble guanylate cyclase and increased production of cyclic GMP. Incubation with glutamate or a glutamate analog does elevate cyclic GMP concentration in the retina [113, 119]. It is therefore reasonable to expect that cyclic GMP may have a role in the glutamate-induced cytotoxicity. However, there has been no evidence for such a role [120]. It may be that nitric oxide causes ganglion cell death through pathways unrelated to soluble guanylate cyclase. Anand *et al.* observed that in the chicken retina glutamate-induced elevation of cyclic GMP concentration is transient, peaking in about 5 min and then slowly returning to baseline [119]. It is possible that following activation by glutamate, soluble cyclase activity is rapidly brought under control and rendered harmless in spite of the continued production of and eventual damage by nitric oxide. Studies on such a control of the cyclase and the putative mechanisms involved are being pursued in other tissues [121–125], but similar efforts have not yet been made in understanding the regulation of soluble cyclases in retina.

In conclusion, there are very good indications that soluble guanylate cyclase plays important roles in the transfer of information between retinal neurons and in the adaptation of retinal networks to light and darkness. Critical control of its activity may be a significant part of preventing retinal cell death following ischemia. It is anticipated that research in the next few years will firmly establish when and how the enzyme is activated in each cell type, how long it remains active, and what mechanisms exist to control its activity under physiological and pathological conditions.

Acknowledgements

I would like to thank Alexander Margulis, Frank J. Giblin, Andrew F.X. Goldberg, Barry S. Winkler, William D. Eldred, and Noga Vardi for reading the manuscript and making helpful suggestions, and Shannon Nagy for preparing the figure. This work was supported by the National Eye Institute grants EY 07158 and EY 05230 (core grant for vision research).

References

1. Koesling D: Studying the structure and regulation of soluble guanylyl cyclase. Methods 19: 485–493, 1999
2. Garbers DL, Lowe DG: Guanylyl cyclase receptors. J Biol Chem 269: 30741–30744, 1994
3. Denninger JW, Marletta MA: Guanylate cyclase and the NO/cGMP signaling pathway. Biochim Biophys Acta 1411: 334–350, 1999
4. Dizhoor AM: Regulation of cGMP synthesis in photoreceptors: Role in signal transduction and congenital diseases of the retina. Cell Signal 12: 711–719, 2000

5. Pugh EN Jr, Duda T, Sitaramayya A, Sharma RK: Photoreceptor guanylate cyclases: A review. Biosci Rep 17: 429–473, 1997
6. Kamisaki Y, Saheki S, Nakane M, Palmieri JA, Kuno T, Chang BY, Waldman SA, Murad F: Soluble guanylate cyclase from rat lung exists as a heterodimer. J Biol Chem 261: 7236–7241, 1986
7. Harteneck C, Koesling D, Soling A, Schultz G, Bohme E: Expression of soluble guanylyl cyclase. Catalytic activity requires two enzyme subunits. FEBS Lett 272: 221–223, 1990
8. Wedel B, Humbert P, Harteneck C, Foerster J, Malkewitz J, Bohme E, Schultz G, Koesling D: Mutation of His-105 in the beta 1 subunit yields a nitric oxide- insensitive form of soluble guanylyl cyclase. Proc Natl Acad Sci USA 91: 2592–2596, 1994
9. Zhao Y, Schelvis JP, Babcock GT, Marletta MA: Identification of histidine 105 in the beta1 subunit of soluble guanylate cyclase as the heme proximal ligand. Biochemistry 37: 4502–4509, 1998
10. Russwurm M, Behrends S, Harteneck C, Koesling D: Functional properties of a naturally occurring isoform of soluble guanylyl cyclase. Biochem J 335: 125–130, 1998
11. Gupta G, Azam M, Yang L, Danziger RS: The beta2 subunit inhibits stimulation of the alpha1/beta1 form of soluble guanylyl cyclase by nitric oxide. Potential relevance to regulation of blood pressure. J Clin Invest 100: 1488–1492, 1997
12. Garthwaite J: Nitric oxide signalling in the nervous system. Semin Neurosci 5: 171–180, 1993
13. Galla HJ: Nitric oxide, NO, an intercellular messenger. Angew Chem Int Ed Engl 32: 378–380, 1993
14. Lancaster JR: Simulation of the diffusion and reaction of endogenously produced nitric oxide. Proc Natl Acad Sci USA 91: 8137–8141, 1994
15. Blute TA, Lee MR, Eldred WD: Direct imaging of NMDA-stimulated nitric oxide production in the retina. Vis Neurosci 17: 557–566, 2000
16. Brandish PE, Buechler W, Marletta MA: Regeneration of the ferrous heme of soluble guanylate cyclase from the nitric oxide complex: Acceleration by thiols and oxyhemoglobin. Biochemistry 37: 16898–16907, 1998
17. Dierks EA, Burstyn JN: The deactivation of soluble guanylyl cyclase by redox-active agents. Arch Biochem Biophys 351: 1–7, 1998
18. Kharitonov VG, Russwurm M, Magde D, Sharma VS, Koesling D: Dissociation of nitric oxide from soluble guanylate cyclase. Biochem Biophys Res Commun 239: 284–286, 1997
19. Margulis A, Sitaramayya A: Rate of deactivation of nitric oxide-stimulated soluble guanylate cyclase: Influence of nitric oxide scavengers and calcium. Biochemistry 39: 1034–1039, 2000
20. Bellamy TC, Garthwaite J: Sub-second kinetics of the nitric oxide receptor, soluble guanylyl cyclase, in intact cerebellar cells. J Biol Chem 276: 4287–4292, 2000
21. Eldred WD: In: H.W.M. Steinbusch, J. De Vente, S.R. Vincent (eds). Nitric Oxide in the Retina, Functional Neuroanatomy of the Nitric Oxide System. Elsevier Science B.V., 2000, pp 111–145
22. Rodieck RW: In: The First Steps in Seeing. Sinauer Associates, Inc., 1998
23. Wassle H, Boycott BB: Functional architecture of the mammalian retina. Physiol Rev 71: 447–480, 1991
24. Sterling P: Vision. Tuning retinal circuits. Nature 377: 676–677, 1995
25. Kolb H, Famiglietti EV: Rod and cone pathways in the inner plexiform layer of cat retina. Science 186: 47–49, 1974
26. Famiglietti EV, Kolb H: A bistratified amacrine cell and synaptic circuitry in the inner plexiform layer of the retina. Brain Res 84: 293–300, 1975
27. Strettoi E, Dacheux RF, Raviola E: Synaptic connections of rod bipolar cells in the inner plexiform layer of the rabbit retina. J Comp Neurol 295: 449–466, 1990
28. Vaney DI: Neuronal coupling in rod-signal pathways of the retina. Invest Ophthalmol Vis Sci 38: 267–273, 1997
29. Pourcho RG: Uptake of [^{3}H]glycine and [^{3}H]GABA by amacrine cells in the cat retina. Brain Res 198: 33–46, 1980
30. Pourcho RG, Goebel DJ: A combined Golgi and autoradiographic study of [^{3}H]glycine-accumulating amacrine cells in the cat retina. J Comp Neurol 233: 473–480, 1985
31. Strettoi E, Dacheux RF, Raviola E: Cone bipolar cells as interneurons in the rod pathway of the rabbit retina. J Comp Neurol 347: 139–149, 1994
32. Nelson R: Cat cones have rod input: A comparison of the response properties of cones and horizontal cell bodies in the retina of the cat. J Comp Neurol 172: 109–135, 1977
33. Wu SM: Synaptic transmission in the outer retina. Annu Rev Physiol 56: 141–168, 1994
34. Troyer EW, Hall IA, Ferrendelli JA: Guanylate cyclases in CNS: Enzymatic characteristics of soluble and particulate enzymes from mouse cerebellum and retina. J Neurochem 31: 825–833, 1978
35. Margulis A, Sharma RK, Sitaramayya A: Nitroprusside-sensitive and insensitive guanylate cyclases in retinal rod outer segments. Biochem Biophys Res Commun 185: 909–914, 1992
36. Ahmad I, Barnstable CJ: Differential laminar expression of particulate and soluble guanylate cyclase genes in rat retina. Exp Eye Res 56: 51–62, 1993
37. Haberecht MF, Schmidt HH, Mills SL, Massey SC, Nakane M, Redburn-Johnson DA: Localization of nitric oxide synthase, NADPH diaphorase and soluble guanylyl cyclase in adult rabbit retina. Vis Neurosci 15: 881–890, 1998
38. Koch KW, Lambrecht HG, Haberecht M, Redburn D, Schmidt HH: Functional coupling of a Ca^{2+}/calmodulin-dependent nitric oxide synthase and a soluble guanylyl cyclase in vertebrate photoreceptor cells. EMBO J 13: 3312–3320, 1994
39. Blute TA, Velasco P, Eldred WD: Functional localization of soluble guanylate cyclase in turtle retina: Modulation of cGMP by nitric oxide donors. Vis Neurosci 15: 485–498, 1998
40. Johansson K, Bruun A, De Vente J, Ehinger B: Immunohistochemical analysis of the developing inner plexiform layer in postnatal rat retina. Invest Ophthalmol Vis Sci 41: 305–313, 2000
41. Gotzes S, De Vente J, Muller F: Nitric oxide modulates cGMP levels in neurons of the inner and outer retina in opposite ways. Vis Neurosci 15: 945–955, 1998
42. Tsuyama Y, Noll GN, Schmidt KF: L-arginine and nicotinamide adenine dinucleotide phosphate alter dark voltage and accelerate light response recovery in isolated retinal rods of the frog (*Rana temporaria*). Neurosci Lett 149: 95–98, 1993
43. Noll GN, Billek M, Pietruck C, Schmidt KF: Inhibition of nitric oxide synthase alters light responses and dark voltage of amphibian photoreceptors. Neuropharmacology 33: 1407–1412, 1994
44. Gold GH, Korenbrot JI: Light-induced calcium release by intact retinal rods. Proc Natl Acad Sci USA 77: 5557–5561, 1980
45. Yau KW, Nakatani K: Light-induced reduction of cytoplasmic free calcium in retinal rod outer segment. Nature 313: 579–582, 1985
46. Gray-Keller MP, Detwiler PB: The calcium feedback signal in the phototransduction cascade of vertebrate rods. Neuron 13: 849–861, 1994
47. Bredt DS, Snyder SH: Isolation of nitric oxide synthetase, a calmodulin-requiring enzyme. Proc Natl Acad Sci USA 87: 682–685, 1990
48. Knowles RG, Palacios M, Palmer RM, Moncada S: Formation of nitric oxide from L-arginine in the central nervous system: A transduction mechanism for stimulation of the soluble guanylate cyclase. Proc Natl Acad Sci USA 86: 5159–5162, 1989
49. Cohen AI, Hall IA, Ferrendelli JA: Calcium and cyclic nucleotide regulation in incubated mouse retinas. J Gen Physiol 71: 595–612, 1978
50. Koch KW, Stryer L: Highly cooperative feedback control of retinal rod guanylate cyclase by calcium ions. Nature 334: 64–66, 1988

51. Miyachi E, Miyakawa A, Murakami M: Modulation of electrical coupling between retinal horizontal cells by intracellular messengers. Neurosci Res 15: S41–S49, 1991
52. Savchenko A, Barnes S, Kramer RH: Cyclic-nucleotide-gated channels mediate synaptic feedback by nitric oxide. Nature 390: 694–698, 1997
53. Rieke F, Schwartz EA: A cGMP-gated current can control exocytosis at cone synapses. Neuron 13: 863–873, 1994
54. Raviola E, Gilula NB: Intramembrane organization of specialized contacts in the outer plexiform layer of the retina. A freeze-fracture study in monkeys and rabbits. J Cell Biol 65: 192–222, 1975
55. Witkovsky P, Owen WG, Woodworth M: Gap junctions among the perikarya, dendrites, and axon terminals of the luminosity-type horizontal cell of the turtle retina. J Comp Neurol 216: 359–368, 1983
56. Qian H, Malchow RP, Ripps H: Gap-junctional properties of electrically coupled skate horizontal cells in culture. Vis Neurosci 10: 287–295, 1993
57. Vaney DI: The coupling pattern of axon-bearing horizontal cells in the mammalian retina. Proc R Soc Lond B Biol Sci 252: 93–101, 1993
58. Tornqvist K, Yang XL, Dowling JE: Modulation of cone horizontal cell activity in the teleost fish retina. III. Effects of prolonged darkness and dopamine on electrical coupling between horizontal cells. J Neurosci 8: 2279–2288, 1988
59. Xin D, Bloomfield SA: Dark- and light-induced changes in coupling between horizontal cells in mammalian retina. J Comp Neurol 405: 75–87, 1999
60. Piccolino M, Neyton J, Gerschenfeld HM: Decrease of gap junction permeability induced by dopamine and cyclic adenosine 3′:5′-monophosphate in horizontal cells of turtle retina. J Neurosci 4: 2477–2488, 1984
61. Lasater EM: Retinal horizontal cell gap junctional conductance is modulated by dopamine through a cyclic AMP-dependent protein kinase. Proc Natl Acad Sci USA 84: 7319–7323, 1987
62. Baldridge WH, Ball AK, Miller RG: Dopaminergic regulation of horizontal cell gap junction particle density in goldfish retina. J Comp Neurol 265: 428–436, 1987
63. McMahon DG, Knapp AG, Dowling JE: Horizontal cell gap junctions: Single-channel conductance and modulation by dopamine. Proc Natl Acad Sci USA 86: 7639–7643, 1989
64. DeVries SH, Schwartz EA: Modulation of an electrical synapse between solitary pairs of catfish horizontal cells by dopamine and second messengers. J Physiol 414: 351–375, 1989
65. Hampson EC, Weiler R, Vaney DI: pH-gated dopaminergic modulation of horizontal cell gap junctions in mammalian retina. Proc R Soc Lond B Biol Sci 255: 67–72, 1994
66. Miyachi E, Murakami M, Nakaki T: Arginine blocks gap junctions between retinal horizontal cells. Neuroreport 1: 107–110, 1990
67. Mills SL, Massey SC: L-Arginine uncouples A-type horizontal cells in rabbit retina. Invest Ophthalmol Vis Sci 34: 1382, 1993
68. Lu C, McMahon DG: Modulation of hybrid bass retinal gap junctional channel gating by nitric oxide. J Physiol (Lond) 499: 689–699, 1997
69. Pottek M, Schultz K, Weiler R: Effects of nitric oxide on the horizontal cell network and dopamine release in the carp retina. Vision Res 37: 1091–1102, 1997
70. Xin D, Bloomfield SA: Effects of nitric oxide on horizontal cells in the rabbit retina. Vis Neurosci 17: 799–811, 2000
71. Qian H, Ripps H: Receptive field properties of rod-driven horizontal cells in the skate retina. J Gen Physiol 100: 457–478, 1992
72. Saez JC, Berthoud VM, Moreno AP, Spray DC: Gap junctions. Multiplicity of controls in differentiated and undifferentiated cells and possible functional implications. Adv Sec Mess Phosphoprot Res 27: 163–198, 1993
73. Goodenough DA, Goliger JA, Paul DL: Connexins, connexons, and intercellular communication. Annu Rev Biochem 65: 475–502, 1996
74. Takens-Kwak BR, Jongsma HJ: Cardiac gap junctions: Three distinct single channel conductances and their modulation by phosphorylating treatments. Pflügers Arch 422: 198–200, 1992
75. Janssen-Bienhold U, Nagel H, Weiler R: *In vitro* phosphorylation in isolated horizontal cells of the fish retina: Effects of the state of light adaptation. Eur J Neurosci 5: 584–593, 1993
76. McMahon DG, Rischert JC, Dowling JE: Protein content and cAMP-dependent phosphorylation of fractionated white perch retina. Brain Res 659: 110–116, 1994
77. McMahon DG, Ponomareva LV: Nitric oxide and cGMP modulate retinal glutamate receptors. J Neurophysiol 76: 2307–2315, 1996
78. McMahon DG, Schmidt KF: Horizontal cell glutamate receptor modulation by NO: Mechanisms and functional implications for the first visual synapse. Vis Neurosci 16: 425–433, 1999
79. Shiells R, Falk G: Retinal on-bipolar cells contain a nitric oxide-sensitive guanylate cyclase. Neuroreport 3: 845–848, 1992
80. Murakami J, Ohtsuka T, Schimazaki H: Effects of aspartate and glutamate on the bipolar cells in the carp retina. Vis Res 15: 456–458, 1975
81. Copenhagen DR, Jahr CE: Release of endogenous excitatory amino acids from turtle photoreceptors. Nature 341: 536–539, 1989
82. Karschin A, Wassle H: Voltage- and transmitter-gated currents in isolated rod bipolar cells of rat retina. J Neurophysiol 63: 860–876, 1990
83. Nawy S, Jahr CE: Suppression by glutamate of cGMP-activated conductance in retinal bipolar cells. Nature 346: 269–271, 1990
84. Nawy S, Jahr CE: cGMP-gated conductance in retinal bipolar cells is suppressed by the photoreceptor transmitter. Neuron 7: 677–683, 1991
85. Shiells RA, Falk G: Glutamate receptors of rod bipolar cells are linked to a cyclic GMP cascade via a G-protein. Proc R Soc Lond B Biol Sci 242: 91–94, 1990
86. Shiells RA, Falk G: Properties of the cGMP-activated channel of retinal on-bipolar cells. Proc R Soc Lond B Biol Sci 247: 21–25, 1992
87. Kikkawa S, Nakagawa M, Iwasa T, Kaneko A, Tsuda M: GTP-binding protein couples with metabotropic glutamate receptor in bovine retinal on-bipolar cell. Biochem Biophys Res Commun 195: 374–379, 1993
88. Dhingra A, Lyubarsky A, Jiang M, Pugh EN, Birnbaumer L, Sterling P, Vardi N: The light response of ON bipolar neurons requires $G\alpha_O$. J Neurosci 20: 9053–9058, 2000
89. Saito T, Kujiraoka T: Physiological and morphological identification of two types of on-bipolar cells in the carp retina. J Comp Neurol 205: 161–170, 1982
90. Muller F, Wassle H, Voigt T: Pharmacological modulation of the rod pathway in the cat retina. J Neurophysiol 59: 1657–1672, 1988
91. Dolan RP, Schiller PH: Evidence for only depolarizing rod bipolar cells in the primate retina. Vis Neurosci 2: 421–424, 1989
92. Daw NW, Jensen RJ, Brunken WJ: Rod pathways in mammalian retinae. Trends Neurosci 13: 110–115, 1990
93. de la Villa P, Kurahashi T, Kaneko A: L-glutamate-induced responses and cGMP-activated channels in three subtypes of retinal bipolar cells dissociated from the cat. J Neurosci 15: 3571–3582, 1995
94. Vardi N, Duvoisin R, Wu G, Sterling P: Localization of mGluR6 to dendrites of ON bipolar cells in primate retina. J Comp Neurol 423: 402–412, 2000
95. Koistinaho J, Swanson RA, De Vente J, Sagar SM: NADPH-diaphorase (nitric oxide synthase)-reactive amacrine cells of rabbit retina: Putative target cells and stimulation by light. Neuroscience 57: 587–597, 1993
96. Mills SL, Massey SC: Differential properties of two gap junctional pathways made by AII amacrine cells. Nature 377: 734–737, 1995

97. Xin D, Bloomfield SA: Comparison of the responses of AII amacrine cells in the dark- and light-adapted rabbit retina. Vis Neurosci 16: 653–665, 1999
98. Cao L, Blute TA, Eldred WD: Localization of heme oxygenase-2 and modulation of cGMP levels by carbon monoxide and/or nitric oxide in the retina (in process citation). Vis Neurosci 17: 319–329, 2000
99. Wexler EM, Stanton PK, Nawy S: Nitric oxide depresses GABAA receptor function via coactivation of cGMP- dependent kinase and phosphodiesterase. J Neurosci 18: 2342–2349, 1998
100. Baldridge WH, Vaney DI, Weiler R: The modulation of intercellular coupling in the retina. Semin Cell Dev Biol 9: 311–318, 1998
101. Cook JE, Becker DL: Gap junctions in the vertebrate retina. Microsc Res Tech 31: 408–419, 1995
102. Bloomfield SA, Xin D, Osborne T: Light-induced modulation of coupling between AII amacrine cells in the rabbit retina. Vis Neurosci 14: 565–576, 1997
103. Ahmad I, Leinders-Zufall T, Kocsis JD, Shepherd GM, Zufall F, Barnstable CJ: Retinal ganglion cells express a cGMP-gated cation conductance activatable by nitric oxide donors. Neuron 12: 155–165, 1994
104. Kawai F, Sterling P: AMPA receptor activates a G-protein that suppresses a cGMP-gated current. J Neurosci 19: 2954–2959, 1999
105. Hirooka K, Kurenny DE, Barnes S: Calcium channel activation facilitated by nitric oxide in retinal ganglion cells. J Neurophysiol 83: 198–206, 2000
106. Kusaka S, Dabin I, Barnstable CJ, Puro DG: cGMP-mediated effects on the physiology of bovine and human retinal Müller (glial) cells. J Physiol 497: 813–824, 1996
107. Liepe BA, Stone C, Koistinaho J, Copenhagen DR: Nitric oxide synthase in Müller cells and neurons of salamander and fish retina. J Neurosci 14: 7641–7654, 1994
108. Kurenny DE, Thurlow GA, Turner RW, Moroz LL, Sharkey KA, Barnes S: Nitric oxide synthase in tiger salamander retina. J Comp Neurol 361: 525–536, 1995
109. Haverkamp S, Kolb H, Cuenca N: Endothelial nitric oxide synthase (eNOS) is localized to Müller cells in all vertebrate retinas. Vision Res 39: 2299–2303, 1999
110. Berkelmans HS, Schipper J, Hudson L, Steinbusch HW, De Vente J: cGMP immunocytochemistry in aorta, kidney, retina and brain tissues of the rat after perfusion with nitroprusside. Histochemistry 93: 143–148, 1989
111. Berger SJ, De Vries GW, Carter JG, Schulz DW, Passonneau PN, Lowry OH, Ferrendelli JA: The distribution of the components of the cyclic GMP cycle in retina. J Biol Chem 255: 3128–3133, 1980
112. Margulis A, Pozdnyakov N, Dang L, Sitaramayya A: Soluble guanylate cyclase and nitric oxide synthase in synaptosomal fractions of bovine retina. Vis Neurosci 15: 867–873, 1998
113. Blute TA, De Grenier J, Eldred WD: Stimulation with N-methyl-D-aspartate or kainic acid increases cyclic guanosine monophosphate-like immunoreactivity in turtle retina: Involvement of nitric oxide synthase. J Comp Neurol 404: 75–85, 1999
114. Zeevalk GD, Nicklas WJ: Chemically induced hypoglycemia and anoxia: Relationship to glutamate receptor-mediated toxicity in retina. J Pharmacol Exp Ther 253: 1285–1292, 1990
115. el Asrar AM, Morse PH, Maimone D, Torczynski E, Reder AT: MK-801 protects retinal neurons from hypoxia and the toxicity of glutamate and aspartate. Invest Ophthalmol Vis Sci 33: 3463–3468, 1992
116. Dreyer EB, Lipton SA: New perspectives on glaucoma. JAMA 281: 306–308, 1999
117. Neufeld AH: Nitric oxide: A potential mediator of retinal ganglion cell damage in glaucoma. Surv Ophthalmol 43(suppl 1): S129–S135, 1999
118. Rothman SM, Olney JW: Excitotoxicity and the NMDA Receptor. TINS 10: 299–302, 1987
119. Anand H, Roberts PJ, Lopez-Colome AM: Excitatory amino acids in the chick retina: Possible involvement of cyclic guanosine monophosphate. Neurosci Lett 58: 31–36, 1985
120. Zeevalk GD, Nicklas WJ: Nitric oxide in retina: Relation to excitatory amino acids and excitotoxicity. Exp Eye Res 58: 343–350, 1994
121. Ujiie K, Hogarth L, Danziger R, Drewett JG, Yuen PS, Pang IH, Star RA: Homologous and heterologous desensitization of a guanylyl cyclase-linked nitric oxide receptor in cultured rat medullary interstitial cells. J Pharmacol Exp Ther 270: 761–767, 1994
122. Davis JP, Vo XT, Sulakhe PV: Altered responsiveness of guanylyl cyclase to nitric oxide following treatment of cardiomyocytes with S-nitroso-D,L-acetylpenicillamine and sodium nitroprusside. Biochem Biophys Res Commun 238: 351–356, 1997
123. Scott WS, Nakayama DK: Sustained nitric oxide exposure decreases soluble guanylate cyclase mRNA and enzyme activity in pulmonary artery smooth muscle. J Surg Res 79: 66–70, 1998
124. Hussain MB, Hobbs AJ, MacAllister RJ: Autoregulation of nitric oxide-soluble guanylate cyclase-cyclic GMP signalling in mouse thoracic aorta. Br J Pharmacol 128: 1082–1088, 1999
125. Bellamy TC, Wood J, Goodwin DA, Garthwaite J: Rapid de-sensitization of the nitric oxide receptor, soluble guanylyl cyclase, underlies diversity of cellular cGMP responses. Proc Natl Acad Sci USA 97: 2928–2933, 2000

Molecular and Cellular Biochemistry **230:** 187–188, 2002.

Index to Volume 230

Developments in Molecular and Cellular Biochemistry

Series Editor: Naranjan S. Dhalla, Ph.D., M.D. (Hon.), FACC

1. V.A. Najjar (ed.): *Biological Effects of Glutamic Acid and Its Derivatives.* 1981 ISBN 90-6193-841-4
2. V.A. Najjar (ed.): *Immunologically Active Peptides.* 1981 ISBN 90-6193-842-2
3. V.A. Najjar (ed.): *Enzyme Induction and Modulation.* 1983 ISBN 0-89838-583-0
4. V.A. Najjar and L. Lorand (eds.): *Transglutaminase.* 1984 ISBN 0-89838-593-8
5. G.J. van der Vusse (ed.): *Lipid Metabolism in Normoxic and Ischemic Heart.* 1989 ISBN 0-7923-0479-9
6. J.F.C. Glatz and G.J. van der Vusse (eds.): *Cellular Fatty Acid-Binding Proteins.* 1990 ISBN 0-7923-0896-4
7. H.E. Morgan (ed.): *Molecular Mechanisms of Cellular Growth.* 1991 ISBN 0-7923-1183-3
8. G.J. van der Vusse and H. Stam (eds.): *Lipid Metabolism in the Health and Diseased Heart.* 1992 ISBN 0-7923-1850-1
9. Y. Yazaki and S. Mochizuki (eds.): *Cellular Function and Metabolism.* 1993 ISBN 0-7923-2158-8
10. J.F.C. Glatz and G.J. van der Vusse (eds.): *Cellular Fatty-Acid-Binding Proteins, II.* 1993 ISBN 0-7923-2395-5
11. R.L. Khandelwal and J.H. Wang (eds.): *Reversible Protein Phosphorylation in Cell Regulation.* 1993 ISBN 0-7923-2637-7
12. J. Moss and P. Zahradka (eds.): *ADP-Ribosylation: Metabolic Effects and Regulatory Functions.* 1994 ISBN 0-7923-2951-1
13. V.A. Saks and R. Ventura-Clapier (eds.): *Cellular Bioenergetics: Role of Coupled Creatine Kinases.* 1994 ISBN 0-7923-2952-X
14. J. Slezák and A. Ziegelhöffer (eds.): *Cellular Interactions in Cardiac Pathophysiology.* 1995 ISBN 0-7923-3573-2
15. J.A. Barnes, H.G. Coore, A.H. Mohammed and R.K. Sharma (eds.): *Signal Transduction Mechanisms.* 1995 ISBN 0-7923-3663-1
16. A.K. Srivastava and J.-L. Chiasson (eds.): *Vanadium Compounds: Biochemical and Therapeutic Applications.* 1995 ISBN 0-7923-3763-8
17. J.M.J. Lamers and P.D. Verdouw (eds.): *Biochemistry of Signal Transduction in Myocardium.* 1996 ISBN 0-7923-4067-1
18. E.-G. Krause and R. Vetter (eds.): *Biochemical Mechanisms in Heart Function.* 1996 ISBN 0-7923-4118-X
19. R. Vetter and E.-G. Krause (eds.): *Biochemical Regulation of Myocardium.* 1996 ISBN 0-7923-4259-3
20. G.N. Pierce and W.C. Claycomb (eds.): *Novel Methods in Molecular and Cellular Biochemistry of Muscle.* 1997 ISBN 0-7923-4387-5
21. F.N. Gellerich and S. Zierz (eds.): *Detection of Mitochondrial Diseases.* 1997 ISBN 0-7923-9925-0
22. P.K. Singal, V. Panagia and G.N. Pierce (eds.): *The Cellular Basis of Cardiovascular Function in Health and Disease.* 1997 ISBN 0-7923-9974-9
23. S. Abdel-aleem and J.E. Lowe (eds.): *Cardiac Metabolism in Health and Disease.* 1998 ISBN 0-7923-8104-1
24. A.K. Srivastava and B. Posner (eds.): *Insulin Action.* 1998 ISBN 0-7923-8113-0
25. V.A. Saks, R. Ventura-Clapier, X. Leverve, A. Rossi and M. Rigoulet (eds.): *Bioenergetics of the Cell: Quantitative Aspects.* 1998 ISBN 0-7923-8118-1
26. G.N. Pierce, H. Rupp, T. Izumi and A. Grynberg (eds.): *Molecular and Cellular Effects of Nutrition on Disease Processes.* 1998 ISBN 0-7923-8171-8
27. K. Ahmed, E. Chambaz and O.G. Issinger (eds.): *Molecular and Cellular View of Protein Kinase CK2.* 1998 ISBN 0-7923-8208-0

Developments in Molecular and Cellular Biochemistry

28. M.V. Cohen, J.M. Downey, R.J. Gelpi and J. Slezák (eds.): *Myocardial Ischemia and Reperfusion*. 1998 ISBN 0-7923-8173-4
29. D.A. Bernlohr and L. Banaszak (eds.): *Lipid Binding Proteins within Molecular and Cellular Biochemistry*. 1998 ISBN 0-7923-8223-4
30. R. Alvarez-Gonzalez (ed.): *ADP-Ribosylation Reactions: From Bacterial Pathogenesis to Cancer*. 1998 ISBN 0-7923-8235-8
31. S. Imai and M. Endo (eds.): *Muscle Physiology and Biochemistry*. 1998 ISBN 0-7923-8264-1
32. D.K. Das (ed.): *Stress Adaptation, Prophylaxis and Treatment*. 1999 ISBN 0-7923-8406-7
33. H. Rupp and B. Maisch (eds.): *Control of Gene Expression by Catecholamines and the Renin-Angiotensin System* ISBN 0-7923-7981-0
34. X. Shi, V. Castranova, V. Vallyathan and W.G. Perry (eds.): *Molecular Mechanisms of Metal Toxicity and Carcinogenesis*. 2001 ISBN 0-7923-7498-3
35. K. Ahmed, J.E. Allende and O.-G. Issinger (eds.): *Protein Kinase CK2 – From Structure to Regulation*. 2001 ISBN 0-7923-7666-8
36. R.K. Sharma (ed.): *Guanylate Cyclase*. 2002 ISBN-0-7923-7682-8

KLUWER ACADEMIC PUBLISHERS – DORDRECHT / BOSTON / LONDON

MIX
Papier aus verantwortungsvollen Quellen
Paper from responsible sources
FSC® C105338

If you have any concerns about our products,
you can contact us on
ProductSafety@springernature.com

In case Publisher is established outside the EU,
the EU authorized representative is:
Springer Nature Customer Service Center GmbH
Europaplatz 3, 69115 Heidelberg, Germany

Printed by Libri Plureos GmbH
in Hamburg, Germany